D1797195

VIRGIL
Eclogues

A COMMENTARY ON

VIRGIL

Eclogues

BY

WENDELL CLAUSEN

CLARENDON PRESS · OXFORD

Oxford University Press, Walton Street, Oxford OX2 6DP
Oxford New York Toronto
Delhi Bombay Calcutta Madras Karachi
Kuala Lumpur Singapore Hong Kong Tokyo
Nairobi Dar es Salaam Cape Town
Melbourne Auckland Madrid
and associated companies in
Berlin Ibadan

Oxford is a trade mark of Oxford University Press

Published in the United States
by Oxford University Press Inc., New York

British Library Cataloguing in Publication Data
Data available

Library of Congress Cataloging in Publication Data
A commentary on Virgil, Eclogues / By Wendell Clausen.
Includes bibliographical references and index.
1. Virgil. Bucolica. 2. Pastoral poetry, Latin—History and criticism.
3. Pastroral poetry, Latin—Translations into English. 4. Country life in literature.
5. Country life—Rome—Poetry. 6. Pastoral poetry, Latin.
I. Virgil. Bucolica. II. Title.
PA6804.B7C54 1994 872'.01—dc20 93–8168
ISBN 0–19–814916–6

3 5 7 9 10 8 6 4 2

Typeset by Latimer Trend & Company Ltd, Plymouth
Printed in Great Britain
on acid-free paper by
Bookcraft (Bath) Ltd.
Midsomer Norton, Avon

In Memoriam
R. A. B. MYNORS

CONTENTS

PREFACE

This commentary was begun, some twenty years ago, at the suggestion of Roger Mynors, who had given up the idea of writing a commentary on the *Eclogues* himself. At the time we planned to publish our commentaries—his on the *Georgics* and mine on the *Eclogues*—together; later, when it seemed that he would be ready before I could, the plan was modified; and, finally, with his death in 1989, dissolved. (His commentary on the *Georgics* was published posthumously, Oxford, 1990.) For more than a decade I enjoyed the benefit of his advice and encouragement, and the memory of those years, of our meetings and conversations, is vivid to me now. I am deeply indebted also to another friend, Guy Lee, who has read, in the various stages of its evolution, and never without improving it, almost all of my commentary. Nor am I unmindful of other debts of gratitude: to Hilary O'Shea for her unfailing interest; to Lenore Parker for her efficiency and marvellous patience; to Julia Budenz for the phrase 'wind-shifted shade' (*E.* 5.5–6 n., from Book Three of her poem *The Gardens of Flora Baum*); to Lucy Gasson for her care and attention; and especially to Leofranc Holford-Strevens for his exact, perspicacious criticism.

The text of the *Eclogues* is that of Roger Mynors (Oxford, 1972). Guy Lee's translation of the *Eclogues* (Liverpool Classical Texts, 1, 1980; Penguin Books, 1984) is used by his permission and that of Penguin Books Ltd.; A. S. F. Gow's translation of Theocritus (Cambridge, 1952) by permission of the Cambridge University Press. Translations of Theocritus not attributed to Gow are my own. The map on p. 234 is reproduced from J. J. Wilkes, *Dalmatia* (Cambridge, Mass., 1969), 18, by permission of the Harvard University Press. My introduction to the Fourth *Eclogue*, as it has now become, was originally conceived as a lecture, 'Virgil's Messianic Eclogue', and published in J. L. Kugel (ed.), *Poetry and Prophecy* (Ithaca, NY, 1990), 65–74; it is used by permission of the Cornell University Press.

W.V.C.

Cambridge, Massachusetts
October 1992

BIBLIOGRAPHY

1. *Texts, Commentaries, and Translation*

J. L. de la Cerda (Cologne, 1647).
J. Martyn, 4th edn. (Oxford, 1820).
C. G. Heyne, 4th edn., rev. G. P. E. Wagner (Leipzig, 1830).
J. H. Voss (Altona, 1830).
J. Conington, 2nd edn. (London, 1865).
A. Forbiger, 4th edn. (Leipzig, 1872).
T. E. Page (London, 1898).
R. Sabbadini (Rome, 1937).
R. A. B. Mynors, corr. edn. (Oxford, 1972).
M. Geymonat (Turin, 1973).
R. Coleman (Cambridge, 1977).
G. Lee (Harmondsworth, 1984).

2. *Abbreviations*

CHCL — *The Cambridge History of Classical Literature*; i, ed. P. E. Easterling and B. M. W. Knox (Cambridge, 1985); ii, ed. E. J. Kenney and W. V. Clausen (Cambridge, 1982).

Enc. Virg. — *Enciclopedia Virgiliana* (Rome, 1984–90).

FPL — *Fragmenta Poetarum Latinorum*, ed. C. Büchner (Leipzig, 1982).

G.–P. — *The Greek Anthology: Hellenistic Epigrams*, ed. A. S. F. Gow and D. L. Page (Cambridge, 1965).

LIMC — *Lexicon Iconographicum Mythologiae Classicae* (Zurich, 1981–).

LSJ — H. G. Liddell and R. Scott, rev. H. S. Jones, *A Greek–English Lexicon* (Oxford, 1940; supplement Oxford, 1968).

New Gallus — R. D. Anderson, P. J. Parsons, and R. G. M. Nisbet, 'Elegiacs by Gallus from Qaşr Ibrîm', *JRS* 69 (1979), 125–55. Text, p. 140 = *FPL*, pp. 129–30, see also now E. Courtney, *The Fragmentary Latin Poets* (Oxford, 1993), 263–8 (Gallus fr. 2).

OCD *The Oxford Classical Dictionary*, ed. N. G. L. Hammond and H. H. Scullard, 2nd edn. (Oxford, 1970).

OLD *Oxford Latin Dictionary*, ed. P. G. W. Glare (Oxford, 1968–82).

RE *Real-Encyclopädie der classischen Altertumswissenschaft* (Stuttgart, 1894–1978).

Suppl. Hell. *Supplementum Hellenisticum*, ed. H. Lloyd-Jones and P. Parsons (Berlin, 1983).

TLL *Thesaurus Linguae Latinae* (Leipzig, 1900–).

3. *Other Books Cited in Abbreviated Form*

E. Abbe, *The Plants of Virgil's Georgics* (Ithaca, NY, 1965).

B. Axelson, *Unpoetische Wörter* (Lund, 1945).

W. Clausen, *Virgil's Aeneid and the Tradition of Hellenistic Poetry* (Berkeley, 1987).

K. J. Dover, *Theocritus* (Basingstoke, 1971).

E. Fraenkel, *Horace* (Oxford, 1957).

M. Gigante (ed.), *Lecturae Vergilianae*, i: *Le Bucoliche* (Naples, 1981).

A. S. F. Gow, *Theocritus* (Cambridge, 1952).

J. B. Hofmann and A. Szantyr, *Lateinische Syntax und Stilistik* (Munich, 1963–4).

A. E. Housman, *The Classical Papers of A. E. Housman*, ed. J. Diggle and F. R. D. Goodyear (Cambridge, 1972).

F. Klingner, *Virgil* (Zurich, 1967).

R. Kühner and C. Stegmann, *Ausführliche Grammatik der lateinischen Sprache*, 3rd edn. (Leverkusen, 1955).

M. Leumann, *Lateinische Laut- und Formenlehre*, 2nd edn. (Munich, 1977).

E. Löfstedt, *Syntactica*, i, 2nd edn. (Lund, 1942), ii (Lund, 1933).

P. Maas, 'Studien zum poetischen Plural bei den Römern', *ALL* 12 (1900–2), 479–550 = *Kleine Schriften* (Munich, 1973), 527–85.

E. Norden, *P. Vergilius Maro: Aeneis Buch VI*, 2nd edn. (Leipzig, 1916).

A. Otto, *Die Sprichwörter und sprichwörtlichen Redensarten der Römer* (Leipzig, 1890).

S. Posch, *Beobachtungen zur Theokritnachwirkung bei Vergil*, (Commentationes Aenipontanae 19; Innsbruck, 1969).

D. O. Ross, Jr., *Style and Tradition in Catullus* (Cambridge, Mass., 1969).

—— *Backgrounds to Augustan Poetry: Gallus, Elegy, and Rome* (Cambridge, 1975).

J. Sargeaunt, *The Trees, Shrubs, and Plants of Virgil* (Oxford, 1920).

J. Soubiran, *L'Élision dans la poésie latine* (Paris, 1966).

H. Tränkle, *Die Sprachkunst des Properz und die Tradition der lateinischen Dichtersprache*, (Hermes Einzelschriften, 15; Wiesbaden, 1960).

F. Ursinus, *Virgilius collatione scriptorum Graecorum illustratus*, ed. L. K. Valckenaer (Leeuwarden, 1747).

J. Wackernagel, *Vorlesungen über Syntax*, 2nd edn. (Basle, 1926–8).

G. Wissowa, *Religion und Kultus der Römer*, 2nd edn. (Munich, 1912).

INTRODUCTION

Pastoral Poetry

PASTORAL as a form of poetry (for pastoral moments occur in earlier Greek poetry)[1] is the invention of Theocritus of Syracuse.[2] Like most Hellenistic poets, belated poets, as they felt themselves to be, in a long and imposing tradition, Theocritus is self-conscious and erudite, a difficult poet, whose pastoral poetry is as complicated, and beautiful, as his other poetry. Ancient pastoral poetry, the poetry of Theocritus and Virgil, is never simple,[3] though it affects to be; and in this affectation of simplicity, the disparity between the meanness of his subject and the refinement of the poet's art, lies the essence of pastoral.

Theocritus was a town-bred poet, and the pleasure he finds in the country is that of a townsman, a pleasure qualified by his own awareness and slightly ironic. He visits the country, or rather, his idea of the country, with a mind closed to all that is harsh and unforgiving there, the endless, weary routine, the small gains, natural disasters; the worst that can befall one of his herdsmen, in this artfully protected landscape, is defeat in a singing-match, a thorn-prick in the foot, or rejection in love.

Occasionally, however, the fiction lapses, or is permitted to lapse, and a Theocritean herdsman bears a striking resemblance to the poet who created him. 'Tityrus, dear friend, feed the goats and drive them to the spring'—so speaks the lovesick goatherd at the beginning of the Third *Idyll*[4] as he goes to serenade Amaryllis, the stony-hearted charmer who lives in a neighbouring cave. He

[1] See A. M. Parry, 'Landscape in Greek Poetry', *YCS* 15 (1957), 3–29 = *The Language of Achilles and Other Papers* (Oxford, 1989), 8–35.

[2] His dates are uncertain, *c.*300–260 BC.

[3] Unlike much of post-Virgilian pastoral poetry, that is, pastoral poetry written in the Renaissance and later (for Virgil's ancient imitators, Calpurnius Siculus and Nemesianus, may be disregarded), which lacks the delicate hardness of Virgil's *Eclogues*; it tends, rather, to be simple and sentimental—Milton's *Lycidas* is a powerful exception—and frequently degenerates into mere prettiness.

[4] See *E.* 9.23–5n. For the word εἰδύλλιον, which has none of the connotations of English 'idyll', see Gow, vol. i, p. lxxi.

worries about his looks, his snub nose, his projecting beard; brings her a present of ten apples, and tomorrow will bring more; wishes he were a buzzing bee that he might slip into her cave through the ivy and fern; and despairs of Love, a cruel god, suckled by a lioness and reared in the wildwood. He thinks of suicide, pauses and reflects, and finally sings, not, however, the naïve love song that might have been expected of him, but a song learnedly alluding, in Hellenistic fashion, to five love-affairs of the mythical past, Hippomenes and Atalanta, Bias and Pero (who is not named), Adonis and Aphrodite, Endymion, Iasion:

> *τὰν ἀγέλαν χὼ μάντις ἀπ' Ὄθρυος ἆγε Μελάμπους*
> *ἐς Πύλον· ἁ δὲ Βίαντος ἐν ἀγκοίναισιν ἐκλίνθη*
> *μάτηρ ἁ χαρίεσσα περίφρονος Ἀλφεσιβοίας.* (3.43–5)

And the seer Melampus drove the herd from Othrys to Pylos; and in the arms of Bias was laid the lovely mother of wise Alphesiboea.

The absurdity of such a song in the mouth of such a singer produces, as Theocritus no doubt intended, an elegantly humorous effect. 'My head aches', says the poor goatherd as he stops singing, 'but you don't care'; he will sing no more, but lie where he has fallen and let the wolves eat him.

Very little is known for certain about the life of Theocritus.[5] Disappointed, it would seem, in an appeal for patronage to Hiero II of Syracuse—*Idyll* 16,[6] a brilliant poem—he emigrated, 16.106–7 *ἐς δὲ καλεύντων* | *θαρσήσας Μοίσαισι σὺν ἁμετέραισιν ἴοιμ' ἄν*, 'but to the houses of those that call me I will take heart and go with my Muses'. Whether he went directly to Alexandria, whose lord, Ptolemy Philadelphus, was renowned for his generosity to poets and men of letters, or first to the island of Cos, Ptolemy's birthplace, and the home of Philitas,[7] Ptolemy's tutor and principal of the Alexandrian school of poetry, is unknown. That Theocritus was well acquainted with both Cos and Alexandria is evident from *Idylls* 7 and 15, and *Idyll* 17 implies that he had gained Ptolemy's favour.[8]

[5] The few facts are discussed by Gow, vol. i, pp. xv–xxix.
[6] Probably composed in 275/4 BC; see Gow, vol. i, p. xvii.
[7] Mentioned in *Idyll* 7.40.
[8] Composed before July 270 BC; see Gow, ii. 326.

A prior residence in Cos may be indicated by Theocritus' knowledge of botany.[9] He refers, mainly in his pastoral *Idylls*, to a large number of plants and trees, far more, in fact, than are found in all Homer; and for the detail and accuracy of his knowledge he is probably indebted to Theophrastus. Cos was famous for its medical school, Theophrastus' *History of Plants* would have been accessible in the medical library, and botany in the mind of Theocritus turned to poetry.

Theocritus had good friends in Cos, Eucritus and Amyntas, the brothers Phrasidamus and Antigenes, and Aratus, all of whom figure in *Idyll* 7, which is laid in Cos; and especially Nicias, the lovesick physician and minor poet to whom *Idyll* 11 is addressed, a poem with a Sicilian subject, the Cyclops in love.[10] The pastoral *Idylls* are thought to be relatively early poems, and *Idyll* 11 may be Theocritus' earliest extant poem; perhaps a poem written in Sicily and then revised in Cos, when the introductory lines to Nicias were added:[11]

> *Οὐδὲν ποττὸν ἔρωτα πεφύκει φάρμακον ἄλλο,*
> *Νικία, οὔτ' ἔγχριστον, ἐμὶν δοκεῖ, οὔτ' ἐπίπαστον,*
> *ἢ ταὶ Πιερίδες* . . . (11. 1–3)

No other cure is there for love, Nicias, neither unguent, in my opinion, nor powder, but the Pierian Muses . . .

It seems likely that Nicias was studying medicine in Cos.

Scientific study, friendships—Theocritus appears to have spent several years in Cos, and may there have composed his pastoral *Idylls*, or most of them, for the subject of *Idyll* 7 is pastoral poetry, 36 *βουκολιασδώμεσθα*, 'let us make country song'. So Simichidas challenges Lycidas, the mysterious goatherd, and his challenge results in two elegant and exemplary songs, the first sung by Lycidas (52–89), the second by Simichidas (96–127).[12] Talk of a young poet in Cos writing a new kind of poetry would eventually

[9] See A. Lindsell, 'Was Theocritus a Botanist?', *G&R* 6 (1937), 78–93, Gow, vol. i, p. xix.

[10] See *E.* 2, Introduction. He retains his Homeric character in *Idyll* 16. 53 *ὀλοοῖο Κύκλωπος*, 'the fell Cyclops', and in *Idyll* 7. 151–3.

[11] See Gow, vol. i, p. xxi.

[12] The subject of Simichidas' song is the passion of his dear friend Aratus, 98 *ὁ τὰ πάντα φιλαίτατος*, for a boy. *Idyll* 6, which resembles *Idyll* 11, the subject of both poems being the Cyclops in love, is addressed to Aratus.

reach the ears of Ptolemy, who cherished his native island, and a passage in *Idyll* 7 suggests that something of the sort did happen.

> *Λυκίδα φίλε, πολλὰ μὲν ἄλλα*
> *Νύμφαι κἠμὲ δίδαξαν ἀν' ὤρεα βουκολέοντα*
> *ἐσθλά, τά που καὶ Ζηνὸς ἐπὶ θρόνον ἄγαγε φάμα.* (7.91–3)

Lycidas my friend, many other songs the Nymphs taught me as I tended my herd upon the hills, good songs, which report has carried, perhaps, even to the throne of Zeus.

The speaker is Simichidas, whom ancient readers took to be Theocritus in pastoral disguise. Yet it is tempting to imagine Theocritus in the great new capital of Egypt, Alexander's city, a city with no relation to the countryside—to imagine him there recalling the herdsmen of his youth and early manhood in a landscape of memory.

Traces of Theocritean presence in Latin poetry before Virgil are slight and elusive. Catullus may have modelled the refrain in his epyllion (64) on the refrain in *Idyll* 1, and he adapted, possibly after Cinna, a line from *Idyll* 15.[13] There is also the remark of Pliny the Elder (*NH* 28.19) to the effect that Catullus, and Virgil after him, imitated Simaetha's song in *Idyll* 2.[14] So erudite a poet and scholar as Parthenius[15] could hardly have been ignorant of Theocritus; perhaps he did not care for pastoral poetry, certainly his first Roman disciples—Cinna, Catullus, Calvus—would not have cared for such poetry.

For Catullus and his friends, young poets concerned to be fashionable, to be urbane, 'of the City', the country possessed no charm; it represented, rather, the very qualities they despised, in poetry as in manners, the inept, the uncouth, the outmoded. Thus the ultimate dispraise of Suffenus, witty and urbane as he is, is this, let him but touch poetry and he seems a goatherd or ditchdigger, coarser than the coarse country (Catull. 22.14 'idem infaceto est infacetior rure'); and of the detestable Volusius, that his *Annals* are

[13] 15.100 *Δέσποιν', ἃ Γολγώς τε καὶ Ἰδάλιον ἐφίλησας*, 'Lady, that dost love Golgi and Idalium'; Catull. 64.96 'quaeque regis Golgos quaeque Idalium frondosum'. Cf. Catull. 36.11–14 and see *CHCL* ii.201.

[14] See *E*. 8, Introduction, p. 238–9.

[15] See *CHCL* ii.185–6, R. Scarcia, *Enc. Virg.* s.v. *Partenio*.

a mass of country coarseness (Catull. 36.19 'pleni ruris et inficetiarum').[16] Not for these poets idealized herdsmen and exquisite pastoral sentiment.

Virgil was different. He was born in the country, near Mantua in the Po valley, and a deep and abiding affection for the country informs his poetry, not only the *Eclogues* and *Georgics* but even, in places, the *Aeneid*.[17] Not that Theocritus was ignorant of the country,[18] he was in fact an accurate observer, especially of plants and trees. But Theocritus keeps his distance; there is little or nothing in his poetry of the nearness and intimacy which characterize Virgil's knowledge of the country—nothing, for example, like the pathos of Meliboeus' new-born kids, so long expected and now left behind on the rocks to die, and the mother-goat so weak and exhausted from her labour (as a countryman would know) that she can hardly be led, much less driven.[19]

In one important respect, however, Theocritus is more strictly pastoral, more realistic—more considerate, that is, of his fiction—than Virgil: very rarely does he allow an extraneous reference to obtrude, an allusion to Ptolemy,[20] the name of a celebrated female citharode,[21] the names of a few friends in Cos.[22] When Theocritus wishes to address Ptolemy, he does so by writing a different kind of poem, unpastoral, an encomium. Yet these few references in Theocritus, together with his own sense of the sufficiency of the country, enabled Virgil, apparently, to include a wider range of experience—politics and politicians, the ravages of civil war,

[16] Contrast Catullus' praise of the delightful young Pollio, 12.8–9 'est enim leporum/differtus puer ac facetiarum'. Horace discerned in the *Eclogues* the wit that Catullus denied to country things, *Serm.* 1.10.44–5 'molle atque facetum/Vergilio adnuerunt gaudentes rure Camenae'.

[17] Virgil is reported to have had the appearance of a countryman, *Vita Donati* 8 'facie rusticana'; but this, like so much else in the *Vita Donati*, has been doubted as being an inference from his poetry. See G. Brugnoli, *Enc. Virg.* s.v. *Vitae Vergilianae*.

[18] As were some post-Virgilian pastoral poets—'men to whom the face of nature was so little known, that they have drawn it only after their own imagination' (Samuel Johnson, *The Rambler*, No. 36, 21 July 1750). Yet it is difficult if not impossible to imagine a poem about country things, *τὰ γεωργικά*, *Georgics*, by Theocritus.

[19] See 1.13n., 14n.

[20] See above, p. xviii.

[21] *Idyll* 4.31.

[22] See above, p. xvii.

religion, poetry, literary criticism—in a pastoral definition. And it was this less stringent definition of pastoral, with its manifold possibilities, that ultimately prevailed.

The Book of Eclogues[23]

From a prefatory epigram it appears that a collection of pastoral poetry was made by Artemidorus of Tarsus, a grammarian active in Alexandria in the first half of the first century BC.[24]

> *Βουκολικαὶ Μοῖσαι σποράδες ποκά, νῦν δ' ἅμα πᾶσαι*
> *ἐντὶ μιᾶς μάνδρας, ἐντὶ μιᾶς ἀγέλας.* (*AP* 9.205)

The bucolic Muses were once scattered, but now are all together in one fold, in one flock.

In all probability, this collected edition was used by Virgil; an edition which included, along with poems attributed to Bion and Moschus, ten poems attributed to Theocritus.[25] Two of these, *Idylls* 8 and 9, are now known to be spurious; but Virgil did not know, and he evidently admired *Idyll* 8. Ten poems 'of Theocritus' varying in quality and casually arranged[26]—Virgil could hardly have attached much importance initially to the number ten; presumably it became important to him as he began to think of arranging his imitations of Theocritus in a book.

Virgil was not the first poet to arrange his poems for publication. Catullus had already done so in his *libellus*,[27] and the artfulness of his arrangement, especially of poems 1–11, might seem to imply a

[23] Virgil called his book of pastoral poems *Bucolica*, as he was to call his next book *Georgica*. The evidence of the capital MSS is unambiguous: P. VERGILI MARONIS BVCOLICA EXPLICIT *G*, P. VERGILI MARONIS BVCOLICON LIBER EXPLICIT *M*, BVCOLICON *P*, VERGILI MARONIS BVCOLICA EXPLIC. FELICITER *R*; cf. Servius, *Prooem.* 'Bucolica, ut ferunt, dicta sunt', Quintilian 8.6.46 (quoting *E.* 9.7–10) 'in Bucolicis'. The word *ecloga* originally meant a selection or excerpt, but had come to mean, by the time of Pliny the Younger, any short poem, *Epist.* 4.14.9 'epigrammata siue idyllia siue eclogas siue, ut multi, poematia'. See N. Horsfall, *BICS* 28 (1981), 108–9, M. Geymonat, *BICS* 29 (1982), 17–18.

[24] See Gow, vol. i, pp. lx-lxii.

[25] *Idylls* 1, 3, 4, 5, 6, 7, 8, 9, 10, 11.

[26] But the primacy of *Idyll* 1 (see *E.* 1, Introduction, n. 1) may imply appreciation of this most beautiful of ancient pastoral poems.

[27] See Clausen, 'Catulli Veronensis Liber', *CP* 71 (1976), 37–43 = *CHCL* ii. 193–7.

precedent.[28] There is a fundamental difference, however, between the books of Catullus and Virgil. Catullus composed his poems, then collected and arranged them in a book;[29] but Virgil composed his book, at least towards the end, while composing and recomposing his poems, when the design of individual poems had become, to some extent, subordinate to the design of the book.[30] Thus, for example, there is a singing-match of forty-eight lines in *Eclogue* 3 and a singing-match of forty-eight lines in *Eclogue* 7, with an odd moment in the singing-match in *Eclogue* 3. 'Bitter is wolf to fold, rain to ripened grain, gales to trees, to me Amaryllis' anger'—so sings Damoetas (3.80–1); to which Menalcas responds, 'Sweet is moisture to crops, arbutus to weaned kids, pliant willow to gravid flock, to me Amyntas only.' Suddenly, into this delicately modulated pastoral fantasy Damoetas introduces Pollio, the unpastoral C. Asinius Pollio, 'Pollio loves my Muse, rustic though she be . . .'.

D. Pollio amat nostram, quamuis est rustica, Musam:
Pierides, uitulam lectori pascite uestro.
M. Pollio et ipse facit noua carmina: pascite taurum,
iam cornu petat et pedibus qui spargat harenam.
D. Qui te, Pollio, amat, ueniat quo te quoque gaudet;
mella fluant illi, ferat et rubus asper amomum.
M. Qui Bauium non odit, amet tua carmina, Meui,
atque idem iungat uulpes et mulgeat hircos. (3.84–91)

These lines, in which Damoetas and Menalcas suddenly become, and as suddenly cease to be, literary critics, bear no relation to what precedes or follows in the singing-match, and were they absent would not be missed. Why, then, are they present? The simplest explanation is that they were added when Virgil was composing his book, when, that is, he decided that *Dic mihi, Damoeta* would be the Third *Eclogue* and *Sicelides Musae*, or rather, *Vltima Cumaei*, the Fourth.[31] Pollio's intrusion in the

[28] Catullus was familiar with Meleager's *Garland*, an anthology of some four dozen poets, each of whom was represented by a flower or a vegetable; see G.–P. on Meleager 1. Whether this elaborate conceit involved the arrangement of individual poems is unknown.

[29] It is of course possible that Catullus composed a few poems with an eye to the shape of his book.

[30] See below, n. 44.

[31] See *E.* 4, Introduction, n. 11.

Third *Eclogue* prepares for the flattering lines addressed to him in the Fourth (11–14)[32], and, more particularly, 3.89 'mella fluant illi, ferat et rubus asper amomum' anticipates the paradisiac landscape of 4.24–5 'occidet et serpens, et fallax herba ueneni / occidet; Assyrium uulgo nascetur amomum'. The rare exotic *amomum* is not found elsewhere in Virgil's poetry, nor in Greek or Latin poetry before Virgil.

'De eclogis multi dubitant, quae licet decem sint, incertum tamen est quo ordine scriptae sint' (Servius, *Prooemium*). The chronology of the *Eclogues* had become a subject of speculation in late antiquity if not before. Three dates, it seemed, could be extracted from the text: 42 BC, the land-confiscations (*Eclogues* 1 and 9); 40 BC, Pollio's consulship (*Eclogue* 4.11–12); and 39 BC, Pollio's campaign against the Parthini (*Eclogue* 8.6–13). The reference in the Eighth *Eclogue*, however, is probably not to Pollio but to Octavian, not therefore to 39 BC but to 35 BC.[33] Individual eclogues were doubtless shown or given to friends as they were written; but since all ten were published together, all ten are, in a sense, contemporaneous, and any attempt to determine the exact order of their composition will prove illusory.[34] Until he finally relinquished his book, Virgil was free to make changes in it—revising, adding, deleting lines—wherever he pleased.[35]

The main design of Virgil's book is evident because Virgil took pains to make it so. (Lesser or partial designs are not necessarily precluded:[36] Virgil is a poet of labyrinthine intricacy.) The book is

[32] Pollio's intrusion also has the effect of making Virgil's two amoebean contests equal in length: 3.60–107, 7.21–68.

[33] See *E.* 8, Introduction.

[34] Here are three hypothetical chronologies: R. Hanslik, *WS* 68 (1955), 8: *E.* 2, 3, 5, 9, 1, 4, 6, 8, 7 (or 7, 8), 10; K. Büchner, *P. Vergilius Maro* (Stuttgart, 1960), 234: *E.* 2, 3, 5, 9, 1, 6, 4, 8, 7, 10; A. La Penna, *Maia*, NS 15 (1963), 491: *E.* 9, 1, 6, 4, 2, 3, 5, 8, 7, 10. E. de Saint-Denis in his edition (Paris, 1942), 5 n. 2 lists nine others.

[35] Cross-reference (so to call it) therefore offers no reliable evidence of priority. The striking phrase—too striking not to be noticed and remembered—in 2.3 'densas, umbrosa cacumina, fagos' is repeated, with a significant modification, in 9.9 'ueteres, iam fracta cacumina, fagos'. The repetition was contrived by Virgil for effect, to enhance the pathos of 9.9, and tells nothing about the temporal relation of the two *Eclogues*.

[36] For a description and criticism of various designs which have been discovered see N. Rudd, *Lines of Enquiry: Studies in Latin Poetry* (Cambridge, 1976), 119–44; also J. Van Sickle, *Enc. Virg.* s.v. *Bucoliche* 549–52.

divided into two halves of five *Eclogues* each: 1–5, 6–10,[37] the first half comprising 420, the second 408 lines; the two longest *Eclogues*, the Third (111 lines) and the Eighth (108 lines)[38] are symmetrically placed, each at the center of its half; as are the two most Theocritean *Eclogues*, the Second and the Seventh, each second in its half and following an un-Theocritean *Eclogue*.[39]

The first word of the First *Eclogue* and the last word of the Fifth is a name, in the vocative case, *Tityre . . . Menalca*, Tityrus and Menalcas, Virgil's occasional personae. Such precision of form cannot be accidental; the two names define the first half of the book as precisely as the name Alexis, in the accusative case, defines the Second *Eclogue*, the first line of which ends, as does the last, with *Alexin*. Did Virgil expect his reader to notice such details? Virgil's Roman reader read aloud, read slowly, and had been trained from boyhood in the discipline of rhetoric. Yet rhetoric need not be noticed to be effective, and even a reader less well trained may experience a certain sense of satisfaction, without knowing quite why, as he reads 'formosum paribus nodis atque aere, Menalca'.

At the end of the Fifth *Eclogue* Menalcas and Mopsus exchange gifts, and Menalcas gives Mopsus his pipe:

> Hac te nos fragili donabimus ante cicuta;
> haec nos 'formosum Corydon ardebat Alexin',
> haec eadem docuit 'cuium pecus? an Meliboei?' (5.85–7)

If Virgil's intention here is—as, obviously, it is—to define the first half of his book, why is there no reference to the First *Eclogue*? Because such a reference would be inappropriate in the mouth of Menalcas; his pipe, Menalcas says, taught him the Second and Third *Eclogues*, Theocritean songs, and the First *Eclogue* is no Theocritean song.[40] Besides, Virgil being Virgil has already referred, subtly and elaborately, to the First *Eclogue* in the preceding lines.

[37] Horace adverts to the design of Virgil's book in his own first book of poems, quite likely published in the same year as Virgil's, *Sermones* 1: ten satires, the first of which begins 'Qui fit, Maecenas' and the sixth 'Non quia, Maecenas'. Apparently it was the impression created by Virgil's book that caused subsequent poets—Horace, Tibullus, Ovid—to compose books of ten (or multiples of five) poems.

[38] See 8.28[a]n.

[39] See H. C. Gotoff, *Philologus*, 111 (1967), 67n. 2.

[40] See *E.* 5, Introduction, pp. 154–5.

T. O Meliboee, *deus* nobis haec *otia* fecit.
namque erit ille mihi semper *deus* . . . (1.6–7)

M. *ipsae* te, Tityre, pinus,
ipsi te fontes, *ipsa* haec *arbusta* uocabant.

T. Quid facerem? neque seruitio me exire licebat
nec tam praesentis alibi cognoscere diuos.
hic illum uidi iuuenem, Meliboee, *quotannis*
bis senos cui nostra dies *altaria* fumant. (1.38–43)

Me. amat bonus *otia*[41] Daphnis.
ipsi laetitia uoces ad sidera iactant
intonsi montes; *ipsae* iam carmina rupes,
ipsa[42] sonant *arbusta: 'deus, deus* ille, Menalca!'
sis bonus o felixque tuis! en quattuor aras:
ecce duas tibi, Daphni, duas *altaria* Phoebo.
pocula bina nouo spumantia lacte *quotannis* . . . (5.61–7)

A reminiscence of the First *Eclogue*, or so these lines would seem to an observant reader, and so no doubt Virgil intended.[43] But the opposite must be the case, for the First *Eclogue*, Virgil's introduction to his book,[44] must be, if not his 'last

[41] *Otia* occurs only in these two places in the *Eclogues*.

[42] The triple polyptoton of *ipse* occurs only in these two places in the *Eclogues*. Klingner 91 n. 1 calls attention to 1.46, 51 'fortunate senex', 5.49 'fortunate puer'.

[43] It might conceivably be argued that the relevant lines in the Fifth *Eclogue* were added or substituted (the songs of Mopsus and Menalcas are parallel in structure) when Virgil was composing his book; but this seems unlikely, since the effect could be obtained more easily as indicated. See also 1. 11 n.

[44] And to the design of his book. Meliboeus begins with five lines, Tityrus answers with five (10), and their succeeding exchanges, though varied in length to suggest the give and take of conversation, are composed almost entirely of multiples of five: M. 8, T. 7 (15); M. 1, T. 9 (10); M. 4, T. 6 (10); M. 13, T. 5 (18, a slight variation to accommodate Octavian precisely at the centre of the poem?); M. 15, T. 5 (20). The Second *Eclogue*, Corydon's 'disordered' utterance (4 'haec incondita'), as if to confirm the design of the First *Eclogue*, begins with a statement of five lines by the poet and ends with Corydon's self-reproach of five lines, which is modelled on the self-reproach of Polyphemus in *Idyll* 11.72–6. *Idyll* 6 may also be compared: it begins and ends with a statement of five lines by the poet and contains two songs, the first by Daphnis of fourteen lines followed by a transitional line (fifteen lines), and the second by Damoetas of twenty lines. Analysis of this sort is not to be confused with the attempt to discover a complicated pattern of numerical correspondences in the book of *Eclogues*, a pattern which attaches undue importance to the Fifth *Eclogue* while virtually ignoring the Tenth; see O. Skutsch, *HSCP* 73 (1969), 153–69, D. West, *CR*, NS 28 (1978), 77–8, O. Skutsch, *BICS* 27 (1980), 95–6, I. M. Le M. DuQuesnay, *LCM* 5 (1980), 245–7, F. Della Corte, *Enc. Virg.* s.v. *Bucoliche* 546–9.

labour',[45] one of his latest, later at any rate than the Fifth.

The opening line of the Sixth *Eclogue*, 'Prima Syracosio dignata est ludere uersu', suggests a new beginning,[46] while l. 8:

agrestem tenui meditabor harundine Musam[47]

recalls l. 2 of the First *Eclogue*:

siluestrem tenui Musam meditaris auena.

Or so a reader would suppose, and so no doubt Virgil intended. But again the opposite must be the case, for 'siluestrem tenui Musam meditaris auena' must be modelled on 'agrestem tenui meditabor harundine Musam', where the literary connotation of *tenui* is immediately intelligible.[48] Virgil's intention is obvious: further to define the first half of his book and, at the same time, to relate it to the second half.

The first half of the book is also related to the second half—the Fifth *Eclogue* to the Tenth—by the figure of Daphnis, the subject of the songs of Mopsus and Menalcas in the Fifth *Eclogue* and present in the Tenth in the imitation of Thyrsis' lament for Daphnis.[49] And the second half is defined by the figure of Gallus, the erudite Alexandrian poet in the Sixth *Eclogue* and the lovesick elegiac poet in the Tenth.[50]

Finally, it was Virgil who introduced shade and shadows into the pastoral landscape,[51] and twice he deploys shadows for an effect of cadence in his book, at the end of the First *Eclogue* and at the end of the Tenth. As the shadows deepen underneath the juniper at the end of the Tenth *Eclogue*, the singer rises:

[45] 'Extremum hunc, Arethusa, mihi concede laborem' (10. 1)—not a statement of fact, as it has been taken to be (G. D'Anna, *Enc. Virg.* s.v. *Cornelio Gallo* 894: 'I critici, concordi nel ritenere l'egloga come scritta per ultima'); see above, n. 34. The reference is not to the date of the poem but to the design of the book. Imagine a previous *Eclogue* beginning with this line.

[46] Some ancient critics, Servius says (*Prooemium*), wished to put the Sixth *Eclogue* first in the book.

[47] Cf. also 1. 10 'ludere quae uellem calamo permisit agresti'. The adjective *agrestis* occurs again only in 10. 24 'agresti capitis Siluanus honore'.

[48] See *E.* 6, Introduction, p. 175.

[49] See 10. 9–12n., 19–20n., 21–3n.

[50] Virgil may have intended a similar if rather vague definition of the first half, with Octavian implied in the First *Eclogue* and Caesar adumbrated in the Fifth; see *E.* 5, Introduction, n. 4.

[51] See 1. 82–3n., *E.* 2, Introduction, n. 9.

surgamus: solet esse grauis cantantibus umbra,
iuniperi grauis umbra, nocent et frugibus umbrae (10.75–6)

—rises, and drives his full-fed goats homeward through the twilight:

ite domum saturae, uenit Hesperus, ite capellae. (10.77)

With this reference to the First *Eclogue*, to an earlier, darker time:

ite meae, felix quondam pecus, ite capellae (1.74)

the Tenth *Eclogue* ends, and the Book of *Eclogues* is complete.

The Composition of a Landscape

The landscapes of Virgil and Theocritus differ in several respects. Virgil's is heavily wooded; his Muse is sylvan, 'siluestrem . . . Musam' (1.2),[52] and sylvan his song, 'si canimus siluas' (4.3). For Virgil, the countryside, all that lies outside the town, is divided into two parts: the woods—first and foremost the woods, 'nobis placeant ante omnia siluae' (2.62)—and the rest of the countryside, 'siluas et cetera rura' (5.58), the small farms below the woods, like the one Tityrus was allowed to keep or the one Meliboeus lost. And in the woods, or rather, in hillside pastures surrounded by the woods, Virgil's herdsmen pass the day, tending their animals and singing to the echoes in the woods.

The woods are a constant presence in Virgil's landscape and intimately related to pastoral life, especially to pastoral song. Tityrus' song teaches the woods to resound Amaryllis' name (1.5); Corydon sings to the woods and hills of his hopeless passion for Alexis (2.5); the woods are in leaf as Damoetas and Menalcas begin their singing-match (3.57); looking for a place in which to sing, Menalcas and Mopsus consider a shady grove of hazels and elms but choose instead a cave overgrown with wild vine (5.1–7); the distraught lover, in Damon's song, bids a last farewell to the woods;[53] and, above all, the woods hear and answer, completely answer, man's song, 'non canimus surdis, respondent omnia siluae' (10.8).

The landscape of Theocritus is sparsely wooded;[54] for the most

[52] Cf. 6.2 'nostra neque erubuit siluas habitare Thalea'.

[53] 8.58 'uiuite siluae'. But the dying Daphnis, in Thyrsis' song, bids farewell to the woods and the thickets, *Idyll* 1.116–17 *οὐκέτ' ἀν' ὕλαν, | οὐκέτ' ἀνὰ δρυμώς, οὐκ ἄλσεα*, 'no more to your woods, to your thickets and groves, no more'.

[54] Cf. A. Cartault, *Étude sur les Bucoliques de Virgile* (Paris, 1897), 454: 'La forêt et les bois, *ὕλα, ἄλσος*, n'occupent dans Théocrite qu'une place restreinte'.

part it is a landscape of the foothills, of the maquis, an odorous tangle of myrtle (*μύρτος*), mastic (*σχῖνος*), bay (*δάφνη*), arbutus (*κόμαρος*), cistus or rock-rose (*κισθός*), and the like;[55] rough country, and very little of it under cultivation.[56] Thus, in *Idyll* 4, there are no trees; the hill on which Aegon's cattle are grazing is overspread with thorns and brambles, and the barefoot Battus gets a thorn under his ankle. In *Idyll* 5, while the oaks and a pine casting its cones modestly commend a *locus amoenus*, the chief delights of the place are (in Comatas' telling) bees humming about the hives, two springs of cold water, birds twittering in the tree, excellent shade, fern and flowering pennyroyal, and goatskins luxuriously soft (45–9, 55–7). In this landscape, in this poetry, trees are incidental, a few oaks, some poplars and elms shading a spring.

From the very beginning, from Homer onwards, poets have described or imagined landscapes. In Book 21 of the *Iliad*, Hephaestus directs his fire against the river Xanthus with devastating effect:

> *καίοντο πτελέαι τε καὶ ἰτέαι ἠδὲ μυρῖκαι,*
> *καίετο δὲ λωτός τε ἰδὲ θρύον ἠδὲ κύπειρον.* (350–1)
>
> Burning were elms and willows and tamarisks,
> Burning was trefoil and rush and galingale

—three trees and three plants arranged in parallel lines. The poet is not describing an observed landscape; he is imagining a landscape, making poetry. It is not necessary that these trees and plants should have been growing there, by the Xanthus in the Trojan plain; it is only necessary that the reader's (or hearer's) sense of decorum not be offended.[57]

While talking with Battus in the Fourth, the least artificial, seemingly,[58] of Theocritus' pastoral *Idylls*, Corydon mentions the Neaethus,[59] 'where all good things grow', *αἰγίπυρος καὶ κνύζα καὶ*

[55] See Lindsell (above, n. 9), 83.

[56] See Cartault (above, n. 54), 459–61.

[57] Homer's trees are properly located; see Theophrastus, *HP* 3. 3. 1.

[58] See *E.* 3, Introduction, n. 3.

[59] A river of southern Italy, now the Neto or Nieto, which rises in the Sila and flows into the sea a few miles north of Croton. Cf. Gow, vol. i, p. xx: 'It also seems reasonable to infer that even if the scenes of *Idd.* 4 and 5 are . . . not Magna Graecia but fairyland, their south-Italian colour implies that their author had some acquaintance with the district'.

εὐώδης μελίτεια,[60] 'restharrow and fleabane and sweet-smelling balm' (4.25)—unliterary plants, which Theocritus sets (were they indeed growing by the Neaethus?) in a line of traditional elegance, for he was thinking of Calypso's cave, *Od.* 5.64 κλήθρη τ' αἴγειρός τε καὶ εὐώδης κυπάρισσος, 'alder and poplar and sweet-smelling cypress'.[61]

Of some sixty-four names of plants and trees found in the pastoral *Idylls*, twenty-four or so are unliterary[62]—are not, that is, found in earlier poetry[63]—and of these, fifteen occur in *Idylls* 4 and 5.[64] Theocritus evidently felt that these humbler names were appropriate to his less elevated style.

Unlike Theocritus, Virgil did not have an ancient and ample poetic tradition to draw upon, and the proportion of unliterary to literary plants and trees in the *Eclogues* is therefore higher than it is in the pastoral *Idylls*.[65] Of some fifty-seven names found in the *Eclogues*,[66] twenty-eight or so, or about half, are unliterary. And of

[60] Fleabane, κόνυζα (κνύζα appears to be Theocritus' modification), is discussed by Theophrastus, *HP* 6.2.6. The form μελίτεια is unique; see Gow ad loc.

[61] Cf. αἰγίπυρος καὶ κνύζα καὶ εὐώδης μελίτεια and κλήθρη τ' αἴγειρος τε καὶ εὐώδης κυπάρισσος; so placed in the hexameter, εὐώδης is not found elsewhere before Theocritus. Cf. also Theocr. *Epigr.* 4.7 δάφναις καὶ μύρτοισι καὶ εὐώδει κυπαρίσσῳ, 'bays and myrtles and sweet-smelling cypress'. The only other line of this shape in the pastoral *Idylls* is 7.68 κνύζᾳ τ' ἀσφοδέλῳ τε πολυγνάμπτῳ τε σελίνῳ, 'fleabane and asphodel and curling celery'. No comparable line is to be found in the *Eclogues*, 4.51 'terrasque tractusque maris caelumque profundum' being a special case (see ad loc.). Somewhat similar is *G.* 2.83–4 'nec fortibus ulmis / nec salici lotoque neque Idaeis cyparissis'; although Theocritus never uses a purely ornamental epithet.

[62] Classification as literary or unliterary is necessarily inexact; all that can be hoped for is a reasonable approximation. No doubt a number of 'literary' names were known to Theocritus (and to Virgil) without reference to literature.

[63] Or, in a very few cases, are found in poetry that Theocritus is not likely to have read.

[64] αἴγιλος, a plant liked by goats, 5.128; αἰγίπυρος, restharrow, 4.25; *ἀνεμώνα, anemone, 5.92 (names with an asterisk occur in Theophrastus); *ἀσπάλαθος, a spinous shrub, 4.57; *ἀτρακτυλλίς, distaff-thistle, 4.52; *κισθός, rock-rose, 5.131; *κνύζα, fleabane, 4.25, 7.68; *κυνόσβατος, wild rose, 5.92; *κύτισος, moon-clover, 5.128, 10.30; *κυκλάμινος, cyclamen, 5.123; μελίτεια, balm, 4.25, 5.130; ὀρομαλίδες, wild apples, 5.94; *ῥάμνος, buckthorn, 4.57; σίον, a marsh plant, 5.125; *σχῖνος, mastic, 5.129.

[65] The pastoral *Idylls* (not counting *Idylls* 8 and 9) number 761 lines, the *Eclogues* 828.

[66] Excepted from this calculation are *amomum*, *baccaris*, *hibiscum*: curious plants.

the literary names, mostly names of trees, nearly half are first attested in Catullus; others in Ennius, Plautus, and Lucretius.

Although Virgil is indebted to Theocritus for the idea of a natural décor, an arranged beauty of trees and plants, he actually borrows rather little from him: acanthus,[67] cytisus,[68] possibly paliurus,[69] and tamarisks. Tamarisks, as observed above, are Homeric; but Virgil's tamarisks are associated with pastoral song, 4.2 'humiles . . . myricae', 6.10 'nostrae . . . myricae'—a symbolism apparently suggested by a casual collocation in Theocritus:

> λῇς ποτὶ τᾶν Νυμφᾶν, λῇς, αἰπόλε, τεῖδε καθίξας,
> ὡς τὸ κάταντες τοῦτο γεώλοφον αἵ τε μυρῖκαι,
> συρίσδεν; (1.12–14)

Will you in the Nymphs' name, will you, goatherd, sit down here, where this sloping hill and the tamarisks are, and pipe?

Theocritus' herdsmen seem to be better botanists than Virgil's; certainly they discourse more readily of plants and shrubs, as in the following exchange between Comatas and Lacon:

> *ΚΟ.* ταὶ μὲν ἐμαὶ κύτισόν τε καὶ αἴγιλον αἶγες ἔδοντι,
> καὶ σχῖνον πατέοντι καὶ ἐν κομάροισι κέονται.
> *ΛΑ.* ταῖσι δ' ἐμαῖς ὀίεσσι πάρεστι μὲν ἁ μελίτεια
> φέρβεσθαι, πολλὸς δὲ καὶ ὡς ῥόδα κισθὸς ἐπανθεῖ. (5.128–31)

Co. Moon-clover and goatwort have my goats for pasture; on mastic they walk, and couch on arbutus.
La. Balm have my sheep to browse, and, like roses, in plenty flowers the cistus.[70]

There is only one passage quite like this in the *Eclogues*, in the Fifth—not, however, an exchange between two singers, though it might be conceived as such,[71] but a passage in the song of Mopsus:

[67] See 3.45n.

[68] See 1.78n.

[69] Paliurus, or Christ's Thorn, a Theocritean plant though not occurring in the pastoral *Idylls*, 24.89–90 ἢ παλιούρου | ἢ βάτου, 'or of paliurus or of bramble', a poem Virgil had read; see *E.* 8.101–2n. Cf., however, Leonidas of Tarentum 18.3–4 G.–P. (= *AP* 7.656) παλιούρου | καὶ βάτου, Theophrastus, *HP* 1.3.1 βάτος παλίουρος, 4.12.4 καὶ ὁ βάτος καὶ ὁ παλίουρος, 6.1.3 παλίουρος βάτος.

[70] Gow's translation.

[71] Two lines (36–7) answered by two lines (38–9), the second 'couplet' being superior to the first.

grandia saepe quibus mandauimus hordea sulcis,
infelix lolium et steriles nascuntur auenae;
pro molli uiola, pro purpureo narcisso
carduus et spinis surgit paliurus acutis. (5.36–9)

The paliurus stands out in this context; an unfamiliar exotic, not a plant that a Roman reader could be expected to know and therefore provided with a descriptive phrase, 'spinis . . . acutis'.[72]

Larger features, too, of the pastoral landscape, mountains and rivers, the seashore, may, like the trees and plants growing by the Neaethus, be imagined. Sometimes Corydon—again in the Fourth *Idyll*—pastures Aegon's calf by the Aesarus, sometimes she frisks in the deep shade of Latymnum. The poem requires that Latymnum, which is not mentioned elsewhere,[73] be a mountain near the Aesarus, a well-attested river of Croton; and the reader imagines the mountain there, near the river.

Virgil is more daring than Theocritus, more personal. Not even Mantua is exempt from his fiction; in the Ninth *Eclogue*, it is imagined lying by the sea.[74] In the first *Eclogue*, reference to the land-confiscations places the scene in the Po valley near Mantua, and yet the landscape, Virgil's native landscape, inhabited now by Tityrus and Meliboeus, is not so much remembered as imagined. Hyblaean bees feed on the willow hedgerow; moon-clover flowers there, where it never grew,[75] for the goats to crop; and evening shadows fall from mountains required by the cadence of the poem.

[72] Rhetorically apt as weighting the second noun in the line and contrasting with 'molli uiola'.

[73] Cf. Gow on l. 19: 'The mountain is not mentioned elsewhere, and if T.'s geographical names are mere ornament (see Introd. p. xx), may belong to quite a different part of the world'.

[74] Some of Virgil's ancient readers were embarrassed by the too-near sea and tried to explain it away; see 9.57–8n.

[75] Cf. Lindsell (above, n.9), 89: 'cytisus, though now fairly common in Sicily, was not introduced till long after Theocritus' day, and it has never grown in the Cisalpine province'.

LATERCVLVS NOTARVM

sub quibus testes in apparatu adlegantur

ps.-Acro	Acronis quae dicuntur scholia in Horatium, ed. O. Keller (Lipsiae, 1902–4)
Char.	Fl. Sosipater Charisius (saec. iv), *Ars grammatica* (*GLK* i); ed. K. Barwick (Lipsiae, 1925)
Cons.	Consentius (saec. v), *Ars* (GLK v)
Diom.	Diomedes (saec. iv), *Ars* (*GLK* i)
Don. ad Ter.	Donatus, *Commentum Terenti*, ed. P. Wessner (Lipsiae, 1902–5)
DSeru.	Seruius qui dicitur Danielis
GLK	*Grammatici Latini*, ex recensione H. Keilii (Lipsiae, 1857–80)
Gramm.	e grammaticis tres pluresue; uide in *GLK* vii indicem scriptorum
Isid.	Isidorus Hispalensis (fl. 600), *Etymologiae*, ed. W. M. Lindsay (Oxonii, 1912)
Lact.	L. Caecilius Firmianus Lactantius (saec. iii/iv)
Macrob.	Macrobius Ambrosius Theodosius (saec. v in.), *Saturnalia*, ed. J. Willis (Lipsiae, 1963)
Non.	Nonius Marcellus (saec. iv), *De compendiosa doctrina*, ed. W. M. Lindsay (Lipsiae, 1903)
Prisc.	Priscianus Caesariensis (saec. v/vi), *Institutio grammatica*, ed. M. J. Hertz (*GLK* ii–iii)
Quint.	M. Fabius Quintilianus (saec. i), *Institutio oratoria*, ed. M. Winterbottom (Oxonii, 1970)
Rufin.	Iulius Rufinianus (saec. iv), *De figuris sententiarum*, ed. C. Halm, *Rhetores Latini minores* (Lipsiae, 1863)
Seru.	Seruius (saec. v in.), in Verg. Commentarii, ed. G. Thilo (Lipsiae, 1881–7)

SIGLA CODICVM

M	Florentinus Laur. xxxix. 1	saec. v
P	Vaticanus Palatinus lat. 1631	saec. iv/v
R	Vaticanus lat. 3867	saec. v
V	fragmenta Veronensia	saec. v
M²P²	corrector aliquis antiquus	

Codices saeculi noni:

a	Bernensis 172 cum Parisino lat. 7929
b	Bernensis 165
c	Bernensis 184
d	Bernensis 255+239
e	Bernensis 167
f	Oxoniensis Bodl. Auct. F. 2. 8
h	Valentianensis 407
r	Parisinus lat. 7926
s	Parisinus lat. 7928
t	Parisinus lat. 13043
v	Vaticanus lat. 1570
ω	consensus horum uel omnium uel quotquot non separatim nominantur
γ	Guelferbytanus Gudianus lat. 2°. 70
def.	deficit (uel mutilus est uel legi non potest)
recc.	codices saec. nono recentiores

BVCOLICA

ECLOGA I

MELIBOEVS TITYRVS

M. TITYRE, tu patulae recubans sub tegmine fagi
siluestrem tenui Musam meditaris auena;
nos patriae finis et dulcia linquimus arua.
nos patriam fugimus; tu, Tityre, lentus in umbra
formosam resonare doces Amaryllida siluas.
T. O Meliboee, deus nobis haec otia fecit.
namque erit ille mihi semper deus, illius aram
saepe tener nostris ab ouilibus imbuet agnus.
ille meas errare boues, ut cernis, et ipsum
ludere quae uellem calamo permisit agresti.
M. Non equidem inuideo, miror magis: undique totis
usque adeo turbatur agris. en ipse capellas
protinus aeger ago; hanc etiam uix, Tityre, duco.
hic inter densas corylos modo namque gemellos,
spem gregis, a! silice in nuda conixa reliquit.
saepe malum hoc nobis, si mens non laeua fuisset,
de caelo tactas memini praedicere quercus.
sed tamen iste deus qui sit, da, Tityre, nobis.
T. Vrbem quam dicunt Romam, Meliboee, putaui
stultus ego huic nostrae similem, quo saepe solemus
pastores ouium teneros depellere fetus.
sic canibus catulos similis, sic matribus haedos
noram, sic paruis componere magna solebam.
uerum haec tantum alias inter caput extulit urbes
quantum lenta solent inter uiburna cupressi.
M. Et quae tanta fuit Romam tibi causa uidendi?
T. Libertas, quae sera tamen respexit inertem,
candidior postquam tondenti barba cadebat,
respexit tamen et longo post tempore uenit,

i 1–29 *PR*   2 siluestrem] agrestem (*E.* vi 8) *Quint.* ix 4. 85   12 turbatur *Quint.* i 4. 28, *Cons.* 372.35, *'uera lectio' iudice Seru.*: turbamur *PRω, agnoscit Seru.*

postquam nos Amaryllis habet, Galatea reliquit.
namque (fatebor enim) dum me Galatea tenebat,
nec spes libertatis erat nec cura peculi.
quamuis multa meis exiret uictima saeptis,
pinguis et ingratae premeretur caseus urbi,
non umquam grauis aere domum mihi dextra redibat.
M. Mirabar quid maesta deos, Amarylli, uocares,
cui pendere sua patereris in arbore poma;
Tityrus hinc aberat. ipsae te, Tityre, pinus,
ipsi te fontes, ipsa haec arbusta uocabant.
T. Quid facerem? neque seruitio me exire licebat
nec tam praesentis alibi cognoscere diuos.
hic illum uidi iuuenem, Meliboee, quotannis
bis senos cui nostra dies altaria fumant.
hic mihi responsum primus dedit ille petenti:
'pascite ut ante boues, pueri; summittite tauros.'
M. Fortunate senex, ergo tua rura manebunt
et tibi magna satis, quamuis lapis omnia nudus
limosoque palus obducat pascua iunco:
non insueta grauis temptabunt pabula fetas,
nec mala uicini pecoris contagia laedent.
fortunate senex, hic inter flumina nota
et fontis sacros frigus captabis opacum;
hinc tibi, quae semper, uicino ab limite saepes
Hyblaeis apibus florem depasta salicti
saepe leui somnum suadebit inire susurro;
hinc alta sub rupe canet frondator ad auras,
nec tamen interea raucae, tua cura, palumbes
nec gemere aëria cessabit turtur ab ulmo.
T. Ante leues ergo pascentur in aethere cerui
et freta destituent nudos in litore piscis,
ante pererratis amborum finibus exsul
aut Ararim Parthus bibet aut Germania Tigrim,
quam nostro illius labatur pectore uultus.
M. At nos hinc alii sitientis ibimus Afros,

30–64 *PR* 37 poma] mala R^1 38 pinus] nobis (*E.* i 18) P^1
59 pascuntur *P* aethere] aequore *Ribbeck e recc.* 63 labantur P^2

pars Scythiam et rapidum cretae ueniemus Oaxen
et penitus toto diuisos orbe Britannos.
en umquam patrios longo post tempore finis
pauperis et tuguri congestum caespite culmen,
post aliquot, mea regna, uidens mirabor aristas?
impius haec tam culta noualia miles habebit,
barbarus has segetes. en quo discordia ciuis
produxit miseros: his nos conseuimus agros!
insere nunc, Meliboee, piros, pone ordine uitis.
ite meae, felix quondam pecus, ite capellae.
non ego uos posthac uiridi proiectus in antro
dumosa pendere procul de rupe uidebo;
carmina nulla canam; non me pascente, capellae,
florentem cytisum et salices carpetis amaras.
T. Hic tamen hanc mecum poteras requiescere noctem
fronde super uiridi: sunt nobis mitia poma,
castaneae molles et pressi copia lactis,
et iam summa procul uillarum culmina fumant
maioresque cadunt altis de montibus umbrae.

ECLOGA II

FORMOSVM pastor Corydon ardebat Alexin,
delicias domini, nec quid speraret habebat.
tantum inter densas, umbrosa cacumina, fagos
adsidue ueniebat. ibi haec incondita solus
montibus et siluis studio iactabat inani:
'O crudelis Alexi, nihil mea carmina curas?
nil nostri miserere? mori me denique cogis?
nunc etiam pecudes umbras et frigora captant,
nunc uiridis etiam occultant spineta lacertos,

i 65–ii 9 *PR* 65 Cretae *nomen proprium agnoscit Seru. ad E.* ii 24 (*ad loc. deest*) 72 perduxit *bdv* his nos *PRb?γ?*: en quis *ω* (en quos *d*) consueuimus agris *R* 74 felix quondam *Rω*: quondam felix *Pf* 78 calices P^{1} 79 hanc . . . noctem P^{1}*Rbdr*: hac . . . nocte P^{2}*ω* 83 de] a P^{1}
ii 1 Alexim *P* 7 coges *Rf* 9 lacertas P^{1}

Thestylis et rapido fessis messoribus aestu
alia serpyllumque herbas contundit olentis.
at mecum raucis, tua dum uestigia lustro,
sole sub ardenti resonant arbusta cicadis.
nonne fuit satius tristis Amaryllidis iras
atque superba pati fastidia? nonne Menalcan,
quamuis ille niger, quamuis tu candidus esses?
o formose puer, nimium ne crede colori:
alba ligustra cadunt, uaccinia nigra leguntur.
despectus tibi sum, nec qui sim quaeris, Alexi,
quam diues pecoris, niuei quam lactis abundans.
mille meae Siculis errant in montibus agnae;
lac mihi non aestate nouum, non frigore defit.
canto quae solitus, si quando armenta uocabat,
Amphion Dircaeus in Actaeo Aracyntho.
nec sum adeo informis: nuper me in litore uidi,
cum placidum uentis staret mare. non ego Daphnin
iudice te metuam, si numquam fallit imago.
o tantum libeat mecum tibi sordida rura
atque humilis habitare casas et figere ceruos,
haedorumque gregem uiridi compellere hibisco!
mecum una in siluis imitabere Pana canendo
(Pan primum calamos cera coniungere pluris
instituit, Pan curat ouis ouiumque magistros),
nec te paeniteat calamo triuisse labellum:
haec eadem ut sciret, quid non faciebat Amyntas?
est mihi disparibus septem compacta cicutis
fistula, Damoetas dono mihi quam dedit olim,
et dixit moriens: 'te nunc habet ista secundum';
dixit Damoetas, inuidit stultus Amyntas.
praeterea duo nec tuta mihi ualle reperti
capreoli, sparsis etiam nunc pellibus albo,
bina die siccant ouis ubera; quos tibi seruo.
iam pridem a me illos abducere Thestylis orat;

10–43 *PR* 12 at ω, *'melius' iudice Seru.*: ad *P* (ad me cum *agnoscit Seru.*): ac *R* 19 quis *cf* 20 pecoris niuei, *'melius est' iudice Seru.* 22 lact *P* desit *de* 27 fallit P^1c: fallat P^2*Rder, Seru.*: fallet *bfγ?* 32 primum *PR*: primus (*E.* viii 24) *ωγ, Seru. ad E.* iii 25, *Isid.* iii 21. 8 41 albo *Pber, Seru.*: ambo *Rω*

et faciet, quoniam sordent tibi munera nostra.
huc ades, o formose puer: tibi lilia plenis
ecce ferunt Nymphae calathis; tibi candida Nais,
pallentis uiolas et summa papauera carpens,
narcissum et florem iungit bene olentis anethi;
tum casia atque aliis intexens suauibus herbis
mollia luteola pingit uaccinia calta.
ipse ego cana legam tenera lanugine mala
castaneasque nuces, mea quas Amaryllis amabat;
addam cerea pruna (honos erit huic quoque pomo),
et uos, o lauri, carpam et te, proxima myrte,
sic positae quoniam suauis miscetis odores.
rusticus es, Corydon; nec munera curat Alexis,
nec, si muneribus certes, concedat Iollas.
heu heu, quid uolui misero mihi? floribus Austrum
perditus et liquidis immisi fontibus apros.
quem fugis, a! demens? habitarunt di quoque siluas
Dardaniusque Paris. Pallas quas condidit arces
ipsa colat; nobis placeant ante omnia siluae.
torua leaena lupum sequitur, lupus ipse capellam,
florentem cytisum sequitur lasciua capella,
te Corydon, o Alexi: trahit sua quemque uoluptas.
aspice, aratra iugo referunt suspensa iuuenci,
et sol crescentis decedens duplicat umbras;
me tamen urit amor: quis enim modus adsit amori?
a, Corydon, Corydon, quae te dementia cepit!
semiputata tibi frondosa uitis in ulmo est:
quin tu aliquid saltem potius, quorum indiget usus,
uiminibus mollique paras detexere iunco?
inuenies alium, si te hic fastidit, Alexin.'

ECLOGA III

MENALCAS DAMOETAS PALAEMON

M. DIC mihi, Damoeta, cuium pecus? an Meliboei?

ii 44–iii 1 *PR* 56 es $P^2\omega$: est P^1R, *nebulo aliquis Pompeianus* (*C.I.L.* iv 1527) 57 certet *R* 61 quae *R* 70 ulmo est $R\omega$: ulmo *P* 73 Alexin *Seru.* (*cf. E.* ii 1, *Mart.* viii 55. 12): Alexim P^1R: Alexis $P^2\omega$

D. Non, uerum Aegonis; nuper mihi tradidit Aegon.
M. Infelix o semper, oues, pecus! ipse Neaeram
dum fouet ac ne me sibi praeferat illa ueretur,
hic alienus ouis custos bis mulget in hora,
et sucus pecori et lac subducitur agnis.
D. Parcius ista uiris tamen obicienda memento.
nouimus et qui te transuersa tuentibus hircis
et quo (sed faciles Nymphae risere) sacello.
M. Tum, credo, cum me arbustum uidere Miconis
atque mala uitis incidere falce nouellas.
D. Aut hic ad ueteres fagos cum Daphnidis arcum
fregisti et calamos: quae tu, peruerse Menalca,
et cum uidisti puero donata, dolebas,
et si non aliqua nocuisses, mortuus esses.
M. Quid domini faciant, audent cum talia fures?
non ego te uidi Damonis, pessime, caprum
excipere insidiis multum latrante Lycisca?
et cum clamarem 'quo nunc se proripit ille?
Tityre, coge pecus', tu post carecta latebas.
D. An mihi cantando uictus non redderet ille,
quem mea carminibus meruisset fistula caprum?
si nescis, meus ille caper fuit; et mihi Damon
ipse fatebatur, sed reddere posse negabat.
M. Cantando tu illum? aut umquam tibi fistula cera
iuncta fuit? non tu in triuiis, indocte, solebas
stridenti miserum stipula disperdere carmen?
D. Vis ergo inter nos quid possit uterque uicissim
experiamur? ego hanc uitulam (ne forte recuses,
bis uenit ad mulctram, binos alit ubere fetus)
depono; tu dic mecum quo pignore certes.
M. De grege non ausim quicquam deponere tecum:
est mihi namque domi pater, est iniusta nouerca,
bisque die numerant ambo pecus, alter et haedos.
uerum, id quod multo tute ipse fatebere maius

2–26 *PR*; 27–35 *PRV*   iii 3 ouis *Pω*   6 lact (*ut E.* ii 22) P^{1} 8 hircis P^{2}*Rω*, *Quint.* ix 3. 59, *Macrob.* iv 6. 21: hircuis P^{1}: hirquis '*alii*' *ap. Seru.*, *quibus auctore Suetonio* 'hirqui *sunt oculorum anguli*'   25 aut] haud *cdv*, haut *e*   26 iuncta *P*, *Rufin.* 42. 3: uincta *Rωγ*, *Prisc.* viii 49 (*ut uid.*)   27 stipula miserum V^{1}

(insanire libet quoniam tibi), pocula ponam
fagina, caelatum diuini opus Alcimedontis,
lenta quibus torno facili superaddita uitis
diffusos hedera uestit pallente corymbos.
in medio duo signa, Conon et–quis fuit alter,
descripsit radio totum qui gentibus orbem,
tempora quae messor, quae curuus arator haberet?
necdum illis labra admoui, sed condita seruo.
D. Et nobis idem Alcimedon duo pocula fecit
et molli circum est ansas amplexus acantho,
Orpheaque in medio posuit siluasque sequentis;
necdum illis labra admoui, sed condita seruo.
si ad uitulam spectas, nihil est quod pocula laudes.
M. Numquam hodie effugies; ueniam quocumque uocaris.
audiat haec tantum–uel qui uenit ecce Palaemon.
efficiam posthac ne quemquam uoce lacessas.
D. Quin age, si quid habes; in me mora non erit ulla,
nec quemquam fugio: tantum, uicine Palaemon,
sensibus haec imis (res est non parua) reponas.
P. Dicite, quandoquidem in molli consedimus herba.
et nunc omnis ager, nunc omnis parturit arbos,
nunc frondent siluae, nunc formosissimus annus.
incipe, Damoeta; tu deinde sequere, Menalca.
alternis dicetis; amant alterna Camenae.
D. Ab Ioue principium Musae: Iouis omnia plena;
ille colit terras, illi mea carmina curae.
M. Et me Phoebus amat; Phoebo sua semper apud me
munera sunt, lauri et suaue rubens hyacinthus.
D. Malo me Galatea petit, lasciua puella,
et fugit ad salices et se cupit ante uideri.
M. At mihi sese offert ultro, meus ignis, Amyntas,
notior ut iam sit canibus non Delia nostris.
D. Parta meae Veneri sunt munera: namque notaui
ipse locum, aëriae quo congessere palumbes.
M. Quod potui, puero siluestri ex arbore lecta
aurea mala decem misi; cras altera mittam.

36–52 *PRV*; 53–71 *PR* 38 facili *agnoscit Seru. hic et ad A.* ii 392: facilis *Vωγ, Donatus ap. Seru.*: fragilis *R (quid P, non liquet)* 60 Musae *uocatiuo casu agnoscit Seru.*

D. O quotiens et quae nobis Galatea locuta est!
partem aliquam, uenti, diuum referatis ad auris!
M. Quid prodest quod me ipse animo non spernis, Amynta,
si, dum tu sectaris apros, ego retia seruo?
D. Phyllida mitte mihi: meus est natalis, Iolla;
cum faciam uitula pro frugibus, ipse uenito.
M. Phyllida amo ante alias; nam me discedere fleuit
et longum 'formose, uale, uale,' inquit, 'Iolla.'
D. Triste lupus stabulis, maturis frugibus imbres,
arboribus uenti, nobis Amaryllidis irae.
M. Dulce satis umor, depulsis arbutus haedis,
lenta salix feto pecori, mihi solus Amyntas.
D. Pollio amat nostram, quamuis est rustica, Musam:
Pierides, uitulam lectori pascite uestro.
M. Pollio et ipse facit noua carmina: pascite taurum,
iam cornu petat et pedibus qui spargat harenam.
D. Qui te, Pollio, amat, ueniat quo te quoque gaudet;
mella fluant illi, ferat et rubus asper amomum.
M. Qui Bauium non odit, amet tua carmina, Maeui,
atque idem iungat uulpes et mulgeat hircos.
D. Qui legitis flores et humi nascentia fraga,
frigidus, o pueri (fugite hinc!), latet anguis in herba.
M. Parcite, oues, nimium procedere: non bene ripae
creditur; ipse aries etiam nunc uellera siccat.
D. Tityre, pascentis a flumine reice capellas:
ipse, ubi tempus erit, omnis in fonte lauabo.
M. Cogite ouis, pueri: si lac praeceperit aestus,
ut nuper, frustra pressabimus ubera palmis.
D. Heu heu, quam pingui macer est mihi taurus in eruo!
idem amor exitium pecori pecorisque magistro.
M. His certe neque amor causa est; uix ossibus haerent;
nescio quis teneros oculus mihi fascinat agnos.
D. Dic quibus in terris (et eris mihi magnus Apollo)

72–104 γ*R*   77 uitula *Macrob.* iii 2. 15, *Seru.*: uitulam *codd., Non.* 313. 31, *Seru. ad E.* v 75   84 est] sit *bcd, Seru.*   91 mulceat *R*   98 aestas *R*   100 eruo *γcfh, Seru. ad A.* ii 69: aruo *Rbderv*   101 est *post* exitium *add. R, post* pecori *γbh, post* magistro *cr*   102 his certe (neque amor causa est) uix *legisse uidetur Don. ad Ter. Eun.* 269, *ut* his *nominatiuus sit* (hi *Stephanus*)

tris pateat caeli spatium non amplius ulnas.
M. Dic quibus in terris inscripti nomina regum
nascantur flores, et Phyllida solus habeto.
P. Non nostrum inter uos tantas componere lites:
et uitula tu dignus et hic, et quisquis amores
aut metuet dulcis aut experietur amaros.
claudite iam riuos, pueri; sat prata biberunt.

ECLOGA IV

SICELIDES Musae, paulo maiora canamus!
non omnis arbusta iuuant humilesque myricae;
si canimus siluas, siluae sint consule dignae.
  Vltima Cumaei uenit iam carminis aetas;
magnus ab integro saeclorum nascitur ordo.
iam redit et Virgo, redeunt Saturnia regna,
iam noua progenies caelo demittitur alto.
tu modo nascenti puero, quo ferrea primum
desinet ac toto surget gens aurea mundo,
casta faue Lucina: tuus iam regnat Apollo.
teque adeo decus hoc aeui, te consule, inibit,
Pollio, et incipient magni procedere menses;
te duce, si qua manent sceleris uestigia nostri,
inrita perpetua soluent formidine terras.
ille deum uitam accipiet diuisque uidebit
permixtos heroas et ipse uidebitur illis,
pacatumque reget patriis uirtutibus orbem.
  At tibi prima, puer, nullo munuscula cultu
errantis hederas passim cum baccare tellus
mixtaque ridenti colocasia fundet acantho.
ipsae lacte domum referent distenta capellae
ubera, nec magnos metuent armenta leones;
ipsa tibi blandos fundent cunabula flores.

iii 105–iv 23 *γR*   105 *et* caeli *et* Caeli (*nomen erit proprium*) *agnoscit Seru.*   107 nascuntur *bdf*   110 aut . . . aut *codd., Seru.*: haut . . . haut *F. W. Graser in Eph. litt. Halens.* 1835, 256   amores *r*   amaros/ aut metuet, dulces aut experietur amores *Peerlkamp*   iv 7 demittitur *γcer*: dimittitur *Rω*   18 ac tibi nulla pater primo *R*   20 fundit *γ*, *Macrob.* vi 6. 18   21 referant *γ*

occidet et serpens, et fallax herba ueneni
occidet; Assyrium uulgo nascetur amomum.
at simul heroum laudes et facta parentis
iam legere et quae sit poteris cognoscere uirtus,
molli paulatim flauescet campus arista
incultisque rubens pendebit sentibus uua
et durae quercus sudabunt roscida mella.
pauca tamen suberunt priscae uestigia fraudis,
quae temptare Thetim ratibus, quae cingere muris
oppida, quae iubeant telluri infindere sulcos.
alter erit tum Tiphys et altera quae uehat Argo
delectos heroas; erunt etiam altera bella
atque iterum ad Troiam magnus mittetur Achilles.
hinc, ubi iam firmata uirum te fecerit aetas,
cedet et ipse mari uector, nec nautica pinus
mutabit merces; omnis feret omnia tellus.
non rastros patietur humus, non uinea falcem;
robustus quoque iam tauris iuga soluet arator.
nec uarios discet mentiri lana colores,
ipse sed in pratis aries iam suaue rubenti
murice, iam croceo mutabit uellera luto;
sponte sua sandyx pascentis uestiet agnos.

'Talia saecla' suis dixerunt 'currite' fusis
concordes stabili fatorum numine Parcae.
adgredere o magnos (aderit iam tempus) honores,
cara deum suboles, magnum Iouis incrementum!
aspice conuexo nutantem pondere mundum,
terrasque tractusque maris caelumque profundum;
aspice, uenturo laetentur ut omnia saeclo!
o mihi tum longae maneat pars ultima uitae,
spiritus et quantum sat erit tua dicere facta!
non me carminibus uincet nec Thracius Orpheus
nec Linus, huic mater quamuis atque huic pater adsit,
Orphei Calliopea, Lino formosus Apollo.
Pan etiam, Arcadia mecum si iudice certet,

24–51 *γR*; 52–8 *PR*   26 ac *R*   parentis *γω*, *Seru.*, *Non.* 331. 34: parentum *R*   28 flauescet *cder*, *Lact. inst.* vii 24?: flauescit *γRbf*   33 tellurem inf. sulco *R*   52 laetantur *R*   53 longe *P*   55 uincat P^1

Pan etiam Arcadia dicat se iudice uictum.
Incipe, parue puer, risu cognoscere matrem
(matri longa decem tulerunt fastidia menses)
incipe, parue puer: qui non risere parenti,
nec deus hunc mensa, dea nec dignata cubili est.

ECLOGA V

MENALCAS MOPSVS

Me. CVR non, Mopse, boni quoniam conuenimus ambo,
tu calamos inflare leuis, ego dicere uersus,
hic corylis mixtas inter consedimus ulmos?
Mo. Tu maior; tibi me est aequum parere, Menalca,
siue sub incertas Zephyris motantibus umbras
siue antro potius succedimus. aspice, ut antrum
siluestris raris sparsit labrusca racemis.
Me. Montibus in nostris solus tibi certat Amyntas.
Mo. Quid, si idem certet Phoebum superare canendo?
Me. Incipe, Mopse, prior, si quos aut Phyllidis ignis
aut Alconis habes laudes aut iurgia Codri.
incipe: pascentis seruabit Tityrus haedos.
Mo. Immo haec, in uiridi nuper quae cortice fagi
carmina descripsi et modulans alterna notaui,
experiar: tu deinde iubeto ut certet Amyntas.
Me. Lenta salix quantum pallenti cedit oliuae,
puniceis humilis quantum saliunca rosetis,
iudicio nostro tantum tibi cedit Amyntas.
sed tu desine plura, puer: successimus antro.
Mo. Exstinctum Nymphae crudeli funere Daphnin
flebant (uos coryli testes et flumina Nymphis),
cum complexa sui corpus miserabile nati
atque deos atque astra uocat crudelia mater.
non ulli pastos illis egere diebus

iv 59–v 24 *PR* 59 Arcadi(a)e $P^{1}R$ dicet *Macrob.* v 14. 6 61 matris P^{2} tulerunt *PRω*: tulerant *de*: abstulerint *'alii' ap. Seru.*: tulerint b^{2} 62 qui *Quint.* ix 3. 8: cui *PRω*, *Seru.*, *Quintiliani codd.* (*corr. Politianus*) parenti *Schrader*: parentes *codd.* v 3 considimus *cde* 8 certet *P* 15 ut *Rb*: *om. Pω*

frigida, Daphni, boues ad flumina; nulla neque amnem
libauit quadripes nec graminis attigit herbam.
Daphni, tuum Poenos etiam ingemuisse leones
interitum montesque feri siluaeque loquuntur.
Daphnis et Armenias curru subiungere tigris
instituit, Daphnis thiasos inducere Bacchi
et foliis lentas intexere mollibus hastas.
uitis ut arboribus decori est, ut uitibus uuae,
ut gregibus tauri, segetes ut pinguibus aruis,
tu decus omne tuis. postquam te fata tulerunt,
ipsa Pales agros atque ipse reliquit Apollo.
grandia saepe quibus mandauimus hordea sulcis,
infelix lolium et steriles nascuntur auenae;
pro molli uiola, pro purpureo narcisso
carduus et spinis surgit paliurus acutis.
spargite humum foliis, inducite fontibus umbras,
pastores (mandat fieri sibi talia Daphnis),
et tumulum facite, et tumulo superaddite carmen:
'Daphnis ego in siluis, hinc usque ad sidera notus,
formosi pecoris custos, formosior ipse.'
Me. Tale tuum carmen nobis, diuine poeta,
quale sopor fessis in gramine, quale per aestum
dulcis aquae saliente sitim restinguere riuo.
nec calamis solum aequiperas, sed uoce magistrum:
fortunate puer, tu nunc eris alter ab illo.
nos tamen haec quocumque modo tibi nostra uicissim
dicemus, Daphninque tuum tollemus ad astra;
Daphnin ad astra feremus: amauit nos quoque Daphnis.
Mo. An quicquam nobis tali sit munere maius?
et puer ipse fuit cantari dignus, et ista
iam pridem Stimichon laudauit carmina nobis.
Me. Candidus insuetum miratur limen Olympi
sub pedibusque uidet nubes et sidera Daphnis.
ergo alacris siluas et cetera rura uoluptas

25–58 *PR*   27 gemuisse *R*   28 ferunt *P*   siluaesque P^{1} feros siluasque *Markland ad Stat. silu.* ii. 5. 13   38 uiola $P^{2}\omega$(uiolae P^{1}): uiola et *R*   purpurea *Diom.* 453. 36   40 umbras] aras *R*   46 fessis] lassis *R*   49 ab illo] Apollo *R*   52 Daphnim *Pcf*

Panaque pastoresque tenet Dryadasque puellas.
nec lupus insidias pecori, nec retia ceruis
ulla dolum meditantur: amat bonus otia Daphnis.
ipsi laetitia uoces ad sidera iactant
intonsi montes; ipsae iam carmina rupes,
ipsa sonant arbusta: 'deus, deus ille, Menalca!'
sis bonus o felixque tuis! en quattuor aras:
ecce duas tibi, Daphni, duas altaria Phoebo.
pocula bina nouo spumantia lacte quotannis
craterasque duo statuam tibi pinguis oliui,
et multo in primis hilarans conuiuia Baccho
(ante focum, si frigus erit; si messis, in umbra)
uina nouum fundam calathis Ariusia nectar.
cantabunt mihi Damoetas et Lyctius Aegon;
saltantis Satyros imitabitur Alphesiboeus.
haec tibi semper erunt, et cum sollemnia uota
reddemus Nymphis, et cum lustrabimus agros.
dum iuga montis aper, fluuios dum piscis amabit,
dumque thymo pascentur apes, dum rore cicadae,
semper honos nomenque tuum laudesque manebunt.
ut Baccho Cererique, tibi sic uota quotannis
agricolae facient: damnabis tu quoque uotis.
Mo. Quae tibi, quae tali reddam pro carmine dona?
nam neque me tantum uenientis sibilus Austri
nec percussa iuuant fluctu tam litora, nec quae
saxosas inter decurrunt flumina uallis.
Me. Hac te nos fragili donabimus ante cicuta;
haec nos 'formosum Corydon ardebat Alexin',
haec eadem docuit 'cuium pecus? an Meliboei?'
Mo. At tu sume pedum, quod, me cum saepe rogaret,
non tulit Antigenes (et erat tum dignus amari),
formosum paribus nodis atque aere, Menalca.

ECLOGA VI

PRIMA Syracosio dignata est ludere uersu

v 59–85 *PR*; v 86–vi 1 *PRV* 68 duos *cfr* 80 uoti *R*
86 Alexim *RVf* 89 tum *RV*ω: nunc P^1: tunc P^2b

nostra neque erubuit siluas habitare Thalea.
cum canerem reges et proelia, Cynthius aurem
uellit et admonuit: 'pastorem, Tityre, pinguis
pascere oportet ouis, deductum dicere carmen.'
nunc ego (namque super tibi erunt qui dicere laudes,
Vare, tuas cupiant et tristia condere bella)
agrestem tenui meditabor harundine Musam:
non iniussa cano. si quis tamen haec quoque, si quis
captus amore leget, te nostrae, Vare, myricae,
te nemus omne canet; nec Phoebo gratior ulla est
quam sibi quae Vari praescripsit pagina nomen.
  Pergite, Pierides. Chromis et Mnasyllos in antro
Silenum pueri somno uidere iacentem,
inflatum hesterno uenas, ut semper, Iaccho;
serta procul tantum capiti delapsa iacebant
et grauis attrita pendebat cantharus ansa.
adgressi (nam saepe senex spe carminis ambo
luserat) iniciunt ipsis ex uincula sertis.
addit se sociam timidisque superuenit Aegle,
Aegle Naiadum pulcherrima, iamque uidenti
sanguineis frontem moris et tempora pingit.
ille dolum ridens 'quo uincula nectitis?' inquit;
'soluite me, pueri; satis est potuisse uideri.
carmina quae uultis cognoscite; carmina uobis,
huic aliud mercedis erit.' simul incipit ipse.
tum uero in numerum Faunosque ferasque uideres
ludere, tum rigidas motare cacumina quercus;
nec tantum Phoebo gaudet Parnasia rupes,
nec tantum Rhodope miratur et Ismarus Orphea.
  Namque canebat uti magnum per inane coacta
semina terrarumque animaeque marisque fuissent
et liquidi simul ignis; ut his ex omnia primis,
omnia et ipse tener mundi concreuerit orbis;

2–20 *PRV*; 21–34 *PR*   vi 2 neque *Pω*: nec *RVf*   siluis *R*   Thalia *ω*, *Seru.*   5 diductum *P*   10 legat *d*, *Prisc.* xviii 87   12 perscripsit *fγ*   23 inridens *P*[2]   30 miratur *Pω* (*cf. A.* ii 317): mirantur *Rde*, *Rufin.* 48. 5   33 ex omnia (*cf. Lucr.* i 61) *P*: exordia *Rω*, Macrob. vi 2. 22   34 omnisa *P*[1]

tum durare solum et discludere Nerea ponto
coeperit et rerum paulatim sumere formas;
iamque nouum terrae stupeant lucescere solem,
altius atque cadant summotis nubibus imbres,
incipiant siluae cum primum surgere cumque
rara per ignaros errent animalia montis.
hinc lapides Pyrrhae iactos, Saturnia regna,
Caucasiasque refert uolucris furtumque Promethei.
his adiungit, Hylan nautae quo fonte relictum
clamassent, ut litus 'Hyla, Hyla' omne sonaret;
et fortunatam, si numquam armenta fuissent,
Pasiphaen niuei solatur amore iuuenci.
a, uirgo infelix, quae te dementia cepit!
Proetides implerunt falsis mugitibus agros,
at non tam turpis pecudum tamen ulla secuta
concubitus, quamuis collo timuisset aratrum
et saepe in leui quaesisset cornua fronte.
a! uirgo infelix, tu nunc in montibus erras:
ille latus niueum molli fultus hyacintho
ilice sub nigra pallentis ruminat herbas
aut aliquam in magno sequitur grege. 'claudite, Nymphae,
Dictaeae Nymphae, nemorum iam claudite saltus,
si qua forte ferant oculis sese obuia nostris
errabunda bouis uestigia; forsitan illum
aut herba captum uiridi aut armenta secutum
perducant aliquae stabula ad Gortynia uaccae.'
tum canit Hesperidum miratam mala puellam;
tum Phaethontiadas musco circumdat amarae
corticis atque solo proceras erigit alnos.
tum canit, errantem Permessi ad flumina Gallum
Aonas in montis ut duxerit una sororum,
utque uiro Phoebi chorus adsurrexerit omnis;
ut Linus haec illi diuino carmine pastor
floribus atque apio crinis ornatus amaro
dixerit: 'hos tibi dant calamos (en accipe) Musae,

35–47 *PR*; 48–69 *MPR* 38 utque *R* 40 ignaros *R*: ignotos *Pω* 41 hic *P* 49 secuta *MP*: secuta est *Rω*, *Macrob*. iv 6. 3 50 timuissent *R* 51 quesissent *P*

Ascraeo quos ante seni, quibus ille solebat
cantando rigidas deducere montibus ornos.
his tibi Grynei nemoris dicatur origo,
ne quis sit lucus quo se plus iactet Apollo.'
  Quid loquar aut Scyllam Nisi, quam fama secuta est
candida succinctam latrantibus inguina monstris
Dulichias uexasse rates et gurgite in alto
a! timidos nautas canibus lacerasse marinis;
aut ut mutatos Terei narrauerit artus,
quas illi Philomela dapes, quae dona pararit,
quo cursu deserta petiuerit et quibus ante
infelix sua tecta super uolitauerit alis?
omnia, quae Phoebo quondam meditante beatus
audiit Eurotas iussitque ediscere lauros,
ille canit, pulsae referunt ad sidera ualles;
cogere donec ouis stabulis numerumque referre
iussit et inuito processit Vesper Olympo.

ECLOGA VII

MELIBOEVS CORYDON THYRSIS

M. FORTE sub arguta consederat ilice Daphnis,
compulerantque greges Corydon et Thyrsis in unum,
Thyrsis ouis, Corydon distentas lacte capellas,
ambo florentes aetatibus, Arcades ambo,
et cantare pares et respondere parati.
huc mihi, dum teneras defendo a frigore myrtos,
uir gregis ipse caper deerrauerat; atque ego Daphnin
aspicio. ille ubi me contra uidet, 'ocius' inquit
'huc ades, o Meliboee; caper tibi saluus et haedi;
et, si quid cessare potes, requiesce sub umbra.
huc ipsi potum uenient per prata iuuenci,
hic uiridis tenera praetexit harundine ripas

vi 70–86 *MPR*; vii 1–11 *MPa*; 12 *MPaV*   73 nec *Rd*   74 ut *R*   79 pararet *Pdf*   81 supra *R*   85 referri M^2P^2c vii 6 huc *Mh*: hic *Paω*   11 ueniunt *a*

Mincius, eque sacra resonant examina quercu.'
quid facerem? neque ego Alcippen nec Phyllida habebam
depulsos a lacte domi quae clauderet agnos,
et certamen erat, Corydon cum Thyrside, magnum;
posthabui tamen illorum mea seria ludo.
alternis igitur contendere uersibus ambo
coepere, alternos Musae meminisse uolebant.
hos Corydon, illos referebat in ordine Thyrsis.
C. Nymphae noster amor Libethrides, aut mihi carmen,
quale meo Codro, concedite (proxima Phoebi
uersibus ille facit) aut, si non possumus omnes,
hic arguta sacra pendebit fistula pinu.
T. Pastores, hedera crescentem ornate poetam,
Arcades, inuidia rumpantur ut ilia Codro;
aut, si ultra placitum laudarit, baccare frontem
cingite, ne uati noceat mala lingua futuro.
C. Saetosi caput hoc apri tibi, Delia, paruus
et ramosa Micon uiuacis cornua cerui.
si proprium hoc fuerit, leui de marmore tota
puniceo stabis suras euincta coturno.
T. Sinum lactis et haec te liba, Priape, quotannis
exspectare sat est: custos es pauperis horti.
nunc te marmoreum pro tempore fecimus; at tu,
si fetura gregem suppleuerit, aureus esto.
C. Nerine Galatea, thymo mihi dulcior Hyblae,
candidior cycnis, hedera formosior alba,
cum primum pasti repetent praesepia tauri,
si qua tui Corydonis habet te cura, uenito.
T. Immo ego Sardoniis uidear tibi amarior herbis,
horridior rusco, proiecta uilior alga,
si mihi non haec lux toto iam longior anno est.
ite domum pasti, si quis pudor, ite iuuenci.
C. Muscosi fontes et somno mollior herba,
et quae uos rara uiridis tegit arbutus umbra,
solstitium pecori defendite: iam uenit aestas

13–37 *MPaV*; 38–47 *MPa* 19 uolebam '*multi*' *ap. Seru.* 22 Phoebo *V* 23 possimus *M[1]P[1]Vcr* 24 pendebis *DSeru.* 25 crescentem *M[2]Paω, Seru. ad E.* iv 19: nascentem *M[1]Vb, Seru. ad loc.* 29 capri '*multi*' *ap. DSeru.*

torrida, iam lento turgent in palmite gemmae.
T. Hic focus et taedae pingues, hic plurimus ignis
semper, et adsidua postes fuligine nigri.
hic tantum Boreae curamus frigora quantum
aut numerum lupus aut torrentia flumina ripas.
C. Stant et iuniperi et castaneae hirsutae,
strata iacent passim sua quaeque sub arbore poma,
omnia nunc rident: at si formosus Alexis
montibus his abeat, uideas et flumina sicca.
T. Aret ager, uitio moriens sitit aëris herba,
Liber pampineas inuidit collibus umbras:
Phyllidis aduentu nostrae nemus omne uirebit,
Iuppiter et laeto descendet plurimus imbri.
C. Populus Alcidae gratissima, uitis Iaccho,
formosae myrtus Veneri, sua laurea Phoebo;
Phyllis amat corylos: illas dum Phyllis amabit,
nec myrtus uincet corylos, nec laurea Phoebi.
T. Fraxinus in siluis pulcherrima, pinus in hortis,
populus in fluuiis, abies in montibus altis:
saepius at si me, Lycida formose, reuisas,
fraxinus in siluis cedat tibi, pinus in hortis.
M. Haec memini, et uictum frustra contendere Thyrsin.
ex illo Corydon Corydon est tempore nobis.

ECLOGA VIII

PASTORVM Musam Damonis et Alphesiboei,
immemor herbarum quos est mirata iuuenca
certantis, quorum stupefactae carmine lynces,
et mutata suos requierunt flumina cursus,
Damonis Musam dicemus et Alphesiboei.
tu mihi, seu magni superas iam saxa Timaui
siue oram Illyrici legis aequoris,–en erit umquam

vii 48–viii 7 *MPa* 48 lento *M^{2}P, interpretari uidetur Seru.*: laeto *M^{1}aω* 51 hinc *'nonnulli' ap. DSeru.* 54 quaque *b^{2}c^2, Bentley ad Manil.* ii 253 56 aberit *P* 64 corylos] Veneris *Hebri exemplar ap. DSeru.* 68 cedet *P, interpretari uidetur Seru.* 69 concedere *a* viii 4 linquerunt *γ* (liqu-*γ^2*) 6 tu] tum *P^1*

ille dies, mihi cum liceat tua dicere facta?
en erit ut liceat totum mihi ferre per orbem
sola Sophocleo tua carmina digna coturno?
a te principium, tibi desinam: accipe iussis
carmina coepta tuis, atque hanc sine tempora circum
inter uictricis hederam tibi serpere lauros.
  Frigida uix caelo noctis decesserat umbra,
cum ros in tenera pecori gratissimus herba:
incumbens tereti Damon sic coepit oliuae.
D. Nascere praeque diem ueniens age, Lucifer, almum,
coniugis indigno Nysae deceptus amore
dum queror et diuos, quamquam nil testibus illis
profeci, extrema moriens tamen adloquor hora.
  incipe Maenalios mecum, mea tibia, uersus.
Maenalus argutumque nemus pinusque loquentis
semper habet, semper pastorum ille audit amores
Panaque, qui primus calamos non passus inertis.
  incipe Maenalios mecum, mea tibia, uersus.
Mopso Nysa datur: quid non speremus amantes?
iungentur iam grypes equis, aeuoque sequenti
cum canibus timidi uenient ad pocula dammae.
  incipe Maenalios mecum, mea tibia, uersus.
Mopse, nouas incide faces: tibi ducitur uxor.
sparge, marite, nuces: tibi deserit Hesperus Oetam.
  incipe Maenalios mecum, mea tibia, uersus.
o digno coniuncta uiro, dum despicis omnis,
dumque tibi est odio mea fistula dumque capellae
hirsutumque supercilium promissaque barba,
nec curare deum credis mortalia quemquam.
  incipe Maenalios mecum, mea tibia, uersus.
saepibus in nostris paruam te roscida mala
(dux ego uester eram) uidi cum matre legentem.
alter ab undecimo tum me iam acceperat annus,

8–18 *MPa*; 19–39 *MPaV*   11 desinam *P*: desinet *Maω*, *DSeru.*: desinit *br*   20 adloquar M^1P^2d 22 pinosque P^2V 24 primum Mb (*ut uidetur, cf. E.* ii 32; *def. V*)   28 timidi $P^2V\omega$, *Quint.* ix 3. 6, *Char.* 269. 2, *Prisc.* v 7, *Seru. hic et ad G.* i 183: timidae *M*, *Seru. ad A.* v 122: timide P^1ac
28a *uersum intercalarem hic habet γ, ita ut u.* 76 *respondeat; om. ceteri*
34 demissaque *P*

iam fragilis poteram a terra contingere ramos:
ut uidi, ut perii, ut me malus abstulit error!
  incipe Maenalios mecum, mea tibia, uersus.
nunc scio quid sit Amor: nudis in cautibus illum
aut Tmaros aut Rhodope aut extremi Garamantes
nec generis nostri puerum nec sanguinis edunt.
  incipe Maenalios mecum, mea tibia, uersus.
saeuus Amor docuit natorum sanguine matrem
commaculare manus; crudelis tu quoque, mater.
crudelis mater magis, an puer improbus ille?
improbus ille puer; crudelis tu quoque, mater.
  incipe Maenalios mecum, mea tibia, uersus.
nunc et ouis ultro fugiat lupus, aurea durae
mala ferant quercus, narcisso floreat alnus,
pinguia corticibus sudent electra myricae,
certent et cycnis ululae, sit Tityrus Orpheus,
Orpheus in siluis, inter delphinas Arion.
  incipe Maenalios mecum, mea tibia, uersus.
omnia uel medium fiat mare. uiuite siluae:
praeceps aërii specula de montis in undas
deferar; extremum hoc munus morientis habeto.
  desine Maenalios, iam desine, tibia, uersus.
  Haec Damon; uos, quae responderit Alphesiboeus,
dicite, Pierides: non omnia possumus omnes.
A. Effer aquam et molli cinge haec altaria uitta
uerbenasque adole pinguis et mascula tura,
coniugis ut magicis sanos auertere sacris
experiar sensus; nihil hic nisi carmina desunt.
  ducite ab urbe domum, mea carmina, ducite Daphnin.
carmina uel caelo possunt deducere lunam,
carminibus Circe socios mutauit Vlixi,
frigidus in pratis cantando rumpitur anguis.
  ducite ab urbe domum, mea carmina, ducite Daphnin.
terna tibi haec primum triplici diuersa colore

40–4 *MPaV*; 45–73 *MPa*   43 qui *a*   nudis *P*[1]*a?b?*: duris (*cf. A.* iv 366) *MP*[2]*Vω*   44 Maros *MP* (*quid a, latet*): Tmarus *V*: Ismarus *Seruii codd.*
58 fiat *MPbf?r, DSeru.*: fiant *aω*   63 possimus *cr* (*cf. E.* vii 23)
70 Vlixis *aω* (Olyxis *titulus Pompeianus, C.I.L.* iv 1982)

licia circumdo, terque haec altaria circum
effigiem duco; numero deus impare gaudet.
  ducite ab urbe domum, mea carmina, ducite Daphnin.
necte tribus nodis ternos, Amarylli, colores;
necte, Amarylli, modo et 'Veneris' dic 'uincula necto'.
  ducite ab urbe domum, mea carmina, ducite Daphnin.
limus ut hic durescit, et haec ut cera liquescit
uno eodemque igni, sic nostro Daphnis amore.
sparge molam et fragilis incende bitumine lauros:
Daphnis me malus urit, ego hanc in Daphnide laurum.
  ducite ab urbe domum, mea carmina, ducite Daphnin.
talis amor Daphnin qualis cum fessa iuuencum
per nemora atque altos quaerendo bucula lucos
propter aquae riuum uiridi procumbit in ulua
perdita, nec serae meminit decedere nocti,
talis amor teneat, nec sit mihi cura mederi.
  ducite ab urbe domum, mea carmina, ducite Daphnin.
has olim exuuias mihi perfidus ille reliquit,
pignora cara sui, quae nunc ego limine in ipso,
Terra, tibi mando; debent haec pignora Daphnin.
  ducite ab urbe domum, mea carmina, ducite Daphnin.
has herbas atque haec Ponto mihi lecta uenena
ipse dedit Moeris (nascuntur plurima Ponto);
his ego saepe lupum fieri et se condere siluis
Moerim, saepe animas imis excire sepulcris,
atque satas alio uidi traducere messis.
  ducite ab urbe domum, mea carmina, ducite Daphnin.
fer cineres, Amarylli, foras riuoque fluenti
transque caput iace, nec respexeris. his ego Daphnin
adgrediar; nihil ille deos, nil carmina curat.
  ducite ab urbe domum, mea carmina, ducite Daphnin.
'aspice: corripuit tremulis altaria flammis
sponte sua, dum ferre moror, cinis ipse. bonum sit!'
nescio quid certe est, et Hylax in limine latrat.
credimus? an, qui amant, ipsi sibi somnia fingunt?
  parcite, ab urbe uenit, iam parcite carmina, Daphnis.

74–109 *MPa*   87 procumbit $MP^2\omega$, *Macrob.* vi 2. 20: concumbit P^1a 107 Hylax *ed. Ascensiana an.* 1500/1: Hylas *codd.*   109 carmina (-ne *c*) parcite *Mcer*

ECLOGA IX

LYCIDAS MOERIS

L. Qvo te, Moeri, pedes? an, quo uia ducit, in urbem?
M. O Lycida, uiui peruenimus, aduena nostri
(quod numquam ueriti sumus) ut possessor agelli
diceret: 'haec mea sunt; ueteres migrate coloni.'
nunc uicti, tristes, quoniam fors omnia uersat,
hos illi (quod nec uertat bene) mittimus haedos.
L. Certe equidem audieram, qua se subducere colles
incipiunt mollique iugum demittere cliuo,
usque ad aquam et ueteres, iam fracta cacumina, fagos,
omnia carminibus uestrum seruasse Menalcan.
M. Audieras, et fama fuit; sed carmina tantum
nostra ualent, Lycida, tela inter Martia quantum
Chaonias dicunt aquila ueniente columbas.
quod nisi me quacumque nouas incidere lites
ante sinistra caua monuisset ab ilice cornix,
nec tuus hic Moeris nec uiueret ipse Menalcas.
L. Heu, cadit in quemquam tantum scelus? heu, tua nobis
paene simul tecum solacia rapta, Menalca!
quis caneret Nymphas? quis humum florentibus herbis
spargeret aut uiridi fontis induceret umbra?
uel quae sublegi tacitus tibi carmina nuper,
cum te ad delicias ferres Amaryllida nostras?
'Tityre, dum redeo (breuis est uia), pasce capellas,
et potum pastas age, Tityre, et inter agendum
occursare capro (cornu ferit ille) caueto.'
M. Immo haec, quae Varo necdum perfecta canebat:
'Vare, tuum nomen, superet modo Mantua nobis,
Mantua uae miserae nimium uicina Cremonae,
cantantes sublime ferent ad sidera cycni.'

ix 1–29*MPa*   1 orbem P^1 6 bene uertat P^2*fhv, Don. ad Ter. Phorm.* 678, *Seru. hic et ad A.* iv 641, *Non.* 348. 25   8 incipiant P^1 dimittere *bcdefr*   9 ueteres . . . fagos *M*: ueteris . . . fagi *Paω, Quint.* viii 6. 46   11 audierat M^1P^1 17 cadit *Maω*: cadet *Pb*   29 ferant P^2

L. Sic tua Cyrneas fugiant examina taxos,
sic cytiso pastae distendant ubera uaccae,
incipe, si quid habes. et me fecere poetam
Pierides, sunt et mihi carmina, me quoque dicunt
uatem pastores; sed non ego credulus illis.
nam neque adhuc Vario uideor nec dicere Cinna
digna, sed argutos inter strepere anser olores.
M. Id quidem ago et tacitus, Lycida, mecum ipse uoluto,
si ualeam meminisse; neque est ignobile carmen.
'huc ades, o Galatea; quis est nam ludus in undis?
hic uer purpureum, uarios hic flumina circum
fundit humus flores, hic candida populus antro
imminet et lentae texunt umbracula uites.
huc ades; insani feriant sine litora fluctus.'
L. Quid, quae te pura solum sub nocte canentem
audieram? numeros memini, si uerba tenerem:
'Daphni, quid antiquos signorum suspicis ortus?
ecce Dionaei processit Caesaris astrum,
astrum quo segetes gauderent frugibus et quo
duceret apricis in collibus uua colorem.
insere, Daphni, piros: carpent tua poma nepotes.'
M. Omnia fert aetas, animum quoque. saepe ego longos
cantando puerum memini me condere soles.
nunc oblita mihi tot carmina, uox quoque Moerim
iam fugit ipsa: lupi Moerim uidere priores.
sed tamen ista satis referet tibi saepe Menalcas.
L. Causando nostros in longum ducis amores.
et nunc omne tibi stratum silet aequor, et omnes,
aspice, uentosi ceciderunt murmuris aurae.
hinc adeo media est nobis uia; namque sepulcrum
incipit apparere Bianoris. hic, ubi densas
agricolae stringunt frondes, hic, Moeri, canamus;
hic haedos depone, tamen ueniemus in urbem.
aut si nox pluuiam ne colligat ante ueremur,

30–63 *MPa* 30 Cyrneas *M*[1]*aω* (Cyrineas *fs*), *Seru. hic et ad G.* iv 47: Grynaeas (Grineas) *M*[2]*Pcer, utrumque Gramm.* 35 Vario *Pa, Seru.*: Varo *Mω, 'alii' ap. DSeru., ps.Acro ad Hor. carm.* i 6. 8 38 nec *a* 45 tenebam *P* 46–50 *Lycidae continuant MP*[1], *Moeridi tribuunt P*[2]*aω* 59 hic *P*

cantantes licet usque (minus uia laedet) eamus;
cantantes ut eamus, ego hoc te fasce leuabo.
M. Desine plura, puer, et quod nunc instat agamus;
carmina tum melius, cum uenerit ipse, canemus.

ECLOGA X

EXTREMVM hunc, Arethusa, mihi concede laborem:
pauca meo Gallo, sed quae legat ipsa Lycoris,
carmina sunt dicenda; neget quis carmina Gallo?
sic tibi, cum fluctus subterlabere Sicanos,
Doris amara suam non intermisceat undam,
incipe: sollicitos Galli dicamus amores,
dum tenera attondent simae uirgulta capellae.
non canimus surdis, respondent omnia siluae.
Quae nemora aut qui uos saltus habuere, puellae
Naides, indigno cum Gallus amore peribat?
nam neque Parnasi uobis iuga, nam neque Pindi
ulla moram fecere, neque Aonie Aganippe.
illum etiam lauri, etiam fleuere myricae,
pinifer illum etiam sola sub rupe iacentem
Maenalus et gelidi fleuerunt saxa Lycaei.
stant et oues circum; nostri nec paenitet illas,
nec te paeniteat pecoris, diuine poeta:
et formosus ouis ad flumina pauit Adonis.
uenit et upilio, tardi uenere subulci,
uuidus hiberna uenit de glande Menalcas.
omnes 'unde amor iste' rogant 'tibi?' uenit Apollo:
'Galle, quid insanis?' inquit. 'tua cura Lycoris
perque niues alium perque horrida castra secuta est.'
uenit et agresti capitis Siluanus honore,

ix 64–x 9 *MPa*; 10–24 *MPR* 64 laedet *Paω*, *Seru.* (*ut uid.*): laedit *M*: laedat *f* 66 et] nunc *cde* x 1 laborum P^{1} 10 peribat $M^{2}PRrt$: periret $M^{1}\omega$ 12 Aonie *ω Seru.* (Aoinie *Rb*): Aoniae *MPder*, *Char.* 13. 33, 14. 23 13 etiam (2°)] illum *R* 19 upilio *MRbt* (ut filio P^{1}), *Seru.*: opilio $P^{2}\omega$ tarde *Pef* 20 umidus *R* 23 castra] saxa P^{1}

florentis ferulas et grandia lilia quassans.
Pan deus Arcadiae uenit, quem uidimus ipsi
sanguineis ebuli bacis minioque rubentem.
'ecquis erit modus?' inquit. 'Amor non talia curat,
nec lacrimis crudelis Amor nec gramina riuis
nec cytiso saturantur apes nec fronde capellae.'
tristis at ille 'tamen cantabitis, Arcades,' inquit
'montibus haec uestris; soli cantare periti
Arcades. o mihi tum quam molliter ossa quiescant,
uestra meos olim si fistula dicat amores!
atque utinam ex uobis unus uestrique fuissem
aut custos gregis aut maturae uinitor uuae!
certe siue mihi Phyllis siue esset Amyntas
seu quicumque furor (quid tum, si fuscus Amyntas?
et nigrae uiolae sunt et uaccinia nigra),
mecum inter salices lenta sub uite iaceret;
serta mihi Phyllis legeret, cantaret Amyntas.
hic gelidi fontes, hic mollia prata, Lycori,
hic nemus; hic ipso tecum consumerer aeuo.
nunc insanus amor duri me Martis in armis
tela inter media atque aduersos detinet hostis.
tu procul a patria (nec sit mihi credere tantum)
Alpinas, a! dura niues et frigora Rheni
me sine sola uides. a, te ne frigora laedant!
a, tibi ne teneras glacies secet aspera plantas!
ibo et Chalcidico quae sunt mihi condita uersu
carmina pastoris Siculi modulabor auena.
certum est in siluis inter spelaea ferarum
malle pati tenerisque meos incidere amores
arboribus: crescent illae, crescetis, amores.
interea mixtis lustrabo Maenala Nymphis
aut acris uenabor apros. non me ulla uetabunt
frigora Parthenios canibus circumdare saltus.
iam mihi per rupes uideor lucosque sonantis
ire, libet Partho torquere Cydonia cornu

25–59 *MPR* 28 ecquis *MP*: et quis *Rω* (*ut fere fit*) non] nec *R*
29 ripis M^1 32 nostris P^1b 40 iaceret $P^2R\omega$: iaceres MP^1
55 Nymphis] siluis *R* 59 Cydonia] Rhodonea M^1

spicula–tamquam haec sit nostri medicina furoris,
aut deus ille malis hominum mitescere discat.
iam neque Hamadryades rursus nec carmina nobis
ipsa placent; ipsae rursus concedite siluae.
non illum nostri possunt mutare labores,
nec si frigoribus mediis Hebrumque bibamus
Sithoniasque niues hiemis subeamus aquosae,
nec si, cum moriens alta liber aret in ulmo,
Aethiopum uersemus ouis sub sidere Cancri.
omnia uincit Amor: et nos cedamus Amori.'
  Haec sat erit, diuae, uestrum cecinisse poetam,
dum sedet et gracili fiscellam texit hibisco,
Pierides: uos haec facietis maxima Gallo,
Gallo, cuius amor tantum mihi crescit in horas
quantum uere nouo uiridis se subicit alnus.
surgamus: solet esse grauis cantantibus umbra,
iuniperi grauis umbra; nocent et frugibus umbrae.
ite domum saturae, uenit Hesperus, ite capellae.

60–77 *MPR* 60 sint *M* 62 neque] nec *Rb* 67 aret Liber *Conington* 69 uincit *Pω, Macrob.* v 14. 5 *et* 16. 7: uincet *M*: uicit *R* 73 hora *P* 74 subducit *R*

COMMENTARY

ECLOGUE 1

Introduction

Meliboeus notices Tityrus reclining beneath the cool, safe canopy of a beech tree and playing on his pipe.

> Tityre, tu patulae recubans sub tegmine fagi
> siluestrem tenui Musam meditaris auena. (1–2)

So begins the First *Eclogue*—so begins Virgil's Book of *Eclogues*—with a Theocritean name and an echo of the First *Idyll*.[1]

> *Ἁδύ τι τὸ ψιθύρισμα καὶ ἁ πίτυς, αἰπόλε, τήνα,*
> *ἁ ποτὶ ταῖς παγαῖσι, μελίσδεται, ἁδὺ δὲ καὶ τύ*
> *συρίσδες.* (1–3)
>
> Sweet is the whispered music of the pine, goatherd,[2]
> The pine there by the springs, and sweet too
> Your piping.

Virgil's idyllic scene, or rather, the suggestion of such a scene, is immediately qualified by Meliboeus' eloquent distress[3] as he speaks of suffering and loss—the harsh reality of the land-

[1] See O. Skutsch, *RhM* 99 (1956), 199–200, V. Pöschl, *Die Hirtendichtung Virgils* (Heidelberg, 1964), 9–11, E. A. Schmidt, *Bukolische Leidenschaft* (Frankfurt am Main, 1987), 29–36. The First *Idyll* stands first in all three families of MSS, as well as in the Antinoe papyrus (*c.* AD 500), and in all probability, therefore, stood first in the edition of Theocritus that Virgil used. See Gow, vol. i, pp. lxvi–lxix.

[2] Tityrus appears to be both a shepherd and a cowherd—cf. 8, 9, 45, 49–50—unlike the herdsmen of Theocritus, who are carefully distinguished in a sort of pastoral hierarchy: cowherd, shepherd, goatherd. See Gow on 1.86, L. E. Rossi, *SIFC* 43 (1971), 6–7.

[3] Note the chiastic arrangement: 'tu (1) . . . nos (3)', 'nos (4) . . . tu (4)', with the repetition of the name: 'Tityre, tu (1) . . . tu, Tityre (4)'. Lines 1–5 are composed of two tricola, the first of three lines, the second of two lines; punctuated rhetorically, the passage would appear thus: 'Tityre tu patulae recubans sub tegmine fagi, siluestrem tenui Musam meditaris auena, nos patriae finis et dulcia linquimus arua. nos patriam fugimus, tu Tityre lentus in umbra, formosam resonare doces Amaryllida siluas'. See Norden 376–7.

confiscations,[4] from which Tityrus, piping and singing of Amaryllis in the shade, is curiously immune.

'O Meliboeus', says Tityrus, 'a god made this peace for me. For to me *he* will ever be a god' (6–7). Despite his own trouble, Meliboeus is not envious; he is surprised, rather, and curious: 'this god of yours, tell me who he is'. But Tityrus tells him instead of Rome, 19 'Vrbem quam dicunt Romam, Meliboee . . .'. Readers ancient and modern have been puzzled by the evasiveness of his answer;[5] a calculated evasiveness, for Virgil intended to defer the moment of revelation and place it at the centre[6] of his poem, 42 'hic illum uidi iuuenem,[7] Meliboee . . .'. Yet Tityrus' answer is emotionally, as well as structurally, justified. Rome, the great city, his journey—these are the thoughts uppermost in his mind, and he attempts to convey, in pastoral terms, something of his wonderment (22–5). Meliboeus, it seems, is not much interested; he simply asks what reason Tityrus had for seeing Rome. 'Libertas', 'freedom', is the emphatic answer (27); and in this ambiguous, charged word lies the principal difficulty of the poem.

Tityrus describes himself as a former slave to whom freedom

[4] After the Battle of Philippi in 42 BC, large tracts of land were confiscated throughout Italy and granted as a reward for service to the soldiers of the victorious Triumvirs. Among the cities so punished were Cremona and, when the territory of Cremona proved insufficient, Mantua (9.28 'Mantua uae miserae nimium uicina Cremonae'). The most vivid account of the confiscations is to be found in Appian, *BC* 5.12 *ἀλλὰ συνιόντες ἀνὰ μέρος ἐς τὴν Ῥώμην οἵ τε νέοι καὶ γέροντες ἢ αἱ γυναῖκες ἅμα τοῖς παιδίοις, ἐς τὴν ἀγορὰν ἢ τὰ ἱερά, ἐθρήνουν, οὐδὲν μὲν ἀδικῆσαι λέγοντες, Ἰταλιῶται δὲ ὄντες ἀνίστασθαι γῆς τε καὶ ἑστίας οἷα δορίληπτοι*, 'They came to Rome in crowds, young and old, women and children, to the forum and the temples, uttering lamentations, saying that they had done no wrong for which they, Italians, should be driven from their fields and their hearthstones, like people conquered in war' (H. White in the Loeb Classical Library). Cf. Prop. 4.1.129–30 'nam tua cum multi uersarent rura iuuenci, / abstulit excultas pertica tristis opes'. It is clear from Appian that Octavian was personally involved in the distribution of the land. L. Keppie, 'Vergil, the Confiscations, and Caesar's Tenth Legion', *CQ*, NS 31 (1981), 367–70, describes the bureaucratic procedure.

[5] Servius ad loc.: 'quaeritur cur de Caesare interrogatus Roman describat. et aut simplicitate utitur rustica, ut ordinem narrationis plenum non teneat, sed per longas ambages ad interrogata descendat . . .'. Servius offers a similar explanation of 3.40. See R. A. Kaster, *Guardians of Language* (Berkeley, 1988), 21 n.27.

[6] The very centre: E. A. Fredricksmeyer, *Hermes*, 94 (1966), 214, R. F. Thomas, *HSCP* 87 (1983), 180 n.16.

[7] See ad loc.

came late—came, that is, after Amaryllis took him in hand, because, he ruefully admits, there had been no hope of freedom while he was attached to the spendthrift Galatea, 32 'nec spes libertatis erat nec cura peculi'.[8] The normal process of manumission is implied by which a slave saved up a certain sum of money, his *peculium*, from activities tolerated or encouraged by his master, and bought his freedom. Tityrus' master may be imagined as residing in the nearby town to which Tityrus brings his lambs for sale on market-days (19–21).

Amaryllis had been sad, calling on the gods, leaving the apples on her tree unpicked. And Meliboeus now understands why: Tityrus was away (36–8). He had gone to Rome, the old slave, to seek redress from the young master of Italy, since nowhere else could he be delivered from slavery, 40 'neque seruitio me exire licebat'.[9] But loss of property did not entail loss of freedom. Why, then, does Tityrus speak of slavery?

'Freedom' (*libertas*) and 'slavery' (*seruitium, seruitus*) were established political metaphors,[10] and *libertas* had acquired a current significance: it was the slogan of Octavian and his party.[11] Virgil deliberately confuses the private with the public sense of *libertas*,[12] and by so doing solves his literary problem, that of expressing gratitude to Octavian in the pastoral mode.[13] Not gratitude for a personal favour—Tityrus has a purely poetic existence—but a

[8] See ad loc.

[9] See ad loc.

[10] E.g. Cic. *Pro Sest.* 118 'sed quid ego populi Romani animum uirtutemque commemoro, libertatem iam ex diuturna seruitute dispicientis?', *Phil.* 5.21 'quid erat aliud nisi denuntiare populo Romano seruitutem?', *Phil.* 6.19 'aliae nationes seruitutem pati possunt, populi Romani est propria libertas', *Ad Brut.* 1.16.9 (Brutus) 'neque desistam abstrahere a seruitio ciuitatem nostram', Sall. *Iug.* 31.11 'uos, Quirites, in imperio nati aequo animo seruitutem toleratis?', 17 'uos pro libertate . . . nonne summa ope nitemini?'

[11] See R. Syme, *The Roman Revolution* (Oxford, 1939), 154–5.

[12] To the confusion of his readers—his later readers, that is. Virgil was writing for his contemporaries.

[13] A problem (if it could have occurred to him) evaded by Theocritus, who praises Hiero and Ptolemy in non-pastoral encomia (*Idylls* 16, 17).

disinterested, larger gratitude, expressed 'Tityri sub persona'[14] as to a god, for the restoration of peace and order.[15]

Now Tityrus can pasture his animals as before. But what of Meliboeus? For him there remains only the dreary prospect of exile.

The second half of the poem consists almost entirely of two long speeches—long in proportion to the poem and to the conversational exchanges preceding—by Meliboeus. In the first of these, he felicitates Tityrus on his great good fortune, small and infertile though his farm may be, and then, repeating his felicitation, imagines the pleasures Tityrus will henceforth enjoy, the familiar streams, cool shade (Meliboeus has been driving his goats in the heat of the day), the susurrus of bees from a neighbour's hedgerow ... Tityrus' response (so to call it) is an elaborate, thankful *adynaton*, in which he uses, somewhat unfeelingly it may seem, though not with reference to Meliboeus, the word 'exsul' (61).

Tityrus' 'impossibility' is a painful reality for Meliboeus; and thus prompted, he conjures up the most remote and barren regions of the earth (64–6). Mainly, however, his second speech is concerned with the pathos of leaving a native place. Will he ever, he wonders, and after how many long years, see his farm,[16] his little kingdom, again? He fears, knows he will not; will never again, stretched at ease in a mossy cavern, watch his goats hang browsing on some tufted crag; will sing no songs.

Virgil's sympathies are usually engaged on the side of defeat and loss; and here, in a poem praising Octavian, it is rather the dispossessed Meliboeus than the complacent Tityrus who more nearly represents Virgil.

[14] Servius on l. 1: 'et hoc loco Tityri sub persona Vergilium debemus accipere; non tamen ubique, sed tantum ubi exigit ratio'. Virgil's family farm may have been confiscated. It does not follow, however, that *E.* 1 is autobiographical, and any attempt to read it as such will fail. Tityrus and Meliboeus are no less imaginary than the landscape in which they are placed.

[15] The First *Eclogue* was probably written after the defeat of Sextus Pompey at Naulochus on 3 Sept. 36 BC; see l. 43n. Among the honours voted to Octavian by the Senate was a golden image with the inscription 'peace, long disturbed, he re-established on land and sea', Appian, *BC* 5.130 *ὅτι τὴν εἰρήνην ἐστασιασμένην ἐκ πολλοῦ συνέστησε κατά τε γῆν καὶ θάλασσαν.*

[16] 67 'patrios ... finis'. Cf. 3 'patriae finis', 4 'patriam'. Love of a native place, profound sorrow for its loss, the pleasure of undisturbed possession—these Roman sentiments are unknown to Theocritus (Jachmann 115).

Yet Tityrus is not without compassion. What little he can he does. He offers Meliboeus a bed for the night and invites him to share his evening meal, apples and chestnuts, fresh cheese—while the landscape gradually darkens around them.[17]

Bibliography

F. Leo, 'Vergils erste und neunte Ekloge', *Hermes*, 38 (1903), 1–18 = *Ausgewählte kleine Schriften*, (Rome, 1960), ii. 11–28.

P. L. Smith, '*Lentus In Umbra*: A Symbolic Pattern in Vergil's *Eclogues*', *Phoenix*, 19 (1965), 298–304.

P. Fedeli, 'Sulla prima bucolica di Virgilio', *GIF*, NS 3 (1972), 273–300.

G. Jachmann, 'Die dichterische Technik in Vergils Bukolika', *N. Jahrb.* 49 (1922), 110–19 = *Ausgewählte Schriften* (Königstein im Taurus, 1981), 312–22.

I. M. Le M. Du Quesnay, 'Vergil's First Eclogue', *Papers of the Liverpool Latin Seminar, Third Volume* (ARCA Classical and Medieval Texts, Papers and Monographs 7; 1981), 29–182.

J. R. G. Wright, 'Virgil's Pastoral Programme: Theocritus, Callimachus and *Eclogue* 1', *PCPS*, NS 29 (1983), 107–60.

A. Traina, 'La chiusa della prima egloga virgiliana (vv. 82–3)', *Lingua e stile* 3 (1986), 45–53 = *Poeti latini (e neolatini): Note e saggi filologici*[2] (Bologna, 1986), i. 175–88.

C. Perkell, 'On *Eclogue* 1.79–83', *TAPA* 120 (1990), 171–81.

1. Tityre: 1.4, 13, 18, 38, 3.20, 96, 5.12, 6.4, 8.55, 9.23, 24. Tityrus appears only twice in Theocritus, in 3.2–4, where he is asked to look after a friend's goats (cf. *E.* 5.12), and in 7.72–82, where he is imagined singing of the woes of Daphnis and the bliss of Comatas. V. tends to exploit Theocritean pastoral names: thus Amaryllis appears in *E.* 1, 2, 3, 8, 9, but only in Theocr. 3 and 4; Damoetas in *E.* 2, 3, 5, but only in Theocr. 6; Menalcas in *E.* 2, 3, 5, 9, 10, but only in [Theocr.] 8 and 9.

[17] In his essay 'On Wordsworth's Poetry', De Quincey maintains that Wordsworth was the first to appreciate the 'abstracting power' of twilight: 'In the dim interspace between day and night all disappears from our earthly scenery . . . which is either mean or inharmonious, or unquiet, or expressive of temporary things' (*The Collected Writings of Thomas De Quincey*, ed. D. Masson, xi (Edinburgh, 1890), 316).

The first line of each *E.* contains, as J. Hubaux, *RBPh* 6 (1927), 603–16, nicely observes, either a Greek name—Theocritean names in *E.* 1, 2, 3, 7, 8—or a name connected with Greek pastoral (4, 6, 10). Cf. Serv. *Prooem.* 'sane sciendum VII eclogas esse meras rusticas, quas Theocritus X habet. hic in tribus a bucolico carmine sed cum excusatione discessit, ut in genethliaco Salonini et in Sileni theologia . . .'. There are additionally the four 'fragments' in *E.* 9, each with a name in the first line: 23 *Tityre*, 27 *Vare* (the Roman name for deliberate contrast; see ad loc.), 39 *Galatea*, 46 *Daphni*. This feature is an imitation of Theocritus: 1.1 *αἰπόλε* (the goatherd remains anonymous throughout the poem), 3.1 *Ἀμαρυλλίδα*, 4.1 *Κορύδων* . . . *Φιλώνδα*, 5.1–2 *Συβαρίταν* . . . *Λάκωνα*, 6.1 *Δαμοίτας καὶ Δάφνις*, 7.1 *Εὔκριτος*, [8.1 *Δάφνιδι*, 9.1 *Δάφνι*], 10.1 *Βουκαῖε*, 11.19 *Γαλάτεια* (the *Idyll* proper begins here). Calpurnius Siculus, the author of the Einsiedeln *Eclogues*, and Nemesianus follow V.'s example. Calp. Sic. 1 appears to be an exception, but Hubaux argues that the poem must originally have begun with what is now l. 4 'cernis ut ecce pater quas tradidit, Ornyte, uaccae'.

tu . . . recubans: cf. Theocr. 7.88–9 *τὺ δ' ὑπὸ δρυσὶν ἢ ὑπὸ πεύκαις | ἁδὺ μελισδόμενος κατεκέκλισο, θεῖε Κομᾶτα,* 'while you lay sweetly singing under the oaks or pines, divine Comatas'. The herdsman is customarily seated while singing or piping; see 3.55n. *Recubo* is a rather unusual verb, here perhaps with a connotation of luxurious ease; cf. Cic. *De or.* 3.63 (Cyrenaic philosophy personified) 'in hortulis quiescet suis, ubi uult, ubi etiam recubans molliter et delicate nos auocat a rostris', Prop. 3.3.1 'molli recubans Heliconis in umbra'.

sub tegmine fagi: apparently modelled on the phrase *sub tegmine caeli*, which first occurs in Cic. *Arat.* 47 'Ales auis, lato sub tegmine caeli / quae uolat', 233, 239 'caeli sub tegmine', 346 'caeli de tegmine'; then in Lucr. 1.988 'sub caeli tegmine', 2.663 'sub tegmine caeli', 5.1016 'caeli sub tegmine'. The grandeur of the metaphor, which owes nothing to Aratus, and its use by Lucretius may indicate that its author was not the youthful Cicero but Ennius (so M. Guendel, *De Ciceronis poetae arte capita tria* (diss. inaug., Leipzig, 1907), 54–5, against Munro on Lucr. 5.619); it will be noticed that Lucretius incorporates it in a passage reminiscent of the high archaic style, 2.661–3 'saepe itaque ex uno tondentes gramina campo / lanigerae pecudes et equorum duellica proles / buceriaeque greges eodem sub tegmine caeli'. *Tegmen* properly refers to any sort of bodily covering or protection: the skin of a wild beast (*A.* 1.275, 11.576), a cap or helmet (*A.* 7.689, 742), armour (*A.* 9.518), a shield (*A.* 10.887); cf. also Cic. *Arat.* 114–15 'foliorum tegmine . . . arbusta ornata', which again owes nothing to Aratus. Like other nouns so formed, *tegmen* is mostly poetic; see J. Perrot, *Les Dérivés latins en* –men *et* –mentum (Paris, 1961), 109–12. V.'s metaphor was parodied by a certain Numitorius, *Vita Donati* 43 'Tityre, si toga calda tibi est, quo tegmine fagi?'—that is 'what point (*quo*) has the phrase *tegmine fagi*?' (Housman, *CR* 49 (1935), 167 = *Class. Papers*, iii. 1245). See 3.1 n. (*cuium pecus*).

fagi: the beech, *Fagus sylvatica* L., a useful tree with its heavy shade (2.3), wood that can be turned into cups (3.36–7), and grey smooth bark in which the notation and words of a song can be carved (5.13–14). It is native to North Italy: 'Charakteristische Bilder wie *patulae recubans sub tegmine fagi* . . . passen noch heute auf die Höhen von Norditalien' (M. C. P. Schmidt, *RE* iii (1897), 972). Placed so prominently at the beginning of the book, the beech seems symbolic or representative; 'beyond all others perhaps, the tree of the Eclogues' (Ross (1975), 72). Not being native to Greece south of Thessaly (see R. Meiggs, *Trees and Timber in the Ancient Mediterranean World* (Oxford, 1982), 112), the beech was not, like the laurel, the plane, the poplar, the cypress, or the pine of Theocritus 1.1, an old and established poetic tree: it is found before V. only in Catull. 64.288–91 'namque ille tulit radicitus altas / fagos ac recto proceras stipite laurus, / non sine nutanti platano lentaque sorore / flammati Phaethontis et aerea cupressu'. Nor is it evenly distributed in the book, being confined to the earlier *Eclogues*: 2.3 (where see note), 3.12, 5.13, 9.9

2. siluestrem . . . Musam: from Lucr. 4.586–9 'cum Pan / pinea semiferi capitis uelamina quassans / unco saepe labro calamos percurrit hiantis, / fistula siluestrem ne cesset fundere Musam'. Muse and music had long since been confused, e.g. *Homeric Hymn to Pan* 15–16 *δονάκων ὕπο μοῦσαν ἀθύρων | νήδυμον*, 'singing a sweet tune to the pipe', Theocr. 1.20 *καὶ τᾶς βουκολικᾶς ἐπὶ τὸ πλέον ἵκεο μοίσας*, 'and you have attained mastery in pastoral song', Meleager 13.2 G.–P. (= *AP* 7.196, to a cicada) *ἀγρονόμαν μέλπεις μοῦσαν ἐρημολάλον*, 'Tu joues une musique champêtre dans la solitude' (Pierre Louÿs). Cf. 6.8.

tenui . . . auena: see above, p. xxv.

meditaris: 'compose'; cf. 6.8, 82–3, Hor. *Epist.* 2.2.76 'i nunc et uersus tecum meditare canoros'. 'Milton endeavours to make the phrase English, cf. Lycidas 66 "and strictly meditate the thankless Muse"' (T. E. Page). See A. Traina, *Enc. Virg.* s.v.

auena: the wild oat, here imagined as a musical instrument; cf. 10.51 'carmina pastoris Siculi modulabor auena'. But what sort of musical instrument? The question is a difficult one: 'not a straw (which would be absurd), but a reed, or perhaps a pipe of reeds, hollow like a straw' (Conington); 'as an oat-straw could not be made into a musical instrument, *avena* must be used for "a reed" or something of the sort' (T. E. Page). This musical absurdity was apparently inspired by the insulting gibe of Comatas in Theocr. 5.7 *καλάμας αὐλόν,* 'an oat-straw pipe', which V. imitates in 3.27. The principal, indeed symbolic, instrument of pastoral music is the vertical flute or pan-pipe (*σῦριγξ*, 'syrinx', *δόνακες*, 'reeds', *fistula, calami*), though the simple pipe (*αὐλός, tibia*) is occasionally mentioned (Theocr. 5.7, 6.43, 10.34; *E.* 8.21, 25, etc.) Whatever the practical objections—for practical objections may be disregarded in the pastoral world—it seems best to take *auena*, like 'calamo . . . agresti' in l. 10, as a collective singular denoting the pan-pipe. So Ovid took it, *Met.* 1.677 'structis cantat auenis', *Trist.* 5.10.25 'pastor iunctis pice cantat auenis'. And V. alludes to 'tenui . . . auena' with another collective singular in 6.8 'tenui . . . harundine', that is, the pan-pipe. For the structure of the pan-pipe see below, 10n. V.'s *auena*, then, will be a metrically useful synonym analogous to Lucretius' *cicuta*, which V. twice employs, in 2.36 (where see n.) and 5.85. See P. L. Smith, 'Vergil's *Avena* and the Pipes of Pastoral Poetry', *TAPA* 101 (1970), 497–510. V.'s *auena* became a convention of pastoral poetry: Calp. Sic. 4.63, Sannazaro, *Ecl. pisc.* 2.28, Spenser's 'oaten reed', Shakespeare's 'oaten straws', Milton's 'oaten flute', Marvell's 'slender oat'. See 2.37n.

3. dulcia . . . arua: cf. *G.* 2.511 'exsilioque domos et dulcia limina mutant', Hom. *Od.* 9.34–5 *ὡς οὐδὲν γλύκιον ἧς πατρίδος οὐδὲ τοκήων | γίγνεται,* 'for nothing is sweeter than one's own country and parents', Ov. *Ex Pont.* 1.3.35–6 'nescioqua natale solum dulcedine cunctos / ducit'.

linquimus: the simple form for the compound is mostly poetic;

see *TLL* s.v., K. F. Smith on Tib. 1.3.44, and Norden on *A.* 6.620. Cf. 5.34n., 9.51n.

4. **patriam fugimus:** Servius quotes Hor. *Carm.* 1.7.21–2 'Teucer Salamina patremque/cum fugeret'. Mostly poetic in this sense; see *TLL* s.v. 1477. 20. Cf. *φεύγω*; see LSJ s.v. III.3.

lentus: easy in the shade, at one with nature. Elsewhere in the *E.*, *lentus* is applied to plants: 1.25, 3.38, 83, 5.31 (the thyrsus simulated a plant or growing thing), 7.48, 9.42, 10.40; an early usage apparently (Plaut. *Asin.* 575 'ulmeis ... lentis uirgis') which survived in poetry; see *TLL* s.v. 1162.43, Tränkle 79.

in umbra: where he had sought shelter from the burning heat; cf. Varro, *RR* 2.2.11 (sheep) 'circiter meridianos aestus, dum deferuescant, sub umbriferas rupes et arbores patulas subigunt quaad refrigeratur'.

5. There may be, as F. Cairns argues (see 2.3n.), an allusion to Callimachus' story of Acontius and Cydippe here. Much more obvious, however, is a connection with Longus 2.7.6 *ἐπῄνουν τὴν Ἠχὼ τὸ Ἀμαρυλλίδος ὄνομα μετ' ἐμὲ καλοῦσαν*, 'I used to praise Echo for calling "Amaryllis" after me'; first noticed by Leo 13n.1, who supposed that Longus was imitating V.—a supposition in the last degree improbable, for Longus nowhere shows the slightest awareness of Latin literature. It is far more probable that V. and Longus drew on a common source in post-Theocritean pastoral. Cf. also Longus 2.5.3 *παρήμην σοι συρίττοντι πρὸς ταῖς φηγοῖς ἐκείναις ἡνίκα ἤρας Ἀμαρυλλίδος*, 'I sat beside you as you played on your pan-pipe under those oaks when you were in love with Amaryllis'.

formosam: V.'s predilection for this adjective in the *E.* is remarkable: sixteen instances; and equally remarkable his avoidance of it thereafter: once in the *G.* (3.219 'formosa iuuenca'; see *E.* 5.44n.), never in the *A.* This distribution is not, however, typical of V.'s practice (Axelson 60–1); of the forty adjectives in *-osus* used by V., only three are peculiar to the *E.* (all in *E.* 7, 29 *saetosus*, 30 *ramosus*, 45 *muscosus*) while no fewer than seventeen are peculiar to the *A.* Here follows a complete list—Ernout's is lacunose—with asterisks indicating those first attested in V.: *animosus* (*G.A.*), *annosus* (*A.*), *aquosus* (*E.G.A.*), **cliuosus* (*G.*), *dumosus* (*E.G.*), *formosus* (*E.G.*), *fragosus* (*A.*), *frondosus* (*E.G.A.*), *fumosus* (*G.*), *generosus* (*G.A.*), *harenosus* (*A.*), *herbosus* (*G.*), *lacrimosus* (*A.*), *lapidosus* (*G.A.*), *latebrosus* (*A.*), *limosus* (*E.A.*), *maculosus* (*G.A.*), *montosus* (*A.* Varro, *montuosus* Cic. Caes.), *muscosus* (*E.*), *nemorosus* (*A.*), *nimbosus* (*A.*), **onerosus* (*A.*), **palmosus* (*A.* 3.705, unique), *piscosus* (*A.*), *ramosus* (*E.*), *religiosus* (*A.*), *rimosus* (*G.A.*), **saetosus* (*E.*), **saxosus* (*E.G.*), **sinuosus* (*G.A.*), *spumosus* (*A.*), *squamosus* (*G.*), **tenebrosus* (*A.*, *tenebricosus* Cic. Catull. Varro), *uadosus* (*A.*), *uentosus* (*E.A.*), **uillosus* (*A.*), *uirosus* (*G.*), *uitiosus* (*G.*), *umbrosus* (*E.G.A.*), **undosus* (*A.*). See A. Ernout, *Les Adjectifs latins en* -osus *et en* -ulentus (Paris, 1948), Tränkle 59–60, Ross (1969), 53–60, and especially P. E. Knox, 'Adjectives in *-osus* and Latin Poetic Diction', *Glotta*, 64 (1986), 90–101. Adjectives in *-osus* are, as R. Syme observes, *Sallust* (Berkeley, 1964), 264n. 149, 'a large and instructive theme'.

resonare: transitive, here for the first time (but cf. Hor. *Serm.* 1.8.41 'resonarint triste et acutum'), as it is in *G.* 3.338 and *A.* 7.12; elsewhere in the *E.* and *G.* and *A.* intransitive.

doces: Tityrus sometimes sings of Amaryllis and sometimes plays on his pan-pipe.

Amaryllida: 1.30, 36, 2.14, 52, 3.81, 8.77, 78, 101, 9.22. Ordinarily, as here, the name is accommodated to the syntax of the sentence; see Fedeli on Prop. 1.18.31 'resonent mihi "Cynthia" siluae', Austin on *A.* 2.769. Cf. 6.44.

Amaryllida siluas: the rhythm of the line (weak caesura in the third foot, strong in the fourth) sets off these two words with their echo-effect: 'Amar*yl*lida si*l*uas'. Cf. 2.13 '*ar*denti resonant *ar*busta', *G.* 1.486 'per noctem resonare *lu*pis *ulul*antibus urbes' (rhythmically identical with 'formosam resonare doces Amaryllida siluas'), 2.328 '*aui*a tum resonant *aui*bus uirgulta canoris' (for the assonance *āui-* *ăui-* and the like see Norden on *A.* 6.204ff.), 3.338 'litoraque *alc*yonen resonant, *acal*anthida dumi' (a rare bird the acalanthis, chosen for the sound of its name?). For this figure of sound V. is indebted to Catullus, 11.3–4 'litus ut longe resonante Eoa / t*und*itur *und*a—a passage that worked powerfully upon his auditory imagination; cf. *G.* 1.358–9 'aut resonantia longe / litora misceri et nemorum increbrescere murmur'. See 6.84n.

siluas: not dense woods, where grazing would be impossible, but partly open hillside with grazing among the trees, 'nemorum ... saltus' (6.56, where see note). Below the woods in the ploughed fields there could be no echoing song; 'The woods shall to me answer, and my Eccho ring' (Spenser, *Epithalamion*).

6. Meliboee: 1.19, 42, 73, 3.1, 5.87, 7.9; a non-Theocritean name modelled, like *Alphesiboeus* (5.73 n.), on the feminine name, which occurs in several Greek legends (*RE* xv. 510). See F. Michelazzo, *Enc. Virg.* s.v. 459–60.

deus ... / ... deus: emphatically repeated so as to suggest a ritual cry; cf. 5.64 'deus, deus ille' and see Norden on A. 6.46 'deus ecce deus'.

otia fecit: the expression seems to be V.'s own. *Otia* is a poetic plural, no doubt recalling the pastoral ease of Lucr. 5.1387 'per loca pastorum deserta atque otia dia'. Cf. 5.61.

7. The boldness of Tityrus' declaration, as Wilamowitz, *Der Glaube der Hellenen* (Berlin, 1956), ii. 422, notices, is at once qualified: 'namque erit ille *mihi* semper deus'. Cf. A. D. Nock, *JHS* 48 (1928), 31 = *Essays on Religion and the Ancient World*, ed. Z. Stewart (Cambridge, Mass., 1972), i. 145: 'Such an individual might be called a god, either unreservedly or with reference to yourself, a *god to you*' (with examples, including this one).

7–9. ille ... illius ... / ... / ille: not simply 'emphatic' (T. E. Page); the repetition implies a style of sacral utterance, Norden's 'Er-Stil', e.g. Lygdamus 6.13–16 (*Liber*) '*ille* facit mites animos deus, *ille* ferocem / contudit et dominae misit in arbitrium; / Armenias tigres et fuluas *ille* leaenas / uicit'; see Norden, *Agnostos Theos* (Berlin, 1913), 163–6. Fedeli 276–7 compares Catull. 51.1–2 '*Ille* mi par esse deo uidetur, / *ille*, si fas est, superare diuos', thus rendering Catullus' addition to Sappho intelligible. Suetonius reports that, while sailing past the harbour of Puteoli, Augustus was saluted as a deity by the passengers and crew of an Alexandrian merchantman, *Aug.* 98.2 'candidati coronatique et tura libantes fausta omina et eximias laudes congesserant: per *illum* se uiuere, per *illum* nauigare, libertate atque fortunis per *illum* frui'.

8. ab ouilibus ... agnus: a phrase like 'a fontibus undae' (*G.* 2.243) or 'pastor ab Amphryso' (*G.* 3.2, where see Mynors); see also Munro on Lucr. 2.51 'fulgorem ... ab auro'. *Ouilibus* is a poetic plural (*TLL* s.v. 1190.22).

9. meas ... boues: one of the earliest instances of the feminine (the text of Plaut. *Truc.* 277 is uncertain); see *TLL* s.v. *bos* 2141.44. Masculine in 5.25.

errare: of animals grazing ('pasci', Serv.) as in 2.21, *G.* 3.139, Hor. *Epod.* 2.11–12 'in reducta ualle mugientium / prospectat errantis greges'.

10. ludere: often of composing light or playful verse, as in Catull. 50.1–2 'Hesterno, Licini, die otiosi / multum lusimus in meis tabellis'; here, explicitly in 6.1 and *G.* 4.565, of pastoral poetry. See Nisbet–Hubbard on Hor. *Carm.* 1.32.2, Lyne on *Ciris* 19–20.

calamo ... agresti: a collective singular; see above, 2n. (*auena*). The pan-pipe consisted of a number of reeds—usually seven—graduated in length and joined together with wax; see 2.36–7n. The player produced a tune by blowing across the upper ends of the reeds, as described by Lucretius (above, 2n.).

11. non equidem: *equidem* occurs only once again in the *E.*, in 9.7, where it is similarly placed in the line and in the poem; that is, following on Moeris' answer, a speech of five lines beginning 'O Lycida', to Lycidas' initial query. To this extent, V. modelled the opening of *E.* 1 on that of *E.* 9, though to a reader of V.'s book, if he noticed the similarity, the opposite would appear to be the case.

magis: 'rather', *potius*, after a negation, as in Catull. 68.30 'non est turpe, magis miserum est', Sall. *Iug.* 96.2 'ab nullo repetere, magis id laborare'; see O. Hey, *ALL* 13 (1902–4), 204–5; *TLL* s.v. 58.63.

12. usque adeo: a Lucretian phrase (37 instances).

turbatur: cf. Lucr. 6.377 'turbatur utrimque'. Wackernagel i.145–6, remarks that Augustan writers, and V. especially, favour these impersonal forms and that here the very indefiniteness heightens the effect. He compares *A.* 4.416 and Livy 2.45.11 'totis castris undique ad consules curritur'. See Austin on *A.* 1.272.

13. protinus: going straight on without rest; only here in the *E.*

ago: cf. 9.37, *A.* 10.675, 12.637, where *ago* is also elided (but not in *A.* 4.534, where see Pease), presumably because the quantity of *-o* was uncertain; see Maas 513n. = 556n.22.

hanc etiam uix ... duco: he drives the other goats ('*ago* autem proprie, nam agi dicuntur pecora', Serv.), the unhappy mother is almost too weak for him even to lead; *Isa.* 40:11 (the shepherd) 'shall gently lead those that are with young'.

14. corylos: hazel-trees, *Corylus avellana* L., the standard undergrowth of V's poetic landscape.

namque: an extreme postposition; cf. Catull. 66.65 'Virginis et saeui contingens namque Leonis' and see Clausen, *HSCP* 74 (1970), 85–6.

gemellos: neoteric in tone (the immediate context being intensely so); cf. Catull. 4.27 'gemelle Castor et gemelle Castoris'; not found elsewhere in V. and generally avoided by later poets except Ovid (*TLL* s.v.). 'Notice the pathos of each word: *gemellos* "twins" heightening the sense of loss; *spem gregis* marking that they were fine ones which, could they have been reared, the flock would have regarded with pride and hope; *silice in nuda* in contrast with the soft bed of litter that would have been provided at home (cf. *G.* 3.297); *conixa* instead of the usual *enixa* emphasising more strongly the pain and effort of the labour; *reliquit* closing the description with the thought of their abandonment' (T. E. Page). The breeding season is almost over (cf. below, 80–1 n.) and the kids Meliboeus has waited so long for he has now lost (Mynors).

15. spem gregis: *spes* is concrete, as in G. 3.473 'spemque gregemque simul', 4.162 'spem gentis'.

a: 10.47n.

silice in nuda: feminine here as elsewhere in V. where its gender can be determined (*A.* 6.471, 602, 8.233, where see Eden), but masculine in Lucretius, and both masculine and feminine in Ovid; see Hofmann–Szantyr 11.

conixa: 'pro eo quod est *enixa*, nam hiatus causa mutauit praepositionem' (Serv.); see Norden 409n.2. Unique in this sense (*TLL* s.v. 319.66). Cf. Ov. *Met.* 11.316 '(namque est enixa gemellos).'

reliquit: cf. Tib. 1.1.31–2 'non agnamue sinu pigeat fetumue capellae / desertum oblita matre referre domum'; 'this is familiar to every shepherd' (K. F. Smith).

16. si mens non laeua fuisset: 'if we had been right-minded' (Lee); reused in *A.* 2.54. In Roman divination the favourable side was the left, in Greek the right; here, however, *laeuus* has its ordinary, non-technical sense of mistaken or stupid. See 9.15n., Mynors on *G.* 4.7.

17. de caelo tactas: the traditional formula for objects struck by lightning; cf. Cic. *De diu.* 1.92 'Etruria autem de caelo tacta scientissume animaduertit, eademque interpretatur quid quibusque ostendatur monstris atque portentis' and see *OLD* s.v. *tango* 4 c.

memini praedicere: the present infinitive, as in old Latin, where the perfect might have been expected (*G.* 4.125–7). Cf. 7.69, 9.52, *A.* 1.619, 7.205–6, 8.157–9, and see Kühner–Stegmann i.703, Hofmann–Szantyr 357.

18. qui: instead of *quis* before *sit* for euphony and with no distinction of meaning; cf. 2.19, *A.* 3.608, 9.146, and see Löfstedt ii. 86–7.

da: instead of *dic*, introducing an appropriate note of gravity; cf. Hor. *Serm.* 2.8.4–5 (mock-solemn) 'da, si graue non est, / quae prima iratum uentrem placauerit esca' (where some MSS have *dic*), Val. Flacc. 5.217–18 'Incipe nunc cantus alios, dea, uisaque uobis / Thessalici da bella ducis'.

19. Vrbem quam dicunt Romam: note the solemn spondaic rhythm: Rome is not mentioned elsewhere in the *E.*

20. stultus: a 'low' word, found also in 2.39 but nowhere else in V.; frequent in comedy, found in Hor. *Serm.* and *Epist.* (but not in the *Carm.*) and in elegy. See Axelson 100.

similem: with *esse* understood as in 3.35, 4.59; 3.109 (*es*), 8.15, 24 (*est*), 5.21 (*estis*), 8.3, 9.53 (*sunt*). For the omission of these words in poetry see F. Leo, *L. Annaei Senecae tragoediae* (Berlin, 1878), i. 184–93.

saepe solemus: on market-days, *nundinae*, every ninth day by Roman reckoning, when they would take their animals and cheese to sell in town (*G.* 3.402 'adit oppida pastor')—in contrast with Tityrus' one great trip. For *saepe* with *soleo* cf. *G.* 2.186, Tib. 1.9.18.

21. depellere: the ancient opinion that *depellere* means *separare a lacte* (Philargyrius) is briefly discussed by most commentators and now accepted by Coleman (with Burman's conjecture *quoi*). But in this sense the verb is always a perfect passive participle; cf. 3.82 'depulsis . . . haedis', 7.15 'depulsos a lacte . . . agnos', *G.* 3.187 'depulsus ab ubere matris', *TLL* s.v. 564.81. Besides, in the vicinity of a town lambs could be delivered to the butcher before being weaned; Columella 7.3.13 (*upilio uilicus*) 'teneros agnos, dum adhuc herbae sunt expertes, lanio tradit, quoniam et paruo sumptu deuehuntur et his submotis fructus lactis ex matribus non minor percipitur'. The flesh of milk-fed or spring lamb is still considered a delicacy in Italy (*abbacchio*).

22. canibus catulos similis: L. P. Wilkinson, *Golden Latin Artistry* (Cambridge, 1963), 82 on *G.* 1.361: 'This rhythm, a succession of three anapaestic words unblurred by synaloepha, is extremely rare'. He cites, in addition to this line, 8.28 (but *cum canibus* is rhythmically a choriamb), *G.* 2.213, 3.165, 410, *A.* 3.259, 606, 4.403, 5.605, 822, 7.479, 9.156, 554, 10.390, 568. See also Norden on *A.* 6.290.

23. paruis componere magna: small things are usually compared to great, as in *G.* 4.176 'si parua licet componere magnis', but cf. Ov. *Trist.* 1.6.28 'grandia si paruis adsimilare licet'. See Otto, no. 1008.

25. uiburna: a species of guelder-rose, the wayfaring-tree, *Viburnum lantana* L.; V. Bertoldi, *Archivum Romanicum*, 15 (1931), 70 n.3, reports a belief current in the Mantuan countryside that the wayfaring-tree (called *antána*) had once been tall but was cursed by the Virgin for providing wood for the Cross and reduced to a shrub (Mynors).

cupressi: the familiar Italian cypress, *Cupressus sempervirens* L., which may attain a height of 150 feet.

26. tanta ... causa: Meliboeus naturally assumes that Tityrus had some compelling reason for such a journey.

27. Libertas: freedom from slavery, purchased with a slave's hard-earned savings, his *peculium*. Leo 4 n.3 compares Plaut. *Stich.* 751 (Stichus) 'Vapulat peculium, actum est'. (Sangarinus) 'Fugit hoc libertas caput'; like Tityrus before Amaryllis took him in hand, Stichus and Sangarinus will squander their *peculium* on a woman. But *Libertas* had also a larger, political reference of which no contemporary reader could be unaware: it was the slogan of Octavian and his party; see Introduction, p.31.

sera tamen: 'though late, yet', (*quamuis*) *sera, tamen*; a fairly common ellipse, e.g. Tib. 1.9.4 'sera tamen tacitis Poena uenit pedibus', Hor. *Carm* 1.15.19–20 'tamen heu serus adulteros / cultus puluere collines'; Cic. *De inuent.* 1.39 'quare iam diu gesta et a memoria nostra remota tamen faciant fidem', *Arat.* 139–41 'exin semotam procul in tutoque locatam / Andromedam tamen explorans fera quaerere Pistrix / pergit', Lucr. 4.952–3 'poplitesque cubanti / saepe tamen summittuntur', *G.* 1.197–8 'uidi lecta diu et multo spectata labore / degenerare tamen'. See Munro on Lucr. 4.952, Housman on Manil. 4.413 and Luc. 1.333, Nisbet–Hubbard on Hor. *Carm.* 1.15.19, Hill on Stat. *Theb.* 1.480.

respexit: though by his own admission Tityrus was shiftless and lazy, Liberty at last looked upon him—as might a deity, and Libertas was a deity in Rome; see Wissowa 138–9, S. Weinstock, *Divus Julius* (Oxford, 1971), 136–7. Cf. Plaut. *Bacch.* 639 'deus respiciet nos aliquis', Cic. *Ad Att.* 7.1.2 'nisi idem deus . . . respexerit rem publicam'; *Exod.* 2:25 'And God looked upon the children of Israel, and God had respect unto them'.

29. respexit tamen: such repetition, artfully varied, is characteristic of pastoral; somewhat similar are 2.37–9, 5.51–2.

30. habet: normally used of the male, e.g. *A.* 9.593–4 (Numanus) 'Turnique minorem /germanam . . . habebat'; see *TLL* s.v. 2408.53. So also in Greek, e.g. *Od.* 4.569 *οὕνεκ' ἔχεις Ἑλένην*, 'for you have Helen to wife'; see LSJ s.v. *ἔχω* A.I.4. Tityrus seems to have been remarkably passive.

Galatea: a beautiful Nereid (Hes. *Theog.* 250 *εὐειδὴς Γαλάτεια*), later localized in Sicily, where she was wooed by Polyphemus; cf. 7.37 'Nerine Galatea', 9.39. The mistress of Dionysius I of Syracuse was named Galatea; see Gow's Preface to Theocr. 6. Galatea and Amaryllis must be, as Tityrus had been, slaves, *conseruae* (cf. 3.64, 72), a brutal reality disguised by the elegance and poetic resonance of their names. Cf. Varro, *RR* 2.10.6–7 'Quod ad feturam humanam pertinet pastorum, qui in fundo perpetuo manent, facile est quod habent conseruam in uilla, nec hac uenus pastoralis longius quid quaerit'. Although a ubiquitous feature of the Italian countryside, slaves hardly appear as such in the *E.*—or in the pastoral *Idylls* of Theocritus (only in 5.5 *δῶλε Σιβύρτα*, 'slave of Sibyrtas', a term of abuse). For the most part, V. leaves the question of status, slave or free, vague so as not to disrupt the harmony of his pastoral landscape, and therefore he does not use the words *seruus* or *serua*; see Axelson 58.

31. namque fatebor: for this rhythm, a trochee followed by an amphibrach, the rhythm of *arma uirumque*, which usually occurs at the end of the line, see C. Weber, *HSCP* 91 (1987), 263 n. 14.

(fatebor enim): again in *A.* 4.20 'Anna (fatebor enim) miseri post fata Sychaei'; cf. *A.* 12.813 'Iuturnam misero (fateor) succurrere fratri'. The parenthesis is colloquial in tone and qualifies a slightly embarrassed or rueful admission, as by Cicero, *Ad Att.* 12.36.1 'hae meae tibi ineptiae (fateor enim) ferendae sunt', or Ovid, *Trist.* 5.6.5 'sarcina sum (fateor)'; see Tränkle 8.

The use of parenthesis to create an effect of liveliness or spontaneity is a feature of Callimachus' style which V. imitates especially in the *E.*; 2.32–3, 53, 3.9, 29–30, 36, 54, 93, 104, 4.48, 61, 5.70, 89, 6.6–7, 69, 7.22–3, 8.38 (cf. Theocr. 11.25–7), 9.3, 6, 23, 25, 64, 10.38–9, 46. Parenthesis is not found in the pastoral poetry of Theocritus (but cf. [Theocr.] 8.50) and very rarely in his other poetry: 12.13, 15.15–16, 22.64. For parenthesis in Callimachus see F. Lapp, *De Callimachi Cyrenaei tropis et figuris* (diss. inaug., Bonn, 1965), 52–3; for Callimachus and V., G. Williams, *Tradition and Originality in Roman Poetry* (Oxford, 1968), 711, 730–1.

Galatea tenebat: cf. Catull. 64.28 'tene Thetis tenuit pulcerrima Nereine?'

32. cura peculi: the *peculium*, as described by M. I. Finley, *The Ancient Economy* (Cambridge, 1973), 64, was 'property (in whatever form) assigned for use, management, and, within limits, disposal to someone who in law lacked the right of property, either a slave or someone in *patria potestas* . . . the possessor normally had a free hand in the management, and, if a slave, he could expect to buy his freedom with the profits'. Cf. Varro, *RR* 1.17.5 'praefectos alacriores faciendum praemiis dandaque opera ut habeant peculium et coniunctas conseruas e quibus habeant filios'. But slaves in the country were rarely if ever emancipated; see K. D. White, *Roman Farming* (London, 1970), 352–3.

33. multa . . . uictima: a collective singular for the unmetrical plural, as in *A.* 1.334 'multa . . . hostia', Tib. 1.3.28 'multa tabella' ('a use characteristic of poetry and post-Augustan prose', K. F. Smith).

exiret: to be sold in town.

34. pinguis . . . caseus: Varro, *RR* 2.11.3 'et etiam est discrimen utrum casei molles ac recentes sint an aridi et ueteres'. In Diocletian's price edict of AD 301 (ed. S. Lauffer, *Diokletians Preisedikt* (Berlin, 1971)), new or soft cheese, *caseus recens*, is listed with fresh farm produce (6.96); hard cheese, *caseus siccus*, with salt fish (5.11). Hard cheese could be kept or exported, but soft cheese had to be consumed within a few days—Columella 7.8.6 'qui recens intra paucos dies absumi debet'—or, as in Tityrus' case, sold in the nearby town. V. clearly distinguishes between the two types of cheese, *G.* 3.400–3 'quod surgente die mulsere horisque diurnis, / nocte premunt; quod iam tenebris et sole cadente, / sub lucem: exportant calathis (adit oppida pastor), / aut parco sale contingunt hiemique reponunt'.

Applied to cheese, *pinguis* is virtually unique (*TLL* s.v. *caseus* 514.35), the usual adjectives being *mollis* and *recens* (ibid. 514.20); and since *pinguis* is commonly applied to animals, especially sacrificial animals (e.g. Plaut. *Capt.* 862 'agnum . . . pinguem', *A.* 11.740 'hostia pinguis'), *multa* and *pinguis* should probably be understood with both nouns: many a fat victim and much fat cheese ('sane *pinguis* melius ad uictimam quam ad caseum refertur', Serv.) Cf. Tib. 2.5.37–8 'fecundi . . . munera ruris, / caseus et . . . agnus'.

et: postponed to the second place as elsewhere in the *E.*: 1.68, 2.10, 3.89 (in the clause), 4.54, 7.60, 8.55; see Norden 402–4.

ingratae . . . urbi: the countryman resents the town, which may refuse the products he brings or, in his opinion, pay him too little ('pretia iniqua', Serv.).

premeretur: cf. below, 81, 'pressi copia lactis'; hand-pressed cheese, the making of which is described by Columella, 7.8.7 'illa uero notissima est ratio faciundi casei, quem dicimus manu pressum . . . feruente aqua perfusus uel manu figuratur uel buxeis formis exprimitur'. Augustus was fond of it (Suet. *Aug.* 76).

35. non umquam grauis aere: unlike the thrifty Simulus, who always returned from town 'ceruice leuis, grauis aere' (*Moretum* 80).

36. maesta: only here in the *E.* and only once in the *G.* (4.515), but thirty-seven times in the *A.* Amaryllis is sad and fearful because Tityrus is away and calls on the gods to protect him.

38. aberat: the final short syllable is lengthened before the caesura, as in 3.97 *erit*, 7.23 *facit*, 9.66 *puer*. See Nettleship's excursus in Conington, iii (London, 1871), 465–70; also Norden 450–2, Austin on *A.* 1.308, Fordyce on *A.* 7.174.

38–9. ipsae . . . / ipsi . . . ipsa: an emphatic but remarkably graceful anaphora with polyptoton and varied ictus; cf. 5.62–4.

39. arbusta: often the equivalent of the unmetrical *arbores*. So probably in 4.2, 5.64, *A.* 10.363. Elsewhere trees planted for some purpose, especially for a vineyard, where branches of the vine were trained in festoons between supporting trees, usually elms; see J. Bradford, *Ancient Landscapes* (London, 1957), 162. Here either meaning is possible, but as trees in general are represented by *pinus* (38) 'vineyards' seems more probable.

uocabant: echoing *uocares* (36). The sympathy of nature, the answering voice, is a fundamental assumption of pastoral; cf. 5.62–4, 10.8.

40. quid facerem?: colloquial, e.g. Ter. *Ad.* 214, *Eun.* 831. In V. only here and in 7.14.

me: understood with *cognoscere* in the next line as *alibi* is with *exire* in this (Leo). See below, 58n.

41. tam praesentis: 'so present to my Pray'r' (Dryden); Ps. 46: 1 'A very present help in trouble'. In this sense *praesens* is old and hallowed: Ter. *Phorm.* 345 'praesentem deum', Cic. *Catil.* 2.19 'deos . . . praesentis auxilium esse laturos', *Tusc. disp.* 1.28 'Hercules tantus et tam praesens habetur deus', *G.* 1.10 'agrestum praesentia numina', Hor. *Carm.* 3.5.2–3 'praesens diuus habebitur / Augustus'; see H. Haffter, *Philologus*, 93 (1938), 137–8. But there may be some Hellenistic colouring here, as in the poem by Hermocles of Cyzicus honouring Demetrius Poliorcetes, in which the presence of the human 'god' is contrasted with the remoteness and indifference of the traditional gods: *Ἄλλοι μὲν ἢ μακρὰν γὰρ ἀπέχουσιν θεοί, | ἢ οὐκ ἔχουσιν ὦτα, | ἢ οὐκ εἰσίν, ἢ οὐ προσέχουσιν ἡμῖν οὐδὲ ἕν, | σὲ δὲ παρόνθ' ὁρῶμεν*, 'For the other gods are far away, or do not have ears, or do not exist, or do not pay attention to us at all, but you we see before us.' See Powell, *Collectanea Alexandrina*, p. 174, ll. 15–18. (I am indebted for this reference and suggestion to R. Renehan.) Cf. Ov. *Trist.* 2.53–4 (Augustus) 'per mare, per terram, per tertia numina iuro, / per te praesentem conspicuumque deum', with Luck's note.

42. uidi: of seeing a god, an epiphany; cf. 10.26n.

iuuenem: Octavian is so styled in *G.* 1.500, Hor. *Serm.* 2.5.62, *Carm.* 1.2.41, but in *G.* 1 and *Carm.* 1.2 he is subsequently addressed by name (Caesar) and in *Serm.* 2.5 his identity is immediately indicated.

quotannis: the young saviour's birthday will be regularly observed with a cult appropriate to a deity; cf. 5.67, 7.33.

43. bis senos . . . dies: twelve days each year (42 'quotannis'). But which twelve? Commentators think that a reference to the Lar familiaris is intended and cite Philargyrius (not, as they suppose, Servius, whose commentary on 1.38–2.10 has been lost; see Thilo on 1.37) 'id est principia mensium uel Idus omnium mensium' and Cato, *De agr.* 143.2 (*uilica*) 'Kalendis, Idibus, Nonis, festus dies cum erit, coronam in focum indat, per eosdemque dies Lari familiari pro copia supplicet'—that is, thirty-six days, not counting special occasions, such as births, marriages, departures, homecomings, and the like, when the Lar was honoured (Wissowa 169). Propertius is occasionally cited as indicating that worship of the Lar was restricted to the Kalends in the Augustan period, 4.3.53–4 'omnia surda tacent, rarisque adsueta Kalendis / uix aperit clausos una puella Lares'; but cited with insufficient regard to the character of his poem. It is an imaginary epistle from a Roman matron to her husband, too often away, as it seems to her, on military campaigns; she is desolate, and the reduced condition of the household reflects her mood; even the indwelling spirit is scanted of his due. Commentators also cite Hor. *Carm.* 4.5.33–5 'te multa prece, te prosequitur mero / defuso pateris, et Laribus tuum / miscet numen'; like Prop. 4.3, a poem written many years after *E.* 1 and unrelated to it. For an additional argument see below (*altaria fumant*).

It is understandable that this interpretation, or rather, this failed interpretation, should appear in commentaries before 1902, when it was refuted by Wissowa, *Hermes*, 37.157–9.; but that it should appear in every commentary since then, with no mention of Wissowa, is not. The reference, as Wissowa demonstrated, is to Hellenistic ruler-cult, to the custom, that is, of celebrating a ruler's birthday every month, κατὰ μῆνα, or twelve times a year. Ptolemy III, Ptolemy V, Attalus II, Antiochus Epiphanes (cf. 2 *Macc.* 6:7 'And in the day of the king's birth every month they were brought by bitter constraint to eat of the sacrifices'), Antiochus I of Commagene—all were so honoured, as was Augustus by the people of Mytilene and Pergamum; see W. Schmidt, *Geburtstag im*

Altertum (Giessen, 1908), 12–16. On 3 September 36 BC, Octavian finally defeated Sextus Pompey at Naulochus and shortly thereafter returned to Rome in triumph. The future now seemed secure. In the grateful municipalities of Italy Octavian's statue was set up beside the statues of the traditional gods; he became an additional god, in Hellenistic style. Cf. Appian, *BC* 5.132 *τοῦτο μὲν δὴ τῶν τότε στάσεων ἐδόκει τέλος εἶναι. καὶ ἦν ὁ Καῖσαρ ἐτῶν ἐς τότε ὀκτὼ καὶ εἴκοσι, καὶ αὐτὸν αἱ πόλεις τοῖς σφετέροις θεοῖς συνίδρυον*, 'This seemed to be the end of the civil dissensions. Octavian was now twenty-eight years of age. Cities joined in placing him among the tutelary gods' (H. White in the Loeb Classical Library). It is unlikely, however, that the peculiar veneration of Tityrus corresponded to any political reality in Rome.

While agreeing with Wissowa, Jachmann 115 n.2 suggests that V. got the idea from Theocritus' encomium of Ptolemy II, 17.126–7 *πολλὰ δὲ πιανθέντα βοῶν ὅγε μηρία καίει | μησὶ περιπλομένοισιν ἐρευθομένων ἐπὶ βωμῶν*, 'And many fat thighs of oxen burns he on the reddening altars as the months come round' (Gow). Both Wissowa and Jachmann had in fact been anticipated by Buecheler, *RhM* 30 (1875), 59 = *Kleine Schriften* (Leipzig, 1927), ii. 19; see Wissowa, *Hermes*, 52 (1917), 101 n.4.

bis senos: a mode of reckoning associated with religion and magic, and often, as here, a metrical convenience; see *TLL* s.v. *bis* 2009.24, Hofmann–Szantyr 212.

altaria fumant: cf. *G.* 2.194 'fumantia ... exta', *A.* 8.106 'tepidusque cruor fumabat ad aras'. The Lar was honoured with flowers and wine and incense, but seldom with a blood-offering unless a death had occurred in the family (Wissowa, *Religion und Kultus*, 169).

44–5. R. Hanslik, *WS* 68 (1955), 16–17, detects an allusion to Hesiod in these lines, *Theog.* 24–6 *τόνδε δέ με πρώτιστα θεαὶ πρὸς μῦθον ἔειπον, | Μοῦσαι Ὀλυμπιάδες, κοῦραι Διὸς αἰγιόχοιο | "ποιμένες ἄγραυλοι ..."*, 'And this was the word the goddesses first spoke to me, the Olympian Muses, daughters of Zeus who holds the aegis, "Shepherds of the field ..."'. Cf. West ad loc. (p. 160): 'it is particularly noteworthy that the Muses deliver this typical address to Hesiod in the plural although he is (presumably) alone ... The plural emphasizes that the addressee belongs to a particular class'.

44. responsum: usually plural in V., but cf. *A.* 6.344 'hoc uno responso animum delusit Apollo'; a formal response to a petitioner from someone in authority or from an oracle (cf. *A.* 7.92 'petens responsa Latinus').

primus: with no suggestion that Tityrus had applied elsewhere; the godlike youth was the first to help him, or who could help him, and is gratefully remembered. Cf. *A.* 7.117–18 'ea uox audita laborum / prima tulit finem' (Conington).

45. Leo 18 n.1 would put a comma after *pueri* instead of a semicolon to indicate that *ut ante* is understood with both imperatives.

summittite: 'raise', 'bring up', as in *G.* 3.73, 159; cf. Varro, *RR* 2.2.18 'quos arietes summittere uolunt, potissimum eligunt ex matribus quae geminos parere solent'.

tauros: used improperly, as elsewhere in V., of plough-oxen, *iuuenci*; see Mynors on *G.* 1.45 ('to plough with bulls is a test of heroic rashness').

46. tua: predicative, 'the land will remain yours'.

47–8. quamuis lapis omnia nudus / limosoque palus obducat pascua iunco: cf. *G.* 1.115–17 'amnis abundans / exit et obducto late tenet omnia limo, / unde cauae tepido sudant umore lacunae', Lucr. 5.206–7 'quod superest arui, tamen id natura sua ui / sentibus obducat', *G.* 2.411 'bis segetem densis obducunt sentibus herbae'.

47. lapis . . . nudus: 'the piles of shingle with which a river in flood leaves some of its late course covered, while it is alluvial marsh in other places, depending on the speed of the water' (Mynors).

omnia: separated from *pascua* and therefore emphatic; cf. *G.* 2.277–8, *A.* 7.573–4, 8.4–5, 604–5, 9.38–9. Jachmann 115 n.3: 'hyperbolisch ausgedrückt. Aber gerade daraus spricht am stärksten die unbedingte Liebe zur heimatlichen Scholle'.

49. grauis . . . fetas: *grauis* in the sense of *grauidus* is first attested here (*TLL* s.v. 2276.73) and seems to be V.'s innovation; cf. *A.* 1.273–4 'regina sacerdos / Marte grauis geminam partu dabit Ilia prolem', 6.515–16 'cum fatalis equus saltu super ardua uenit /

Pergama et armatum peditem grauis attulit aluo' (Enn. *Sc.* 76–7 V.[2] = 72–3 J. 'nam maximo saltu superabit grauidus armatis equus, / qui suo partu ardua perdat Pergama'). The tautology 'grauis ... fetas' may be owing to V.'s wish to support and define *grauis* in this new sense.

50. contagia: poetic plural for the unmetrical *contagio*, like *obliuia* for *obliuio*, both of which are found in Lucretius; see Norden on *A.* 6.715. Used once again by V. in *G.* 3.469; cf. Hor. *Epod.* 16.61 'nulla nocent pecori contagia'.

51. flumina nota: cf. *G.* 4.266 'pabula nota', *A.* 2.256, 3.657 'litora nota', 6.221 'uelamina nota', 7.491 'limina nota', 11.195 'munera nota'. See 2. 44n.

52. frigus captabis opacum: reads like a refinement of 2.8 'umbras et frigora captant'; the verb does not occur elsewhere in the *E.* See 3.18n.

opacum: 'opaca uocantur umbrosa' (Festus, p. 200 L.). For the relative frequency of these two adjectives in poetry see *TLL* s.v. 657.25. Generally speaking, *opacus* belongs to the higher, *umbrosus* to the lower style of poetry, with the notable exception of Catull. 37.19 'Egnati, opaca quem bonum facit barba', where G. Lee suspects parody; cf. Pacuvius 362 R.[3] 'nunc primum opacat flora lanugo genas'. Thus *opacus* occurs only here in the *E.* and only once in the *G.* (1.156), but seventeen times in the *A.* (also *G.* 2.55 *opacant, A.* 6.195 *opacat,* where see Norden); *umbrosus* once in the *E.* (2.3), twice in the *G.* (2.66, 3.331), and twice in *A.* (8.34, 242).

53. hinc: corresponds to *hinc* in 56, 'on this side', 'on that'; each adverb being further defined by a prepositional phrase, 'uicino ab limite', 'alta sub rupe'. Cf. 3.12, *G.* 4.423, *A.* 3.616–17, 6.305, where see Norden, 385, 7.209, 10.656–7, and see Wagner here. The construction is as old as Homer, *Od.* 10.511 *αὐτοῦ ... ἐπ' Ὠκεανῷ βαθυδίνῃ*, 'there ... by the deep-eddying Ocean', 11.69 *ἐνθένδε ... δόμου ἐξ Ἀΐδαο*, 'hence ... from the house of Hades'.

quae semper: *suasit* or *suadebat* is to be supplied from *suadebit* in l. 55.

uicino ab limite: the *limes* was a strip of land left unploughed to mark the boundary and used as a highway.

ab: for V.'s use of *ab* before consonants see Wagner, *Quaest. Virg.* 1.1; also *TLL* s.v. *a, ab, abs* 2.57.

saepes: the boundary hedge of willows where the bees feed. The willow was a variously useful tree; it provided leafage for the flock, shade for shepherds, hedges for the crops, and food for bees (*G.* 2.434–6).

54. Hyblaeis: not an established ornamental epithet, like *Chaonius* (9.13) or *Cydonius* (10.59), but as such V.'s own invention; this, or rather 7.37, being the first reference in poetry to the honey of Hybla. The city Megara Hyblaea, situated a few miles north of Syracuse, had long since ceased to exist, but the older name of the place remained, Strabo says, because of the excellence of its honey, 6.2.2 *τὸ δὲ τῆς Ὕβλης ὄνομα συμμένει διὰ τὴν ἀρετὴν τοῦ 'Υβλαίου μέλιτος*. In the opinion of Varro, *RR* 3.16.14, Sicilian honey is the best 'quod ibi thymum bonum frequens est', and Columella 9.14.19 remarks that thyme blooms longer at Hybla than anywhere else in Sicily (cf. *E.* 7.37 'thymo mihi dulcior Hyblae'). V. may already have had personal knowledge of this region, which he evidently knew later: *Vita Donati* 13 'quamquam secessu Campaniae Siciliaeque plurimum uteretur'; the description of the coast near Syracuse in *A.* 3.688–98 suggests autopsy.

The honey of Hybla was to enjoy a long success in poetry: Ovid, *Ars* 2.517, 3.150, *Trist.* 5.6.38, 13.22, *Ex Pont.* 2.7.26, 4.15.10, Columella 10.170, Lucan 9.291, Seneca, *Oed.* 601, Calp. Sic. 4.63, Statius, *Silu.* 2.1.48, 3.118, *Ach.* 1.557, Martial 2.46.1–2, 5.39.3, 7.88.8, 9.26.4, 13.105.1, Silius 14.200, *Peruigilium Veneris* 51, 52, Claudian, *De rapt. Pros.* 2.125, Shakespeare, *1 Henry IV*, I.ii.42 'the honey of Hybla', *Julius Caesar* V.i.34 'but for your words, they rob the Hybla bees', Drummond of Hawthornden, *Madrigals*, 5 'Hybla's hills', Collins, *Ode to Simplicity*, 14 'Hybla's thymy shore', Bridges, *Testament of Beauty*, 2.334 'Hybla's renown'.

florem depasta salicti: to be distinguished from the accusative of respect, e.g. *G.* 4.181 'crura thymo plenae' or *A.* 1.320 'nuda genu', a Graecism and not found in the *E.* The past participle had originally a medial force and was therefore capable of assuming a direct object; for this construction and its development, under Greek influence, in Latin poetry see G. Landgraf, *ALL* 10 (1896–8), 215–24, Löfstedt ii.421–2, Kühner–Stegmann i.288–92, Hofmann–Szantyr 36–8. Of the seven examples in the *E.*, four occur in *E*. 6: 1.54, 3.106, 6.15, 53, 68, 75, 7.32, 8.4. For the verb cf. 5.77, Lucr. 3.11–12 'floriferis ut apes in saltibus omnia libant, / omnia nos itidem depascimur aurea dicta'.

55. leui: 'soft', cf. Catull. 64.273 'leuiterque sonant plangore cachinni' and see Fedeli on Prop. 1.3.43 'interdum leuiter mecum deserta querebar'.

suadebit inire: strong caesura in the third foot and weak in the fourth; for lines so constructed see Norden 427–9.

susurro: here only in V., and here only of bees (but cf. *G*. 4.260 *susurrant*); the hedge is drowsy with their soft murmuring.

56. alta sub rupe canet: so Gallus sings, 10.14 'sola sub rupe', and Orpheus, *G*. 4.508 'rupe sub aëria'; perhaps suggested by [Theocr.] 8.55 ἀλλ' ὑπὸ τᾷ πέτρᾳ τᾷδ' ᾄσομαι, 'Rather beneath this rock will I sit and sing' (Gow).

frondator: it was his task to prune the vine and the elm on which the vine depended so as to let in sufficient light; cf. 2.70, *G*. 1.156–7, 2.362–70, Catull. 64.41 'non falx attenuat frondatorum arboris umbram'. The leaves might be used for fodder; see 9.61 n. He sings to pass the time, as solitary workers will, 'sweetning his labour with a cheareful song' (Francis Quarles, *On the Ploughman*). Cf. *G*. 1.293–4 'interea longum cantu solata laborem / arguto coniunx percurrit pectine telas', Tib. 2.1.65–6 'atque aliqua adsiduae textrix operata Mineruae / cantat', where see K. F. Smith, *Moretum* 29–30 (Simulus grinding grain) 'modo rustica carmina cantat / agrestique suum solatur uoce laborem'. See also Pease on Cic. *De nat. deor*. 2.89, to whose note Kenney on *Moretum* 29–30 adds Quintil. 1.10.16.

57. raucae, tua cura, palumbes: for the word-order, parenthetic apposition, see J. B. Solodow, *HSCP* 90 (1986), 129–53. V. uses this construction five times in the *E*.: here, 2.3, 3.3, 7.21, 9.9; three times in the *G*.: 2.146–7, 4.168, 246 'dirum tiniae genus', all three of this simple type; and only once (but cf. *A*. 3.305, 537, 7.717) in the *A*., 6.842–3 'geminos, duo fulmina belli, / Scipiadas', where the pattern is disguised by extension into the next line (cf. Stat. *Theb*. 3.237–8 'turpis, primordia belli, / insidias') and the reader's attention distracted by Ennian sonorities (see Norden ad loc.). Evidently V. felt that parenthetic apposition was appropriate to the *E*., marginally so to the *G*., but hardly to the *A*. Strictly speaking, the only precedent in Hellenistic poetry is Hedylus 2.5

G.–P. (= *AP* 5.199). Cf., however, Meleager 31.3–4 G.–P. (= *AP* 5.144) ἡ φιλέραστος, ἐν ἄνθεσιν ὥριμον ἄνθος, | Ζηνοφίλα, 'the lovely, the spring flower of flowers, Zenophila', the structure of which bears a strange resemblance to that of *A.* 6.842–3.

O. Skutsch, *RhM* 99 (1956), 198–9, citing Prop. 3.3.31 'et Veneris dominae uolucres, mea turba, columbae', detects the presence of Gallus here; see 10.22n.

palumbes: wood-pigeons, whose song consists of a triple phrase with a sigh at the close, Pliny, *NH* 10.106 'cantus ... trino conficitur uersu praeterque in clausula gemitu'. The Romans kept pigeons and turtle-doves on their farms (Varro, *RR* 3.7–8).

58. gemere: here first of bird-sound (*TLL* s.v. 1762.35); cf. Theocr. 7.141 ἔστενε τρυγών, 'the turtle-dove moaned'. *Gemere* and *cessabunt* are understood with *palumbes*; for this, the ἀπὸ κοινοῦ construction, see Norden on *A.* 6.471, E. J. Kenney, *CQ*, NS 8 (1958), 55.

aeria ... ulmo: cf. Catull. 64.291 'aeria cupressu', *A.* 3.680 'aeriae quercus', *E.* 3.69 'aeriae ... palumbes', Hor. *Carm.* 1.2.9–10 'summa ... ulmo, / nota quae sedes fuerat columbis'. Pigeons seek high places, Varro says (*RR* 3.7.1), because of their natural shyness.

59–62: highly stylized 'impossibilities', ἀδύνατα, found in both Greek and Latin poetry, but more often in Latin; nearly 200 examples were collected and classified by H. V. Canter, *AJP* 51 (1930), 32–41. See also K. F. Smith on Tib. 1.4.65–6, E. Dutoit, *Le Thème de l'Adynaton dans la poésie antique* (Paris, 1936), with the review by H. Herter, *Gnomon*, 15 (1939), 205–11, and G. O. Rowe, *AJP* 86 (1965), 387–96. This is the most elaborate passage of its kind in the *E.* (59 *ante*, 61 *ante*, 63 *quam*); cf. 3.90–1, 8.26–8, 52–6.

59. leues: 'on the wing', as in *G.* 4.55, *A.* 5.838; 'auium ante cerui uolabunt more' (Philargyrius 1).

in aethere: Hellenistic poets use ἀήρ and αἰθήρ indifferently (Mineur on Callim. *Hymn* 4.176), but *in aere* is usual in Latin: Plaut. *Asin.* 99–100 'iubeas una opera me piscari in aere, / uenari autem rete iaculo in medio mari' (where see Leo), Cic. *De nat. deor.* 2.42 'in aqua, in aere', Varro, *LL* 5.75 'animalia in tribus locis quod sunt, in aere, in aqua, in terra', *G.* 1.404 'apparet liquido sublimis in aere (M: *aethere* R) Nisus'. But V. was thinking here of Lucr. 3.784–5 'denique in aethere non arbor, non aequore in alto / nubes esse queunt nec pisces uiuere in aruis' (cf. Lucr. 5. 128–9).

61. exsul: only here in V., with melancholy emphasis; grammar notwithstanding, V.'s reader will sooner think of Meliboeus and his fellow-exiles than of the warlike Parthians and Germans. Tityrus' impossibly distant places (here first in an *adynaton*; see A. Manzo, *Enc. Virg.* s.v. *Arar*) are only too real for Meliboeus—as his answer demonstrates (64–6).

62. Ararim: strict reciprocity would have the Parthians drinking from the Rhine, the German river; cf. Luc. 7.433 'libertas ultra Tigrim Rhenumque recessit'. But the Arar, now the Saône, in its upper reaches is not far removed from the Rhine, and while V. was still a boy the Germans crossed the Rhine and occupied much of the territory between it and the Arar, which their king, Ariovistus, arrogantly claimed as 'his Gaul' (Caes. *BG* 1.34.4 'in sua Gallia'). Read *aut Rhenum Parthus* (K. Wellesley, *CP* 63 (1968), 139–41) and fled is the music. V.'s ear seems to have been engaged by the sound or rhyming effect of *Ararim* . . . *Tigrim*, which is set off by the repetition *aut* . . . *aut* and chiastic word-order. So J. Aymard, *Latomus*, 14 (1955), 120–2, who also suggests another reason for opposing these two rivers: the Arar was known to be sluggish, Caes. *BG* 1.12.1 'flumen est Arar . . . incredibili lenitate, ita ut oculis in utram partem fluat iudicari non possit', but the Tigris terribly swift, Varro, *LL* 5.100 'uehementissimum flumen' (see *E.* 5.29n.). Poets from Homer onward sometimes identify a people by naming the river from which they drink; cf. *G.* 1.509 'hinc mouet Euphrates, illinc Germania bellum', 4.211–12 'nec populi Parthorum aut Medus Hydaspes / obseruant', and see Nisbet–Hubbard on Hor. *Carm.* 2.20.20, Tarrant on Sen. *Ag.* 318ff.

Parthus: for the use of the collective singular to designate an enemy ('the Hun') see Löfstedt i.22–3.

63. The *adynata* end with an oath or asseveration of fidelity, as in Prop. 2.15.31–6.

64. ibimus Afros: the name of a people instead of the place where they live, a construction as old as Homer, *Od.* 4.84 *Αἰθίοπάς θ' ἱκόμην καὶ Σιδονίους καὶ Ἐρεμβούς*, 'and I came to the Ethiopians and the Sidonians and the Erembi'; see Denniston on Eur. *El.* 917, Hofmann–Szantyr 50. Cf. *A.* 3.254 'ibitis Italiam'.

64–5. nos . . . / pars: for the distributive apposition cf. *A.* 1.423–5 'Tyrii: pars . . . / . . . / pars', 12.277–8 'fratres . . . / pars . . . pars', Sall. *Iug.* 38.5 'milites Romani . . . alii, alii . . ., pars', and see *TLL* s.v. *pars* 454.48, Kühner–Stegmann ii.72–3.

65. rapidum cretae: idiosyncratic but not unintelligible Latin ('which has yet to be supported by examples', Conington). *Rapidus* is a standing epithet of rivers, with *rapax* as a metrical variant: Lucr. 1.15 'rapidos . . . amnis', 17 'fluuiosque rapaces' (cf. *G.* 3.142 'fluuiosque . . . rapacis'); and *rapax*, like other such adjectives (Kühner–Stegmann i.451), may assume an objective genitive; see 4.24n. 'The verbal force, dominant in *rapax*, was not far below the surface in *rapidus*; and this perhaps made it easier for V. to write *rapidum cretae* in *E.* 1.65' (Mynors on *G.* 4.425). Cf. Plaut. *Men.* 64–5 'ingressus fluuium rapidum ab urbe haud longule, / rapidus raptori pueri subduxit pedes'.

Oaxen: context and proportion alike require that this be the name of a great river in the East, apparently the Oxus, now the Amu-darya, which flows into the Aral Sea; a remote and fabulous river of which V. can have had only the vaguest notion, and which he calls, for reasons not entirely clear, the Oaxes. Pliny states that the Oxus rises in Lake Oaxus, *NH* 6.48 'Oxus amnis, ortus in lacu Oaxo'; and H. Myśliwiec, *RE* Suppl. xi. 1022–30, concludes, after a dense argument, that *Oaxos/-es* is a by-form of *Oxos*. Arrian describes the Oxus as the greatest of the rivers crossed by Alexander except those in India, with a sandy bottom and a current so swift that piles sunk by the royal engineers were instantly swept away, 3.29.3 *ὡς τὰ καταπηγνύμενα πρὸς αὐτοῦ τοῦ ῥοῦ ἐκστρέφεσθαι ἐκ τῆς γῆς οὐ χαλεπῶς*; cf. Curtius Rufus 7.10.13 'quia limum uehit, turbidus semper'.

The old opinion that V. is here referring to Crete should be mentioned only to be rejected. There was no such river in Crete, nor would a beautiful and fertile island washed round by the soft Mediterranean (*Od.* 19.173 *καλὴ καὶ πίειρα, περίρρυτος*) be compatible as a place of exile with the African desert, distant Britain, or the frozen North. See Clausen, *AJP* 76 (1955), 60–1, adding Gratt. *Cyn.* 132 'Eois . . . Sabaeis'; see also G. Funaioli, *Esegesi virgiliana antica* (Milan, 1930), 309–10.

66. penitus toto diuisos orbe Britannos: *penitus* is to be taken with *diuisos*, as it is with *diuersa* in *A*. 9.1 'Atque ea diuersa penitus dum parte geruntur' (Conington). Cf. Catull. 11.11–12 'ultimosque Britannos', 29.4 'ultima Britannia', and see Pease on Cic. *De nat. deor*. 2.88 'in Scythiam aut in Britanniam'.

67–9. Cf. *G*. 3.474–7, noting 'post tanto' and 'desertaque regna / pastorum'.

67. en umquam: asking a passionate rhetorical question, as in Plaut. *Trin*. 589–90 'o pater, / en umquam aspiciam te?' Cf. 8.7 and see A. Köhler, *ALL* 6 (1889), 26–7.

68. tuguri: only here in V. Cf. Varro, *RR* 3.1.3 'quod tempus si referas ad illud principium, quo agri coli sunt coepti atque in casis et tuguriis habitabant nec murus et porta quid esset sciebant, immani numero annorum urbanos agricolae praestant'.

69. post: preposition or adverb? The former, a temporal preposition, according to the ancient grammarians: 'quasi rusticus per aristas numerat annos' (Philargyrius 1; see V. Schindel, *Hermes*, 97 (1969), 472–89); so also Klingmer 21: 'mein Reich nach vielen (?) Ernten sehend'. This interpretation is defended on the analogy of *πόα*, which occasionally means 'year' in Hellenistic poetry (references in Schindel 481 n.4); but 'post aliquot aristas' in this sense would contradict 'longo post tempore' in 67 (Leo 11 n.1). And what would there be for Meliboeus to see and marvel at? Leo (ibid.) takes *post* as a local preposition and compares 3.20 'tu post carecta latebas'; an interpretation hesitantly accepted by V. Pöschl, *Die Hirtendichtung Virgils* (Heidelberg, 1964), 59: 'hinter ein paar Ähren mein Reich schauend'. Is Meliboeus to be imagined creeping up to his old homestead (so as not to be seen by the brutal soldier?) and peering at it from behind a few ears of grain? And again, what would there be for him to marvel at? *Post* must be an adverb repeating, pathetically repeating, 'longo post tempore'; so Germanus (Antwerp, 1575), La Cerda, Heyne, and others, but Conington finds the repetition 'very awkward', as does Pöschl. Germanus compares *G*. 2.259–62 'his animaduersis terram *multo ante* memento / excoquere et magnos scrobibus concidere montis, / *ante* supinatas Aquiloni ostendere glaebas / quam laetum infodias uitis genus'. It will be noticed that *ante*, like *post* (cf. *A*. 1.612,

2.216, 12.185), is followed by an accusative which it does not govern.

aliquot: a prosaic word found only here in V. (*TLL* s.v.).

mea regna: so Scaevola refers to Crassus' Tusculan estate, Cic. *De or.* 1.41 'nisi hic in tuo regno essemus'. Meliboeus dreams of returning one day, of viewing his 'kingdom' and marvelling at its sadly impoverished state.

70. Note the coincidence of ictus and accent; such lines are rare: 5.52, 7.33, 8.80.

impius: as having taken part in a civil war; cf. Hor. *Epod.* 16.9 'impia ... aetas', *Carm.* 2.1.30 'impia proelia', and see H. Fugier, *Recherches sur l'expression du sacré dans la langue latine* (Paris, 1963), 382–3.

noualia: fallow land left unseeded in alternate years; cf. *G.* 1.71–2, Pliny, *NH* 18.176 'nouale est quod alternis annis seritur'.

71. barbarus: not a foreign mercenary, who could hardly be termed *impius*; the contrast is between soldier—the brutal, blood-stained soldier—and civilian.

discordia: domestic strife, civil war; cf. *G.* 2.496 'infidos agitans discordia fratres', Cic. *Phil.* 7.25 'omnia ... plena odiorum, plena discordiarum, ex quibus oriuntur bella ciuilia'. The collocation 'discordia ciuis' is striking; repeated in *A.* 12.583 'trepidos inter discordia ciuis' and imitated by Propertius 1.22.5 'cum Romana suos egit discordia ciuis'.

73. insere ... piros: 'graft your pear-shoots'; cf. Varro, *RR* 1.40.5 'si in pirum siluaticam inserueris pirum quamuis bonam, non fore tam iucundam quam si in eam quae siluestris non sit'. Pear can also be grafted upon apple; see A. S. Pease, *TAPA* 64 (1933), 66–76.

nunc: sarcastic, usually *i nunc*; see *TLL* s.v. *eo* 632.37, E. B. Lease, *AJP* 19 (1898), 59–69. In 9.50 the advice is seriously intended.

74. ite meae, felix quondam pecus, ite capellae: cf. 7.44, 10.77.

75. posthac: cf. 3.51; not found elsewhere in V.

antro: ἄντρῳ, cf. Theocr. 3.6, 13, 6.28, 7.137, 149, [8.72], [9.15], 11.44; first attested in the *E.*, 5.6, 19, 6.13, 9.41. See

Norden on *A*. 6.10, Mynors on *G*. 2.469. *Spelunca*, the ordinary Latin word, is not found in the *E*. but occurs in the *G*. and *A*. Theocritus' Amaryllis lives in a cave (3.6 ἄντρον) as do post-Theocritean herdsmen, [9.15–16] κἠγὼ καλὸν ἄντρον ἐνοικέω | κοίλαις ἐν πέτραισιν, 'I too dwell in a fine cave among the hollow rocks'. In the *E*., however, caves or grottoes provide only a temporary shelter—for Meliboeus as he watches his flock, for Mopsus and Menalcas while they are singing (5.6–7). V.'s herdsmen live in cabins or huts, to which they return in the evening with their animals (2.29, 3.33, 7.44, 49–50, 10.77).

76. dumosa: cf. *G*. 3.314–15 (goats) 'pascuntur uero siluas et summa Lycaei / horrentisque rubos et amantis ardua dumos'. For the adjective see above, 5n.

pendere: an image admired by Wordsworth, who in the preface to the 1815 edition of his poems compares Shakespeare's 'half way down / *Hangs* one who gathers samphire' and observes that 'the apparently perilous situation of the goat, hanging upon the shaggy precipice, is contrasted with that of the shepherd contemplating it from the seclusion of the cavern in which he lies stretched at ease and in security'. Ovid would recall this passage in exile, *Ex Pont*. 1.8.51–2 'ipse ego pendentis, liceat modo, rupe capellas, / ipse uelim baculo pascere nixus ouis'.

78. cytisum: *citiso virgiliano*, moon-trefoil, *Medicago arborea* L., mentioned twice by Theocritus, 5.128 and 10.30, as goat-food; 'an inhabitant of Greece and southern Italy' (Abbe 118). Cf. 2.64, 9.31 (cows), 10.30 (bees).

79. poteras: in effect a polite invitation; cf. Ov. *Met*. 1.678–9 'at tu, / quisquis es, hoc poteras mecum considere saxo'. Possibly V. was thinking of Polyphemus' invitation to Galatea, Theocr. 11.44 ἅδιον ἐν τὤντρῳ παρ' ἐμὶν τὰν νύκτα διαξεῖς, 'you will pass the night more pleasantly in the cave with me'.

80–1. Apples, chestnuts, and cheese—simple rustic fare set out in an elegant tricolon.

mitia: 'matura, quae non remordent cum mordentur' (Philargyrius 1), an unexpected touch of verbal wit. Cf. Hor. *Epod*. 2.17 'mitibus pomis', Mart. 10.48.18 'mitia poma'.

castaneae: the sweet chestnut, *Castanea sativa* Mill.; see Abbe 89–90 and cf. 2.52, 7.53.

molles: soft and mealy, perhaps roasted over a slow fire like those of Martial 5.78.15 'lento castaneae uapore tostae'. The sweet chestnut ripens in late autumn (Calp. Sic. 2.82–3).

82–3. Cf. 2.66–7, Ap. Rhod. 1.451–2 *αἱ δὲ νέον σκοπέλοισιν ὑποσκιόωνται ἄρουραι,* | *δειελινὸν κλίνοντος ὑπὸ ζόφον ἠελίοιο,* 'and now the fields are overshadowed by the rocks as the sun descends towards evening and darkness', Hor. *Carm.* 3.6.41–3 'sol ubi montium / mutaret umbras et iuga demeret / bobus fatigatis', and see J. Nováková, *Umbra: Ein Beitrag zur dichterischen Semantik* (Berlin, 1964), 35.

83. altis de montibus: a Lucretian phrase slightly modified, 4.1020 'de montibus altis', 5.492, 663, 6.735. Cf. 7.66.

ECLOGUE 2

Introduction

The Second *Eclogue* begins with a studied simplicity:

> Formosum pastor Corydon ardebat Alexin,
> delicias domini.

Virgil is imitating a device, a configuration—the subject of passion and its object together with a verb—which he had observed in Hellenistic poetry. So, in effect, begins the story of Polyphemus and Galatea:[1]

> *ὡρχαῖος Πολύφαμος, ὅκ' ἤρατο τᾶς Γαλατείας.* (Theocr. 11.8)

Polyphemus of old, when he loved Galatea.

So also the story of Acontius and Cydippe:

> *Αὐτὸς Ἔρως ἐδίδαξεν Ἀκόντιον, ὁππότε καλῇ*
> *ἤθετο Κυδίππῃ παῖς ἐπὶ παρθενικῇ,*
> *τέχνην.* (Callim. *Aet.* fr. 67.1–3 Pf.)

Love himself taught Acontius the art, when the boy was burning for the beautiful girl Cydippe.

And so the story of Orpheus and Calaïs:[2]

> *Ἢ ὡς Οἰάγροιο πάϊς Θρηΐκιος Ὀρφεὺς*
> *ἐκ θυμοῦ Κάλαϊν στέρξε Βορηϊάδην.*
> (Phanocles, *Ἔρωτες ἢ Καλοί* fr. 1.1–2 Powell)

Or how Oeagrus' son, Thracian Orpheus, loved Calaïs, the son of Boreas, with all his heart.

Virgil's imitation is evident, as he intended it to be, but not inert. Even in his earliest work—and the Second *Eclogue* may be his

[1] See Gow on Theocr. 11.13.

[2] Cf. also [Theocr.] 23.1–2 *Ἀνήρ τις πολύφιλτρος ἀπηνέος ἤρατ' ἐφάβω, | τὰν μορφὰν ἀγαθῶ, τὸν δὲ τρόπον οὐκέθ' ὁμοίω*, 'A passionate lover longed for a cruel youth, whose form was goodly but his ways unlike thereto' (Gow), Parthenius, *Ἐρωτικὰ παθήματα* 4.1 *Ἀλέξανδρος ὁ Πριάμου βουκολῶν κατὰ τὴν Ἴδην ἠράσθη τῆς Κεβρῆνος θυγατρὸς Οἰνώνης*, 'Alexander, Priam's son, while tending cattle on Mt. Ida, fell in love with Oenone, the daughter of Cebren', 10.1, 24.1, 25.1, 26.1, Ov. *Met.* 4.55–62 (Pyramus and Thisbe). See Du Quesnay 48–9, 214 n.130.

earliest published work[3]—Virgil shows himself incapable of mere appropriation: the elegance of his first line is a Latin elegance.[4]

With 'Formosum pastor' Virgil insinuates a contrast between town and country—the uncouth country, 'sordida rura' (28), for which Corydon pleadingly apologizes—and with 'delicias domini' involves his poor shepherd in a hopelessly unequal 'triangle': Corydon's rival for the affections of Alexis[5] is their master. A typical landowner, Iollas[6] (named, almost casually, in l. 57; here it is not the name but the position that matters) lives in town,[7] where for his pleasure he keeps Alexis.

Frustrated, on fire with passion, Corydon retires to the usual grove, and there, in the dense shade of the beeches, utters his artless complaint[8] to the woods and hills, singing in the sultry afternoon, singing until the shadows lengthen and the slow-moving oxen return from the fields. With the fading light Corydon's song draws to a close,[9] and the madness of his unquenched desire is brought home to him, 'a, Corydon, Corydon, quae te dementia cepit!' (69). He reproaches himself for his neglected chores; he will find another, a less fastidious, Alexis.

The Second *Eclogue* is modelled on Theocritus' Eleventh *Idyll*,[10] the Cyclops in love. No longer Homer's monster but Theocritus' fellow-countryman, Polyphemus of old has become a sentimental rustic, grotesquely in love with Galatea, his unattainable Galatea, who lives not in town but in the sea. (The opposition

[3] See *E.* 8, Introduction, p. 238.

[4] See ad loc.

[5] For the connotation of the name see below, 1 n. Cf. Tac. *Ann.* 14.42.1 'Haud multo post praefectum urbis Pedanium Secundum seruus ipsius interfecit, seu negata libertate, cui pretium pepigerat, siue amore exoleti incensus et dominum aemulum non tolerans'.

[6] An unpastoral name; see ad loc.

[7] Cf. *E.* 9.1–6, Longus 3.31.4, 4.5.1.

[8] 'Haec incondita' (4)—a suitable fiction.

[9] The equation of song with the passage of time, so that song (or conversation) ends as the day ends, is a characteristic of Virgilian pastoral; perhaps suggested by Theocr. 11.12–15: Polyphemus would sing of Galatea from daybreak until his sheep returned of their own accord in the evening. So ends the First *Eclogue* (with a hint of autumn also) and the Sixth, and so ends Virgil's pastoral song (10.75–7). In Theocritus, the preferred time of day is noon, 'the bright severity of noon' (Thomson, *Summer* 503): 1.15 τὸ μεσαμβρινόν, 'at noon', 6.4 θέρεος μέσῳ ἄματι, 'in summer at midday,' 7.21 μεσαμέριον, 'at noon'.

[10] With some reference to his Third *Idyll*; see Du Quesnay 43 ff.

between town and country, latent in all pastoral, never surfaces in Theocritus.[11])

In both Theocritus and Virgil, the complaint of the distressed lover is introduced with a description of his case: a concise introduction of five lines in Virgil, of eighteen in Theocritus, addressed to his friend the physician and poet Nicias, on lovesickness, the only remedy for which, as Nicias knows and as the experience of Polyphemus proves, is song.

The song of Polyphemus is divided into unequal sections which occur, as his thoughts or changing moods may be supposed to occur, in no very orderly sequence. Why does she flee from him? For himself, he remembers, it was love at first sight when she came with his mother, and he was their guide, to gather hyacinths on the hill. He knows he is repulsive . . . At last, realizing the hopelessness of his case, Polyphemus reproaches himself, 'O Cyclops, Cyclops'.

Each section begins with an apostrophe, an entreaty, or a regret, which momentarily interrupts and defines his song: 19 *Ὦ λευκὰ Γαλάτεια,* 'O white Galatea', 30 *γινώσκω, χαρίεσσα κόρα,* 'I know, fair maid', 42 *ἀλλ' ἀφίκευσο ποθ' ἁμέ,* 'Do come to me', 50 *αἰ δέ τοι αὐτὸς ἐγὼν,* 'But if I myself', 54 *ὤμοι,* 'Ah me', 63 *ἐξένθοις, Γαλάτεια,* Come forth, Galatea', 72 *ὦ Κύκλωψ Κύκλωψ,* 'O Cyclops, Cyclops'.

Corydon's song is similarly disjointed (the course of passion, Bentley remarks on *Paradise Lost* 1.107, loves asyndeta): 6 'O crudelis Alexis', 19 'Alexi', 28 'o tantum libeat mecum tibi', 45 'huc ades, o formose puer', 56 'rusticus es, Corydon', 69 'a, Corydon, Corydon'. And yet Corydon seems different, not so naïve, so simply true as Polyphemus, nor so passionate; he seems self-conscious, as though aware—as Virgil expected his reader to be aware—of Polyphemus. Corydon's is a more composed passion.

Bibliography

Posch 29–53.

I. M. Le M. Du Quesnay, 'From Polyphemus to Corydon', in D. West and T. Woodman (eds.), *Creative Imitation and Latin Literature* (Cambridge, 1979), 35–69, 206–21.

[11] See Dover, p. lvii: 'Theokritos's countrymen, although (unlike characters in the pastoral poetry of later times) they do not moralize about the superiority of their life to that of the city-dweller, are intensely conscious of its beauties and comforts and extremely fluent in describing these in terms appropriate to a man who has come out on a fine day after long confinement in a city'.

M. Geymonat, 'Lettura della seconda bucolica,' in Gigante 107–27.

E. J. Kenney, 'Virgil and the Elegiac Sensibility', *ICS* 8 (1983), 49–52.

1. Formosum pastor Corydon ardebat Alexin: adjective at the beginning of the line, noun at the end, *formosum* . . . *Alexin*, a neoteric arrangement (see Norden 391, C. Conrad, *HSCP* 69 (1965), 225–9, Ross (1969), 132–4); modifiers and names disposed in a chiastic relation, *formosum pastor, Corydon* . . . *Alexin*, with the names joined, or separated, by an old verb used in a new way—a complicated, artful line, which yet contains the (seemingly) simple statement, 'Corydon ardebat Alexin'.

Corydon: the cowherd with musical pretensions in Theocr. 4.29–33, the victor in the singing-match in *E.* 7.

ardebat: first attested here (the text of Ter. *Phorm.* 82 is uncertain) with the accusative, a construction V. was not to repeat, except in 'quoting' this line (5.86); see *TLL* s.v. 486.74. Cf. Moschus fr. 2.3 Gow *ὡς Ἀχὼ τὸν Πᾶνα, τόσον Σάτυρος φλέγεν Ἀχώ*, 'As Echo for Pan, so Satyr burned for Echo', Prop. 1.13.23 'nec sic caelestem flagrans amor Herculis Heben', where see Fedeli. See also Tränkle 69–70. The metaphor itself is banal, and in Latin as old as Ter. *Eun.* 72 'amore ardeo'. Here Corydon's passion for Alexis seems to be associated with the noonday heat; cf. Meleager 79.1–4 G.–P. (=*AP* 12.127).

Alexin: the traditional name of a catamite; cf. Anacreon 394 (b) Page, Plato, *AP* 7.100.1, Meleager 79.1 G.–P. (=*AP* 12.127), 80.3 G.–P. (=*AP* 12.164), Straton, *AP* 12.229.1. But the identification of Alexis as a slave-boy loved by V. seems to have been well established by the time of Martial, who however makes Maecenas the donor, 5.16.11–12, 6.68.6, 7.29.7–8, 8.55.12–20, 73.9–10; cf. Apuleius, *Apol.* 10 'quanto modestius tandem Mantuanus poeta, qui itidem ut ego puerum amici sui Pollionis bucolico ludicro laudans et abstinens nominum sese quidem Corydonem, puerum uero Alexin uocat', *Vita Donati* 9 'maxime dilexit Cebetem et Alexandrum, quem secunda Bucolicorum ecloga Alexim appellat, donatum sibi ab Asinio Pollione', Philargyrius 1 (Servius is lacking here) '*Alexin*, quem dicunt Alexandrum, fuit seruus Asinii Pollionis. Vergilius, rogatus ad prandium, cum uidisset in ministerio nimium pulcherrimum, dilexit eumque dono accepit'.

2. domini: Iollas (below, 57).

nec quid speraret habebat: 'nor had he anything to hope for'; the normal prose construction, e.g. Cic. *Verr.* 2.2.69 'neque ... quid responderet habuit' (Milton, *Paradise Regained* 4.2 'the Tempter stood, nor had what to reply'), *Phil.* 2.62 'quo se uerteret non habebat'.

3. densas, umbrosa cacumina, fagos: so Acontius sang of his passion for Cydippe (see below, *fagos*); Orpheus of his passion for Calaïs, Phanocles, Ἔρωτες ἢ Καλοί fr. 1.3–4 Powell πολλάκι δὲ σκιεροῖσιν ἐν ἄλσεσιν ἕζετ' ἀείδων | ὃν πόθον, 'often he would sit in the shady groves and sing of his desire'; and Propertius of his passion for Cynthia, 1.18.1–3 'Haec certe deserta loca et taciturna querenti, / et uacuum Zephyri possidet aura nemus. / hic licet occultos proferre impune dolores', where see Fedeli (pp. 416–19). Cf. 9.9 and, for the parenthetic apposition, see 1.57n.

umbrosa: 1.52n.

fagos: in Callimachus, as paraphrased by the fifth-century epistolographer Aristaenetus 1.10.57 Mazal (see Pfeiffer on Callim. *Aet.* fr. 73), Acontius utters his complaint sitting under the oaks or the poplars, φηγοῖς ὑποκαθήμενος ἢ πτελέαις, and φηγοῖς may have put V. in mind of the familiar *fagus* (1.1n.); cf. Varro (Charisius 165.17 Barwick) 'fagus quas Graeci φηγούς uocant'. See A. La Penna, *Maia*, NS 15 (1963), 488, F. Cairns, 'Propertius 1.18 and Callimachus, *Acontius and Cydippe*', *CR*, NS 19 (1969), 133, Ross (1975), 71–2. Since V. would have known that *fagus* and φηγός, the Valonia oak, *Quercus aegilops* L., were not the same tree, *fagus* may be considered a 'learned catachresis' (Kenney 50 n.23).

4. incondita: 'rude', 'artless'; cf. Varro, *Men.* 363 Buecheler 'homines rusticos in uindemia incondita cantare', Livy 4.53.11 'alternis inconditi uersus militari licentia iactati', and see *TLL* s.v. 1002.11.

4–5 solus / montibus et siluis: in Euripides, a female speaker, alone on the stage, acquaints heaven and earth with her disquiet or misery (*Med.* 57, *Andr.* 91, *Ion* 870, *El.* 59, *IT* 42), as does the lover in New Comedy; see F. Leo, *Plautinische Forschungen*[2] (Berlin, 1912), 151, on Plaut. *Merc.* 3–5 'non ego item facio ut alios in comoediis / ui uidi amoris facere, qui aut nocti aut die / aut soli aut lunae miserias narrant suas', and R. F. Thomas, *HSCP* 83 (1979), 198; also Page on Eur. *Med.* 57–8, Stevens on Eur. *Andr.*

93, and Pfeiffer on Callim. fr. 714.3–4 ὅτε κωφαῖς | ἄλγεα μαψαύραις ἔσχατον ἐξερύγῃ, 'when finally a man pours out his troubles to the unheeding gusts of wind', adding Catull. 64.164 (Ariadne) 'sed quid ego ignaris nequiquam conqueror auris?' Under the influence of Callimachus and other Hellenistic poets, V. adapted this motif to pastoral; cf. Theocr. 11.17–18 (Polyphemus) καθεζόμενος δ' ἐπὶ πέτρας | ὑψηλᾶς ἐς πόντον ὁρῶν ἄειδε τοιαῦτα, 'seated on some high rock would gaze seaward and sing thus' (Gow).

5. iactabat: like *ueniebat* in the preceding line, an imperfect of repeated action; cf. Theocr. 11.18 ἄειδε, Phanocles (above, 3n.) ἕζετ' ἀείδων.

6. O crudelis Alexi: cf. Theocr. 3.6 Ὦ χαρίεσσ' Ἀμαρυλλί, 11.19 Ὦ λευκὰ Γαλάτεια. Greek name and Greek verse-technique, as below, 24, 4.16, 34, 57, 5.52, 9.60, 10.12; that is, a weak caesura in the third foot unsustained by strong caesurae in the second and fourth feet (Norden 431–2).

7. mori me denique cogis: Theocr. 3.9 ἀπάγξασθαί με ποησεῖς, 'you will make me hang myself', would seem to support *coges*, but *cogis* is the reading of the better MS, less obvious, and more dramatic—'quasi iam sit mors in oculis' (La Cerda). In any case, Virgil was not merely translating Theocritus.

8. umbras et frigora: shade and coolness, the cool shade, a hendiadys. See 1.52n.

9. nunc uiridis etiam occultant spineta lacertos: cf. Theocr. 7.22 (μεσαμέριον) ἁνίκα δὴ καὶ σαῦρος ἐν αἱμασιαῖσι καθεύδει, (noon) 'when even the lizard sleeps in the wall'. It is so hot that even the sun-loving lizards seek shelter, those of Theocritus in walls of loose stone, V.'s under thorn-bushes (there are no stone walls in V.'s pastoral landscape). But the scholiast on Theocr. 1.47 (where see Gow) thought that αἱμασιαῖσι meant a thorn hedge, and V.'s *spineta* suggests that he thought so too. See 10.9n.

lacertos: the scribe of the Palatinus, generally a better MS than the Romanus, wrote *lacertas*, perhaps thinking of Hor. *Carm.* 1.23.6–7 'uirides rubum / dimouere lacertae'. Other poets use the feminine (see *TLL.* s.v. *lacerta*), but V. uses the masculine in *G.* 4.13 'picti squalentia terga lacerti'. Cf. schol. Juv. 3.231 *lacertae*: 'cum Vergilius masculino dixerit *lacertos*'.

10–11. There may be a distant echo here of Homer, *Il.* 18.559–60 αἱ δὲ γυναῖκες | δεῖπνον ἐρίθοισιν λεύκ' ἄλφιτα πολλὰ πάλυνον, 'Meanwhile the women scattered, for the workmen to eat, abundant white barley' (Lattimore). But cf. Longus 2.1.3 ἡ δὲ τροφὴν παρεσκεύαζε τοῖς τρυγῶσι, 'and Chloe prepared food for the vintagers'.

10. Thestylis: the name of Simaetha's servant in Theocr. 2, who assists her in preparing a love charm. The connection seems to be that in both Theocritus and V. Thestylis brays herbs, magic herbs in Theocritus (2.59). See Posch 35–6.

rapido ... aestu: the fierce scorching heat of the sun; cf. *G.* 1.92 'rapidiue potentia solis' and see Mynors on *G.* 4.425.

11. alia: Thestylis prepares lunch, a *moretum*, for the weary harvesters. The *moretum*, a traditional dish, was compounded of several ingredients—garlic and other herbs, salt, hard cheese, and oil—in a mortar; see the pseudo-Virgilian *Moretum* 85–116, with Kenney's notes. Cf. Hor. *Epod.* 3.3–4 'cicutis alium nocentius. / o dura messorum ilia!'

serpyllum: wild thyme, *Thymus serpyllum* L.; see Abbe 170. Not an ingredient in the *Moretum*.

serpyllum ... herbas ... olentis: for this mannered apposition, which is as old as Homer, *Il.* 19.30–1 ἄγρια φῦλα, | μυίας, 'fierce things, flies', cf. *G.* 1.10 'agrestum praesentia numina, Fauni', *A.* 6.7–8 'densa ferarum / tecta ... siluas', with Norden's note, 10–11 'secreta Sibyllae, / antrum immane', 10.601 'latebras animae pectus'; see also Nisbet–Hubbard on Hor. *Carm.* 1.3.20.

12–13. at mecum raucis ... / ... cicadis: only Corydon and the cicadas disturb the midday peace; '*cicadis* is of course the real subject, though *arbusta* is made the grammatical subject' (Conington). The song of the cicada was traditionally melodious (see West on Hes. *Works and Days* 583), but, as with the swan (9.29n.), V. knew better, *G.* 3.327–8 'inde ubi quarta sitim caeli collegerit hora / et cantu querulae rumpent arbusta cicadae'. Cf. C. T. Ramage, *The Nooks and By-ways of Italy* (Liverpool, 1868), 236: 'the cicada, whose sound resembles nothing so much as the scream of the corn-craik in Scotland; and if you can imagine thousands of such birds assembled in a small plain, and all screaming at the same moment, and continually from ten in the morning till sunset, you will have some idea of the disturbance caused by this small insect'. For other, similarly irritated descriptions of the cicada's song see M. Davies and J. Kathirithamby, *Greek Insects* (Oxford, 1986), 116–17.

12. uestigia lustro: many of Acontius' suitors, in an excess of passion, would place their feet in his footprints: Aristaenetus 1.10.13–14 Mazal πολλοί γε διὰ τὸ λίαν ἐρωτικὸν τοῖς ἴχνεσι τοῦ μειρακίου τοὺς ἑαυτῶν ἐφήρμοζον πόδας. Cf. Meleager 114.5 G.–P. (= *AP* 12.84) and see J. Hubaux, *Le Réalisme dans les Bucoliques de Virgile* (Liège, 1927), 51n.2. The phrase recurs in *A.* 11.763, of Arruns tracking Camilla.

13. sole sub ardenti: cf. Catull. 64.353–4 'namque uelut densas praecerpens messor aristas / sole sub ardenti flauentia demetit arua' (Ursinus).

arbusta: 1.39n.

14–15. nonne fuit satius ... / ... pati: this construction, colloquial in tone, occurs a number of times in Plautus and Terence, once in Lucretius, 5.1127 'ut satius multo iam sit parere quietum', and once again in V., *A.* 10.59 (Venus speaking to Jupiter) 'non satius cineres patriae insedisse supremos ...?'

nonne ... / ... nonne: repeated as in Ter. *Andr.* 238–9. *Nonne* occurs only here in the *E.*, three times in the *G.* (in the Lucretian formula *nonne uides*), but never in the *A.*; see Axelson 89–90.

15. atque superba: cf. 3.11, 5.23, 6.38, 63, 8.99; in his later poetry V., like the other Augustan poets (Horace excepted), usually avoids placing *atque* before a consonant. Although the numbers are small, the difference in V.'s practice in the *E.* is striking. In the *E.* the proportion of *atque* elided to *atque* unelided is 15:6 (=28.57% unelided), while in the *G.* it is (assuming the accuracy of Axelson's figures; his figures for the *E.* are inaccurate) 98:9 (=8.41% unelided), and in the *A.* 294:35 (=10.64% unelided), with most instances of *atque* unelided occurring in Books 7–12. See Axelson 84, Kenney on *Moretum* 44.

Menalcan: in [Theocr.] 8 an 'extremely shadowy' figure (Gow on l. 2) who challenges Daphnis to a singing-match and loses. An important figure in *E.* 5 and 9, here merely a convenient pastoral name.

16. niger: a traditional contrast, with the beloved's dark complexion requiring some apology ('I am black, but comely ...'); see

Nisbet–Hubbard on Hor. *Carm.* 2.4.3, Howell on Mart. 1.115. Menalcas is a field-hand, sunburnt like Bombyca in Theocr. 10.27 ἁλιόκαυστον; cf. Hor. *Epod.* 2.41–2 'perusta solibus / pernicis uxor Apuli'.

candidus: an adjective usually applied to women, fair-complexioned from staying indoors; more generally of feminine beauty, as in Catull. 13.4, 35.8, Hor. *Epod.* 11.27 (*TLL* s.v. 241.10). The effeminate house-slave performs no labour out of doors.

17. o formose puer: Servius reports approvingly a curious, oversubtle interpretation of this phrase by Donatus: 'bene dicit Donatus suspendendum "o" et sic dicendum "formose puer", ut intellegamus per iram eum aliud uoluisse dicere sed in haec uerba, ne Alexin offenderet, amoris necessitate compulsum'.

nimium ne crede colori: an admonition suggesting the 'gather ye rosebuds' theme, for which see K. F. Smith on Tib. 1.4.29. But the point here is rather different: the beautiful boy is like a flower to which another flower not conventionally beautiful may be preferred.

18. alba ligustra cadunt, uaccinia nigra leguntur: 'et rustice et amatorie ex floribus facit comparationem' (Serv.). Cf. 10.38–9, Theocr. 10.28–9 (no matter that the charming Bombyca is sunburnt) καὶ τὸ ἴον μέλαν ἐστί, καὶ ἁ γραπτὰ ὑάκινθος· | ἀλλ' ἔμπας ἐν τοῖς στεφάνοις τὰ πρᾶτα λέγονται, 'Dark is the violet and the lettered hyacinth, yet in garlands these are accounted first' (Gow). The structure of this line was noticed in antiquity: Diomedes, *GLK* 1.499.18 'aequidici (uersus) sunt qui singulis propositionibus antithetas apparant dictiones, ut ... albis enim nigra opposuit, ligustris autem uaccinia tribuit et cadentibus legenda adsignauit'.

ligustra: of uncertain identity, perhaps privet (Howell on Mart. 1.115.3).

cadunt: fall unregarded, as autumn leaves fall, *A.* 6.309–10 'autumni frigore primo / lapsa cadunt folia'.

uaccinia: some ancient readers identified the *uaccinium* with the hyacinth (DServ. on *G.* 4.183). Whether V. did so is a question; see Ernout–Meillet, *Dict. étym.* s.v., A. W. Pickard-Cambridge, *JRS* 8 (1918), 204.

19. qui sim: 1.18n.

20. diues: here first with the genitive (Hofmann–Szantyr 77).

niuei . . . lactis: despite Homer's *γάλα λευκόν*, 'white milk', Servius thinks that *niuei* is better taken with *pecoris*, and is followed by La Cerda and Voss. But rhetoric—a chiastic arrangement with the second noun weighted—and sense are decisive; cf. Ov. *Met.* 13.829 (Polyphemus speaking) 'lac mihi semper adest niueum', Theocr. 5.53 *λευκοῖο γάλακτος*.

abundans: normally with the ablative, as in *G.* 4.139–40.

21. mille meae Siculis errant in montibus agnae: cf. Theocr. 11.34 *βοτὰ χίλια βόσκω*, 'I tend a thousand cattle', 3.1–2 *ταὶ δέ μοι αἶγες | βόσκονται κατ' ὄρος*, 'and my goats graze on the hill'. Corydon far outdoes Polyphemus in boasting of his pastoral wealth: a thousand ewe–lambs, implying a thousand ewes (so, explicitly, Calp. Sic. 2.68–9).

mille meae: here speaks the pride of ownership. Polyphemus tends and milks (11.35 *ἀμελγόμενος*) his animals himself, but Corydon's sheep are far away, grazing ('errant') on the mountainside and looked after by others. Being a slave, Corydon cannot be the legal owner, the *dominus*, of this absurdly large flock; it is his *peculium*, the fruit of his industry, which he naturally speaks of as his own. See 1.32n., J. Van Sickle, '"Shepheard Slave": Civil Status and Bucolic Conceit in Virgil, *Eclogue* 2', *QUCC*, NS 27 (1987), 127–9.

22. lac mihi non aestate nouum, non frigore defit: cf. Theocr. 11.36–7 *τυρὸς δ' οὐ λείπει μ' οὔτ' ἐν θέρει οὔτ' ἐν ὀπώρᾳ, | οὐ χειμῶνος ἄκρω*, 'Cheese I never lack, neither in summer, nor in autumn, nor at winter's end', with Dover's note. Corydon has an abundance of fresh milk even in the most difficult seasons of the year, in summer, when milk turns sour with the heat, and in winter, when fodder is scarce. Cf. Serv. 'multo melius quam Theocritus; ille enim ait *τυρὸς δ' οὐ λείπει μ' οὔτ' ἐν θέρει οὔτ' ἐν ὀπώρᾳ*. sed caseus seruari potest, nec mirum est si quouis tempore quis habeat caseum; hoc uero laudabile est, si quis habeat lac nouum . . . sane hunc uersum male distinguens Vergiliomastix uituperat "lac mihi non aestate nouum, non frigore: defit", id est, semper mihi deest'. See 1.1n. (*sub tegmine fagi*).

defit: a verb found in Plautus and Terence and the early dramatists but rarely in poetry (or prose) thereafter; see *TLL* s.v. *deficio* 326.47, Tränkle 43–4, Fedeli on Prop. 1.1.34.

23–4. Being a Virgilian shepherd, Corydon is of course an accomplished musician (like the fabled Amphion) with exquisite literary taste, while Polyphemus claims only to be the best piper among the Cyclopes and sings a simple song of love to solace his sleepless nights, Theocr. 11.38–40 *συρίσδεν δ' ὡς οὔτις ἐπίσταμαι ὧδε Κυκλώπων, | τίν, τὸ φίλον γλυκύμαλον, ἁμᾷ κἠμαυτὸν ἀείδων | πολλάκι νυκτὸς ἀωρί*, 'I can pipe as no other Cyclops here, and often at dead of night I sing of you, my sweet-apple, and of myself too'. For the tale of Amphion and Zethus, their long-suffering mother Antiope, wicked Dirce, who was turned into, or whose bones were thrown into, the spring that bore her name, and the 'song-built' walls of Thebes, see *OCD* s.v. *Amphion*.

24. Amphion Dircaeus in Actaeo Aracyntho: a verse of the most precious Alexandrian sort. Since Aracynthus cannot be the famous mountain in Acarnania, which had no connection with Amphion, it must be another mountain of the same name, largely unknown to fame, in Boeotia, but situated on the border of, or partly in, Attica. (It was not uncommon for mountains to bear the same name: eighteen Olympuses besides the famous one are identified in *RE* xviii.310–15.) The evidence is as follows: Steph. Byz. s.v. *Ἀράκυνθος, ὄρος Βοιωτίας, ἀφ' οὗ ἡ Ἀθηνᾶ Ἀρακυνθιάς*, 'Aracynthus, a mountain of Boeotia, whence Athena Aracynthias'; Sext. Empir. *Adu. math.* 1.257 *Ἀράκυνθος τῆς Ἀττικῆς ἐστιν ὄρος*, 'Aracynthus is a mountain of Attica'; Prop. 3.15.41–2 'prata cruentantur Zethi, uictorque canebat / paeana Amphion rupe, Aracynthe, tua'; Stat. *Theb.* 2.239 (Artemis and Athena) 'illa suas Cyntho comites agat, haec Aracyntho'. It was noticed by Heyne that this line is essentially Greek: 'totus versus ex Graeco factus est: *Ἀμφίων Διρκαῖος ἐν Ἀκταίῳ Ἀρακύνθῳ*, ut adeo manifestum sit poetam Graecum esse expressum'. Quite possibly (see, however, Thomas on *G.* 1.138); there is another line in V., *G.* 1.437 'Glauco et Panopeae et Inoo Melicertae', identified by both Gellius and Macrobius as rendering a line of Parthenius (see Mynors ad loc.). May not this, too, be a line of Parthenius, whose typically Alexandrian delight in learned geographical allusion is evident in his fragments? See M. Geymonat, *MCr* 13–14 (1978–9), 375–6.

Dircaeus: not simply, as in later Latin poetry, and especially Statius, equivalent to *Thebanus*, for Amphion was intimately associated with Dirce: he and his brother Zethus killed her. Διρκαῖος is found in Pindar and the tragedians, but, with one exception, only of the spring or the stream flowing from it (the stream in fact had several sources, *RE* iv. 1169): Pind. *Pyth.* 9.88 Διρκαίων ὑδάτων, 'the waters of Dirce', Aesch. *Sept.* 307 ὕδωρ . . . Διρκαῖον, Soph. *Ant.* 104–5 Διρκαίων . . . ῥεέθρων, 'Dirce's stream', 844 Διρκαῖαι κρῆναι, 'Dirce's spring', Eur. *Supp.* 637 ῥεῦμα Διρκαῖον, 'Dirce's stream', *Phoen.* 730 Διρκαῖος . . . πόρος, 'the Dircaean ford', fr. 819.4 N.[2] Διρκαῖον . . . πέδον, 'the Dircaean plain'. So also in the one Latin instance before V., Accius 602 R.[3] 'Dircaeo fonte' (*TLL* Onomast. s.v. *Dirce* 188.23). In Greek poetry after Euripides, Διρκαῖος occurs only once, in Nonnus, *Dionys.* 2.671 Ἄρεα . . . Διρκαῖον; the dragon guarding the spring that would one day be Dirce's was Ares' son, 'Dircaean Ares'.

Actaeo: 'Attic'; the adjective Ἀκταῖος is not attested before Callimachus (Hollis on *Hec.* fr. 1). Here first in Latin and here only in V. (*TLL* s.v. *Acte* 436.23). It should perhaps be noted that Parthenius was a strict Callimachean; see Clausen 6–8.

Actaeo Aracyntho: cf. 8.44 'extremi Garamantes', 10.12 'Aonie Aganippe'. This rhythmical pattern, Hellenistic in character though as old as Homer, e.g. *Il.* 21.553, *Od.* 6.4, is first attested in Latin poetry where it might have been expected, in Catull. 64.141 'optatos hymenaeos'. See Hopkinson on Callim. *Hymn* 6.84 and cf. Theocr. 15.102, 16.31, 17.79, 22.141. See also 7.53n.

25. nec sum adeo informis: cf. Theocr. 6.34 (Damoetas impersonating Polyphemus) καὶ γάρ θην οὐδ' εἶδος ἔχω κακόν, 'For really I am not even bad-looking'; Soubiran 636: 'Ce n'est pas un hasard si trois élisions, dont deux de monosyllabes, se rencontrent dans cet hexamètre: elles glosent, sur le plan du rythme, l'adjectif *informis*'. Cf. *A.* 3.658 (Polyphemus) 'monstrum horrendum, informe, ingens'.

25–6. nuper me in litore uidi, / cum placidum uentis staret mare: cf. Theocr. 6.35 ('Polyphemus') ἦ γὰρ πρᾶν ἐς πόντον ἐσέβλεπον, ἦς δὲ γαλάνα, 'for lately I looked in the sea, and there was a calm'; also Callim. *Hymn* 5.19–20 οὔτ' ἐς ὀρείχαλκον μεγάλα θεὸς οὔτε Σιμοῦντος | ἔβλεψεν δίναν ἐς διαφαινομέναν, 'the great goddess looked neither in her mirror of orichalc nor in the clear waters of the Simois' (Ursinus). By 'in litore' V. means the shallows along the shore, as in *A.* 6.362 'nunc me fluctus habet uersantque in litore uenti', where Norden cites Lucan 8.698–9 'litora Pompeium feriunt truncusque uadosis / huc illuc iactatur aquis'; see E. Wistrand, *Nach innen oder nach aussen?* (Göteborgs Högskolas Årsskrift 52, 1946), 36–42, V. Skånland, '*Litus*: The Mirror of the Sea. Vergil, *ecl.* 2, 25', *SO* 42 (1968), 93–101. V. has caused his commentators some distress: Serv. 'negatur hoc per rerum naturam posse fieri'; La Cerda (betraying his exasperation): 'adeant mare, qui hoc negant'; Heyne: 'Verum in eo commode haec de Cyclope dici poterant; non aeque nunc de pastore, qui melius fonte, rivo, amni, pro speculo utitur'; Conington: 'It is just possible that a Mediterranean cove might be calm enough to mirror a giant, not possible that it should be calm enough to mirror Corydon' (but in Theocr. 6.11–12 the waves mirror Polyphemus' dog as it runs along the gently murmuring beach, *τὰ δέ νιν καλὰ κύματα φαίνει | ἅσυχα καχλάζοντος ἐπ' αἰγιαλοῖο θέοισαν*). V.'s imitators reduce the seashore to a fountain: Calp. Sic. 2.88–9, Nemes. 2.74, Tasso, *Aminta* 1.90; and Marvell, most ingeniously, to the blade of a scythe, *Damon the Mower*, 57–8 'Nor am I so deform'd to sight, / If in my Sithe I looked right'.

26. placidum . . . mare: cf. Enn. *Ann.* 377 Skutsch 'uerrunt extemplo placidum mare: marmore flauo', Catull. 64.269 'hic, qualis flatu placidum mare matutino'.

uentis staret: cf. *G.* 4.484 'Ixionii uento rota constitit orbis'. The sea, like Ixion's whirling wheel, is stilled by the wind. Cf. *A.* 1.66, 3.69–70, 5.763 'placidi strauerunt aequora uenti', and see Nisbet–Hubbard on Hor. *Carm.* 1.3.16.

non ego Daphnin: with the maturing of his style, V. tended to avoid this older cadence, i.e. a pyrrhic preceded by a monosyllable in the fifth foot; for a discussion see Norden 446–8 and Mynors on *G.* 2.153. Norden's figure for the *E.*, however, is very inaccurate; there are twenty-nine (not thirteen) instances of this cadence in the *E.*, distributed as follows: 2 (5), 3 (5), 9 (5, counting 51 'saepe ego longos'), 8 (4), 7 (3, counting 7 'atque ego Daphnin'), 4 (2), 5 (2), 6 (2), 10 (1), 1 (0). If instances with the enclitic *quoque* are not counted (so Mynors), the number shrinks to nineteen; and if instances with elision are not counted, to seventeen.

Daphnin: the ideal shepherd, the beloved of Pan.

26–7. non ego Daphnin / iudice te metuam: cf. the New Gallus 8–9 'non ego, Visce, / . . . Kato, iudice te uereor'; also 4.58–9.

27. si numquam fallit imago: entirely unlike Theocr. 6.37 ('Polyphemus') *ὡς παρ' ἐμὶν κέκριται*, 'as my judgment goes' (Gow). There is a moment of delicate irony here; cf. below, 65n. In Epicurean dogma, the senses are infallible and error arises from the viewer's opinion or inference; cf. Lucr. 4.464–6 'pars horum maxima fallit / propter opinatus animi quos addimus ipsi, / pro uisis ut sint quae non sunt sensibus uisa'. It was not the image but his own conceit that deceived Corydon. See A. Traina, *A&R* 10 (1965), 72–8 = *Poeti latini (e neolatini)*[2] (Bologna, 1986), i. 163–74. As Dryden remarks, Preface to *Sylvae* (1685): 'Virgil's Shepherds are too well read in the Philosophy of Epicurus'.

numquam: colloquial for *non*; cf. 3.49.

28–30. o tantum libeat . . .: *o* with the optative subjunctive expressing a passionate wish; cf. 4.53, *G.* 2.488 (*o qui*), *A.* 4.578, 8.78, 560 (*o si*), 579 'nunc, nunc o liceat crudelem abrumpere uitam', and see Fraenkel, *Horace*, 242n. 1. So Polyphemus urges Galatea to leave the sea and share his humble existence, Theocr. 11.65–6 *ποιμαίνειν δ' ἐθέλοις σὺν ἐμὶν ἅμα καὶ γάλ' ἀμέλγειν | καὶ τυρὸν πᾶξαι τάμισον δριμεῖαν ἐνεῖσα*, 'Come be a shepherd with me, and milk the flock, and pour sharp rennet in to set the cheese'.

28. tibi: with both *libeat* and *sordida*.

sordida: in the eyes of the fastidious, city-bred Alexis; see below, 44n. Martial often applies the adjective to country things; see Howell on Mart. 1.49.28.

29. habitare: transitive by analogy with *colo*; cf. Cic. *Verr.* 2.4.119 'coliturque ea pars et habitatur' (Wackernagel i. 142). Cf. below, 60, 6.2.

casas: hut or cottage, a word that connotes a primitive, idyllic life. Cf. Livy 5.53.8 'in casis ritu pastorum agrestiumque habitare', Tib. 2.5.25–6 'sed tunc pascebant herbosa Palatia uaccae / et stabant humiles in Iouis arce casae', Ov. *Met.* 8.632–3 (Philemon and Baucis) 'illa / consenuere casa', and see *TLL* s.v., Perutelli on *Moretum* 60.

figere ceruos: cf. *G.* 1.307–8 'tum gruibus pedicas et retia ponere ceruis / auritosque sequi lepores, tum figere dammas'. Hunting is one of the pleasures of pastoral life in V., *E.* 3.12–13, 75, 5.60–1, 7.29–30, 8.28, 10.55–60; a pleasure virtually unknown to the herdsmen of Theocritus, who mentions hunting deer once, in an *adynaton* (1.135). Cf. Theocr. 1.110 (Adonis), 5.106–7, and see Dover, p. lix.

30. haedorumque gregem uiridi compellere hibisco: either of two interpretations seems possible: 'to drive the kids to the green *hibiscum*' (so Serv. 'ad hibiscum compellere . . . et sic dixit *hibisco* ad hibiscum, ut *it clamor caelo*, id est ad caelum') or 'to drive the kids with a switch of green *hibiscum*' (so Philargyrius 2 'genus uirgulti, quo pastores flagellant pecus'). Hor. *Carm.* 1.24.16–18 'quam uirga semel horrida /. . ./ nigro compulerit Mercurius gregi' (invariably if partially cited here) can be used to support either interpretation. Destination is usually expressed with *compellere*; cf. 7.2, Calp. Sic. 5.57 'ad fontem compelle gregem', *TLL* s.v. 2029.67 (cf., however, Cic. *In Pis.* 87 'omni totius prouinciae pecore compulso'). Was the *hibiscum*, a malvaceous plant (*RE* i.1694–6), edible? Calpurnius Siculus considered it a poor substitute for human food, 4.32 'uiridique famem solarer hibisco' (doubtful evidence); there is no evidence for it as a forage-plant. Theophrastus, *HP* 1.3.2, has a μαλάχη, apparently a sort of wild mallow like the *hibiscum* (Pliny, *NH* 20.29 'hibiscum, quod molochen agrian uocant'), whose stalk grows to the size of a spear in length and thickness, so that men use it as a walking-stick, and in Lucian, *Fugitiui* 33, runaway slaves are beaten with it (Mynors). The plant used for small basketwork in 10.71 'gracili . . . hibisco' is probably another species; see A. W. Pickard-Cambridge, *JRS* 8 (1918), 203.

compellere hibisco: for the elision cf. 4.11, 5.6, 6.76, 8.92, 10.53, and see G. Eskuche, *RhM* 45 (1890), 249, Soubiran 551–2.

31. imitabere Pana canendo: cf. Longus 3.23.5 (the buried though still singing limbs of Echo) μιμεῖται καὶ αὐτὸν συρίττοντα τὸν Πᾶνα, 'They even imitate Pan himself when he plays on the pipe', 2.37.1 (dancing) ὁ Δάφνις Πᾶνα ἐμιμεῖτο, τὴν Σύριγγα Χλόη, 'Daphnis imitated Pan, and Chloe Syrinx'. Cf. 5.73.

canendo: here 'piping', as the context shows; see 5.9n.

32–3. Pan primum calamos cera coniungere pluris / instituit: cf. 5.29–30 'Daphnis et Armenias curru subiungere tigris / instituit', *G.* 1.147–8 'prima Ceres ferro mortalis uertere terram / instituit', *A.* 6.142–3 'hoc sibi pulchra suum ferri Proserpina munus / instituit'. In each of these remarkably similar passages the institution of a custom or practice is described. The Ancients believed, or pretended to believe, that most things in the world had been invented by someone, god or man, a πρῶτος εὑρετής or *primus inuentor*; for discussion and bibliography see Nisbet–Hubbard on Hor. *Carm.* 1.3.12.

33. Pan curat ouis: *G.* 1.17 'Pan, ouium custos', Πὰν νόμιος; see Nisbet–Hubbard on Hor. *Carm.* 1.17.3.

ouis ouiumque magistros: cf. 3.101.

35. Amyntas: cf. 3.66n.

36–7. est mihi disparibus septem compacta cicutis / fistula: 1.10n. Cf. Tib. 2.5.31–2 'fistula, cui semper decrescit harundinis ordo, / nam calamus cera iungitur usque minor'. A sadder and wiser Pan constructed his pipe (σῦριγξ) of unequal reeds because his love for Syrinx had not been matched (Longus 2.34.3). Cf. Ov. *Met.* 1.689–712.

36. compacta: cf. Theocr. 1.128–9 τάνδε φέρευ πακτοῖο μελίπνουν | ἐκ κηρῶ σύριγγα, 'take this pipe, breathing honey from its compacted wax'.

cicutis: the hemlock, a ferule-like plant with a hollow stem; a metrical variant for *calami* invented by Lucretius, 5.1382–3 'zephyri, caua per calamorum, sibila primum / agrestis docuere cauas inflare cicutas', and twice appropriated by V., here and in 5.85. See 1.2n. (*auena*).

37. fistula: the pan-pipe, Theocritus' σῦριγξ; cf. *E.* 3.22, 25, 7.24, 8.33, 10.34. Theocritus uses σῦριγξ in the nominative, dative, and accusative singular, but V. can use *fistula* only in the nominative singular; in the oblique cases he uses *auena* (1.2, 10.51), *calamus* (1.10, 2.34, 5.2, 48, 6.69), *cicuta* (5.85), *harundo* (6.8), and *stipula* (3.27).

Damoetas: the singer who impersonates Polyphemus in Theocr. 6; one of the two singers in *E.* 3, mentioned in 5.72.

dono mihi quam dedit: cf. Theocr. 4.30 (the pan-pipe) δῶρον ἐμοί νιν ἔλειπεν, 'he left it to me as a present', 5.8 τάν μοι ἔδωκε Λύκων, 'which Lycon gave to me'; the Latin is a little more formal.

38. et dixit moriens: cf. Longus 1.29.3 (the dying shepherd Dorcon gives his pan-pipe to Chloe) χαρίζομαι δέ σοι καὶ τὴν σύριγγα αὐτήν, ᾗ πολλοὺς ἐρίζων καὶ βουκόλους ἐνίκησα καὶ αἰπόλους, 'and I give you this pipe, with which in contests I defeated many a cowherd and many a goatherd'. See 5.49n.

secundum: cf. Longus 2.37.3 (Philetas) *τὴν σύριγγα χαρίζεται φιλήσας καὶ εὔχεται καὶ Δάφνιν καταλιπεῖν αὐτὴν ὁμοίῳ διαδόχῳ*, 'he gives Daphnis his pipe with a kiss and begs that he too will leave it to a successor like himself'.

39. stultus: 1.20n.

40–1. duo . . . / capreoli: roe-deer fawns, which could be captured and suckled by a ewe, *capreoli* being used here as though it were a diminutive (see W. Schulze, *Kl. Schriften*[2] (Göttingen, 1966), 76–7); rustic love-tokens, like the eleven fawns and four bear-cubs Polyphemus is raising for Galatea (Theocr. 11.40–1), their value enhanced by the dangerous circumstances in which they were found (Serv. 'commendat a difficultate'). Ovid imitates both V. and V.'s model, *Met.* 13.834–6 'inueni geminos, qui tecum ludere possint, / inter se similes, uix ut dinoscere possis, / uillosae catulos in summis montibus ursae' (La Cerda); see 8.41n.

40. nec tuta: i.e. *et non tuta*, as often in poetry and prose; see Kühner–Stegmann ii.39–40, D. R. Shackleton Bailey, *Propertiana* (Cambridge, 1956), 279.

41. sparsis etiam nunc pellibus albo: cf. *G.* 3.56 'maculis insignis et albo'. They lose their white markings before the end of the first year (hence 'etiam nunc'), as Serv. seems to have known. Pretty things, and Alexis is hardly more than a child; see J. Bodel, 'Trimalchio's Coming of Age', *Phoenix*, 43 (1989), 72–4.

42. bina die: cf. 3.34 'bisque die', Hor. *Serm.* 2.1.4 'mille die'; *die* without *in* in such phrases is first attested in Varro (Hofmann–Szantyr 148). Twice a day, mornings and evenings, when the ewes are let out to pasture and when they return (Varro, *RR* 2.2.15).

siccant ouis ubera: in plain Latin, *sugunt ouis mammam*; cf. Serv. '*siccant*: sugunt', Varro, *RR* 2.1.20 (*agni*) 'matris sugant mammam'; see J. N. Adams, *BICS* 27 (1980), 58. *Sicco* in this sense is first attested here; cf. Hor. *Epod.* 2.46 'distenta siccet ubera'.

42–4. quos tibi seruo . . .: cf. Theocr. 3.34–6 *ἦ μάν τοι λευκὰν διδυματόκον αἶγα φυλάσσω, | τάν με καὶ ἁ Μέρμνωνος ἐριθακὶς ἁ μελανόχρως | αἰτεῖ· καὶ δωσῶ οἱ, ἐπεὶ τύ μοι ἐνδιαθρύπτῃ*, 'Truly I keep for you a white she-goat with twin kids, which Mermnon's swarthy serving-girl begs of me. And give it to her I will, since you are so prudish with me'.

43. orat: here first with the infinitive, 'orat ut abducat' (Serv.); cf. *A.* 6.313 and see *TLL* s.v. 1040.76.

44. faciet: replacing the particular verb (*abducet*); an old and common usage, e.g. Ter. *Andr.* 775, *Eun.* 620, *Heaut.* 416.

sordent: found three times in Plautus, *Poen.* 1179, *Truc.* 379, 381, and once in Accius 23 R.[3], but nowhere else in poetry and never in prose of the Republican period. In effect, introduced into poetry by Catullus, 61.129–30 'sordebant tibi uilicae, / concubine', Prop. 2.32.11, Hor. *Epist.* 1.11.4, 18.18. See Tränkle 137–8 and cf. Livy 4.25.11 (only here in Livy) 'se suis etiam sordere nec a plebe minus quam a patribus contemni', with Ogilvie's note: 'It is a good touch for plebeians to use coarse and plebeian language'. *Sordeo* is found occasionally in later poetry and prose.

munera nostra: noun and adjective with the same ending, a rare cadence found only once elsewhere in the *E.*, where however *nigra* is predicative, 10.39 'uaccinia nigra'—unless 'flumina nota' (1.51) should also be counted. See S. J. Harrison, *CQ*, NS 41 (1991), 142.

45–6. huc ades, o formose puer: tibi . . . / . . . tibi: the beautiful boy is invoked in a style reminiscent of prayer, Norden's 'Du-Stil'; cf. *G.* 2.2–8 and see Norden, *Agnostos Theos* (Berlin, 1913), 143–63—a minor deity, as it were, to whom the Nymphs bring offerings of flowers.

45. huc ades: cf. 7.9, 9.39, 43.

lilia plenis: a cadence that lingered in V.'s memory, *A.* 6.883 'manibus date lilia plenis'. For the plural *lilia* see Kühner–Stegmann i.68, Maas 507–8 = 550–1.

46. The Nymphs, water-spirits who cause flowers to grow (see H. Herter, *RE* xvii. 1548), bring baskets full of lilies as from a garden. Cf. Ap. Rhod. 4.1143–5 (the marriage of Jason and Medea) *ἄνθεα δέ σφι | νύμφαι ἀμεργόμεναι λευκοῖς ἐνὶ ποικίλα κόλποις | ἐσφόρεον*, 'the Nymphs gathered and brought flowers of various hues in their white bosoms'.

calathis: a Greek word, here first in Latin, though Catullus 64.319 has *calathiscus*; Tränkle 115 notes that both *κάλαθος* and καλαθίσκος belong to the vocabulary of Hellenistic poetry. Cf. 5.71 and see Mynors on *G.* 3.402.

46–50. So Meleager 46.1–4 G.–P. (=*AP* 5.147) weaves for Heliodora a garland of white stocks, narcissus, lilies, crocus, hyacinths, and roses—a literary garland, for Meleager gathered his flowers from poetry; cf. *Cypria* fr. 4.3–6 Davies, *Homeric Hymn to Demeter* 6–8, where see Richardson, Theocr. 7.63–4, Moschus, *Eur.* 65–70, where see Bühler. In composing their garlands and bouquets the poets did not bother about seasonal consistency, although Polyphemus, not wishing to seem ungallant, naïvely explains to Galatea that he might have brought her a bouquet of snowdrops and scarlet poppies but one flower grows in summer, the other in winter (Theocr. 11.56–9).

46. candida Nais: cf. [Theocr.] 8.43 *καλὰ Ναΐς*; neither the singular nor the plural is found in Theocritus. Cf. 6.21, 10.10.

47. pallentis uiolas: Pliny, *NH* 21.27 'plurima sunt genera, purpureae, luteae, albae'; pale yellow therefore, for the pallor of an Italian is not white or ashen but yellowish; cf. Hor. *Epod.* 10.16 'pallor luteus', *Carm.* 3.10.14 'tinctus uiola pallor amantium' (Serv.), Tib. 1.8.52 'sed nimius luto corpora tingit amor', with K. F. Smith's note. Here first in the plural; see Kühner–Stegmann i.68, Maas 516=558.

papauera: here first in the plural; see Kühner–Stegmann i.69, *TLL* s.v. 250.40.

48. narcissum: the Pheasant's Eye or Poet's Narcissus, *N. poeticus* L., which is again joined with the violet in 5.38. Not the late-flowering narcissus, *N. serotinus* L., of *G.* 4.122–3; Abbe 68: 'Since *N. serotinus* is rare except in localities from Naples southward, it was probably unknown to Virgil before he moved to his villa in the Neapolitan area'.

florem . . . bene olentis anethi: dill, *Anethum graveolens* L.; cf. Catull. 61.6–7 'floribus / suaue olentis amaraci', *G.* 4.270 'graue olentia centaurea', *A.* 6.201 'graue olentis Auerni'. V. knew the adjective βαρύοδμος, 'heavy-scented', from Nicander, *Ther.* 51. Dill is a garland-flower in Theocr. 7.63–4 and so already in Sappho fr. 81 (b).2 L.–P. and Alcaeus fr. 362.1 L.–P.

49. casia: a species of daphne; see Mynors on *G.* 4.30, Abbe 142–4. Listed as a garland-flower by Pliny, *NH* 21.53.

intexens: entwining 'hyacinths' (*uaccinia*) with casia and other fragrant herbs. Cf. Theocr. 3.23 ἀμπλέξας ('entwining ivy with the other plants named', Dover), *E.* 5.31, *A.* 7.488.

suauibus herbis: *suauis* is found in comedy (oftener in Plautus than in Terence), in Lucretius (13 times and *suauiter* twice), in Catullus (3 times, on which see Ross (1969), 79), in Horace (not in the *Carm.*, 8 times in the *Serm.*, and *suauiter* once in the *Serm.* and in the *Epist.*); but after V. and Horace *suauis* virtually disappears from poetry, its place being taken by the metrically equivalent *dulcis*, which already predominates in Lucretius (35 times) and Catullus (27 times). See Axelson 35–6, A. J. Mamoojee, *Phoenix*, 35 (1981), 220–3. Apart from *G.* 4.200 'e suauibus herbis', V. uses *suauis* only in the *E.*: here, 55 below, 3.63 'suaue rubens hyacinthus' (cf. Nemes. 2.45 and 48 'dulce rubens hyacinthus'), and 4.43–4 'suaue rubenti / murice'): only in his earliest *E.*, that is, apparently under the influence of Catullus. In *A.* 1.693–4, thinking of Catull. 61.6–7 'floribus / suaue olentis amaraci', V. wrote 'amaracus illum / floribus et dulci aspirans complectitur umbra' and thus produced a phrase, 'dulci . . . umbra', virtually unique (*TLL* s.v. *dulcis* 2190.77).

50. mollia luteola pingit uaccinia calta: a verse 'which they call golden, or two Substantives and two Adjectives, with a Verb betwixt to keep the peace' (Dryden, Preface to *Sylvae* (1685)). Disregarding incidental words—*quamuis, et, non, iam, tibi, quae, per, hic, ante, ab, de*—which hardly impair the basic pattern, there are in the *E.* sixteen golden lines, in eleven of which the word-order is interlocking (abVAB): 1.33, 34, 49, *2.50, 4.23, 29, *5.7, 6.17, 40, 7.24, 8.4; in five, concentric (abVBA): 4.4, *6.8, 7.12, 9.15, 10.20. Three purely golden lines are indicated by asterisks. See Norden 393–8, Thomas on *G.* 1.117, Hopkinson on Callim. *Hymn* 6.9.

mollia: of other flowers also: *hyacinthus*, 6.53, *G.* 4.137; *uiola*, 5.38, *A.* 11.69. By implication, the texture of the *calta* should be coarse or uneven to the touch.

luteola: the diminutive first occurs here, then in Columella 9.4.4 (roses), 12.49.9 (the olive). Cf. Columella 10.307 'pressaque flammeola rumpatur fiscina calta'; *flammeola* is unique.

pingit: she sets off the dark hyacinths (18 'uaccinia nigra') with yellow marigolds; 'ex colorum diuersitate quaerit ornatum' (Serv.).

calta: of uncertain identity; Pliny, *NH* 21.28, compares it with the violet and mentions its strong scent; cf. Ov. *Ex Pont.* 2.4.28 (an *adynaton*) 'caltaque Paestanas uincet odore rosas'. Perhaps the common marigold (Sargeaunt 24–5).

51–5. Corydon offers fruit and nuts, decorated with sprays of laurel and myrtle, from his garden; a simple gift in contrast to the complicated garland (46–50).

51. tenera lanugine mala: 'quinces with tender down'; the variety indicated according to B. W. Boyd, '*Cydonea Mala*: Virgilian Word-Play and Allusion', *HSCP* 87 (1983), 169–74, is not the Cydonian quince, which has no down, but the sparrow-quince, *malum strutheum*. This is the first extant occurrence of *lanugo* in a metaphorical sense (*TLL* s.v. 937.43). V. was thinking of Lucr. 5.888–9 'puerili aeuo florente iuuentas / occipit et molli uestit lanugine malas'; cf. *A.* 10.324–5 'tu quoque, flauentem prima lanugine malas / dum sequeris Clytium infelix, noua gaudia, Cydon'.

52. castaneasque nuces: 'bene speciem addidit dicens *castaneas*' (Serv.); see Dodds on Eur. *Bacch.* 1024–6, R. Renehan, *CP* 75 (1980), 348 and 80 (1985), 148–9.

mea . . . Amaryllis: an affectionate reference (7.22n.); Corydon naïvely imagines that what pleased his rustic sweetheart will also please Alexis. Cf. Ov. *Ars* 2.267–8 'aut uuas aut, quas Amaryllis amabat, / at nunc, castaneas, non amat illa, nuces'.

53. cerea pruna: plums with a lustrous waxy yellow skin. Diaeresis in the second foot and weak caesura in the third, a rare rhythm: 5.19, *G.* 4.448, *A.* 3.697, 4.486.

pruna (honos . . .): the hiatus is unparalleled in V., but does occur in Theocritus; see Gow on 7.8. See also *E.* 8.11n.

54. o lauri . . . et te, proxima myrte: laurel (properly the bay, *Laurus nobilis* L.) and myrtle are perpetually joined by the poets, e.g. Theocr. *Epigr.* 4.7 Gow *δάφναις καὶ μύρτοισι*, *E.* 7.62, Hor. *Carm.* 3.4.19 'lauroque collataque myrto', Petrarch, *Canzoniere* 7.9 'Qual vaghezza di lauro, qual di mirto?', Milton, *Lycidas*, 1–2 'Yet once more, o ye Laurels, and once more, / Ye Myrtles brown . . .', *Paradise Lost*, 4.693–4 'inwoven shade / Laurel and Myrtle'. See E. A. Schmidt, *Bukolische Leidenschaft* (Frankfurt am Main, 1987), 139, 151n.1.

55. suauis . . . odores: cf. Catull. 64.87–9 (Ariadne) 'quam suauis exspirans castus odores / lectulus in molli complexu matris alebat, / quales Eurotae progignunt flumina myrtus'.

57. Iollas: the *dominus* of l. 2. A non-Theocritean name which recurs in 3.76 and 79; in *A.* 11.640 a nondescript Trojan who is named only to be killed.

58. heu heu, quid uolui misero mihi?: something may be owed here to Theocr. 3.24 *ὤμοι ἐγών, τί πάθω, τί ὁ δύσσοος;*, 'Alack, what is to become of me, poor wretch?' (Gow); but form and tone are essentially Latin, e.g. Plaut. *Aul.* 721 'heu me miserum, misere perii, / male perditus', *Merc.* 661 'heu misero mihi'. The double interjection is found here and in 3.100, and seems to be V.'s imitation of Theocr. 4.26–7 *φεῦ φεῦ βασεῦνται καὶ ταὶ βόες, ὦ τάλαν Αἴγων, | εἰς Ἀΐδαν*, 'Oh! oh! poor Aegon, your cows too will go to the devil'; cf. also Theocr. 5.86, the two *Idylls* V. was most concerned with while writing *E.* 3. And it is probably no coincidence that *heu heu*, like *φεῦ φεῦ* in the *Idylls*, occurs only twice in the *E.*; see Clausen 165n.64.

Austrum: the South Wind or sirocco, which blows with devastating effect in winter and spring; cf. Stat. *Silu.* 3.3.129 'pubentesque rosae primos moriuntur ad Austros'. Corydon realizes that his peace of mind has been destroyed.

60. quem fugis?: an old motif; cf. Sappho 1.21 L.–P. (but see A. Giacomelli, *TAPA* 110 (1980), 135–42), Anacreon 417.1–2 Page, Theocr. 6.17, 11.75, Callim. *Epigr.* 31.5–6 Pf., *A.* 4.314 (Dido) 'mene fugis?' In the Underworld their roles are reversed and Aeneas puts the question, 6.466 'quem fugis?' Cf. Milton, *Paradise Lost* 4.481–2 'Thou following cried'st aloud, "Return fair Eve, / Whom fli'st thou?"' See Clausen 145n.39.

a: 10.47n.

61. Dardanius: cf. *A.* 10.92 'me duce Dardanius Spartam expugnauit adulter?' Young Paris herded his father's cattle on the slopes of Mt. Ida, that is, in Dardania (cf. *Il.* 20.215–18 and see Kirk on *Il.* 2.218–19); and there, on being confronted by Hera, Athena, and Aphrodite, he judged Aphrodite the most beautiful of the three, receiving Helen, the most beautiful of women, as a gift from the grateful goddess. For the connotation of this oblique reference to Paris as herdsman cf. Eur. *IA* 180–1 *Πάρις ὁ βουκόλος ἂν ἔλαβε |*

δῶρον τᾶς Ἀφροδίτας, 'whom (Helen) the herdsman Paris seized, the gift of Aphrodite', [Bion] 2.10 Gow ἅρπασε τὰν Ἑλέναν πόθ' ὁ βωκόλος, ἆγε δ' ἐς Ἴδαν, 'once upon a time the herdsman ravished Helen, and led her away to Ida', *A.* 7.363–4 'at non sic Phrygius penetrat Lacedaemona pastor, / Ledaeamque Helenam Troianas uexit ad urbes?', and see Nisbet–Hubbard on Hor. *Carm.* 1.15.1 'Pastor cum traheret'.

Paris. Pallas: a piquant juxtaposition.

61–2. Pallas quas condidit arces / ipsa colat: let the chaste goddess frequent the citadel she founded, i.e. Athens; speaking is a Graeco-Roman herdsman.

61. condidit arces: V.'s variation of *urbem condere*; cf. *A.* 7.61 (Latinus) 'conderet arces' and see *TLL* s.v. *condo* 142.75.

63–5. torua leaena lupum sequitur, lupus ipse capellam, / florentem cytisum sequitur lasciua capella, / te Corydon, o Alexi: 'admodum iucunda gradatio' (Ursinus); see W. R. Race, *The Classical Priamel from Homer to Boethius* (Leiden, 1982), 24. 'Theocriti sunt isti uersus' (Serv.); true, though not as shaped by V., Theocr. 10.30–1 ἁ αἶξ τὰν κύτισον, ὁ λύκος τὰν αἶγα διώκει, | ἁ γέρανος τὤροτρον· ἐγὼ δ' ἐπὶ τὶν μεμάνημαι, 'The goat pursues the moon-clover, the wolf the goat, the crane the plough, and I am mad for you.' The trope is very old and seems usually to have had a homosexual reference, as in Plato, *Phaedr.* 241D ὡς λύκοι ἄρν' ἀγαπῶσ', ὣς παῖδα φιλοῦσιν ἐρασταί, 'as wolves crave a lamb, so lovers desire a boy'; see G. J. de Vries (Amsterdam, 1969), ad loc., M. Haupt, *Opuscula* (Leipzig, 1876, repr. Hildesheim, 1967), ii. 404–5, G. Luck, 'Kids and Wolves', *CQ*, NS 9 (1959), 34–7.

63. leaena: possibly suggested by Catull. 60. 1 'Num te leaena montibus Libystinis', but λέαινα occurs three times in Theocritus, 2.68, 3.15 (see *E.* 8.43n.), 26.21.

64. cytisum: 1.78n.

65. o Alexi: the interjection *o* is shortened twice in Theocritus, 1.115 and 15.123, both times as the third in a sequence, but only here in V.

trahit sua quemque uoluptas: critics objected, says Servius, 'quod hanc sententiam dederit rustico supra bucolici carminis legem aut possibilitatem'. Cf. Lucr. 2.258 'quo ducit quemque uoluptas', as V. seems to have read.

66. aratra iugo referunt suspensa: 'It was customary when returning home to turn the plough over (cf. Hor. Epod. 2.63 *videre fessos vomerem inversum boves / collo trahentes languido*), so as to prevent the share catching in the ground; the main body of the plough would thus seem to be "hanging" in the air' (T. E. Page). Cf. also Ov. *Fast.* 5.497 'uersa iugo referuntur aratra'.

iuuenci: 1.45n.

68. me tamen urit amor: Corydon is still burning with desire for Alexis ('ardebat'). The turbulence of human emotion as contrasted with nature's tranquillity is an old theme of poetry; see Clausen 62–3, and cf. Theocr. 2.38–40 *ἠνίδε σιγῇ μὲν πόντος, σιγῶντι δ' ἀῆται· | ἁ δ' ἐμὰ οὐ σιγῇ στέρνων ἔντοσθεν ἀνία,| ἀλλ' ἐπὶ τήνῳ πᾶσα καταίθομαι,* 'Look, quiet is the sea, and quiet the winds; yet the torment in my breast is not quiet, but I am all on fire for him'.

quis enim modus adsit amori: cf. 10.28 and, for the shape and balance of this line, 10.69.

69–72. Corydon abandons his usual tasks, a symptom of lovesickness in Hellenistic literature; cf. Theocr. 11.72–4 *ὦ Κύκλωψ Κύκλωψ, πᾷ τὰς φρένας ἐκπεπότασαι; | αἴ κ' ἐνθὼν ταλάρως τε πλέκοις καὶ θαλλὸν ἀμάσας | ταῖς ἄρνεσσι φέροις, τάχα κα πολὺ μᾶλλον ἔχοις νῶν,* 'O Cyclops, Cyclops, whither have your wits fled? You would show much more sense were you to go and weave cheese-crates and gather foliage for your lambs'. In Longus 1.13.5, Chloe pays no attention to her flock; and in Theocr. 10.2, Bucaeus, now for some eleven days in love with Bombyca, has left the plot before his door unhoed. The significance of *A.* 4.86–9—the building of Dido's new city languishes—was understood by R. Heinze, *Virgils epische Technik*[3] (Leipzig, 1915), 130, but hardly by Pease and Austin.

69. a: 10.47n.

quae te dementia cepit!: 6.47; cf. *G.* 4.488, *A.* 5.465. No doubt V. had noticed that *πᾷ τὰς φρένας ἐκπεπότασαι;* occurs twice in Theocritus. For a similar subtlety see above, 58n.

70. Spring pruning (Pliny, *NH* 17.190 'pampinatio uerna') when surplus shoots were removed and foliage on the vine and supporting tree thinned out to let the sun get at the grapes; cf. *G.* 2.362–70. The normal time for this was May.

semiputata: such compounds seem mostly to have been invented by the poets: Catull. 50.15 *semimortuus*, 54.2 *semilautus*, 59.5

semirasus, 61.213 *semihians*, *A.* 3.578 *semustus* (cf. Ap. Rhod. 4.598 ἡμιδαής, with Livrea's note), Ov. *Her.* 1.55 *semisepultus*, 7.176 *semirefectus*, *Am.* 1.6.4 *semiadapertus*, where see McKeown, *Ars* 2.614 *semireductus, Ibis* 634 *semicrematus*, Mart. 5.14.9 *semifultus*. See 3.38n.

71. quorum: for the ellipse of the antecedent cf. *G.* 1.104, 111, *A.* 11.81, 172, and see Munro on Lucr. 1.883, Pease on *A.* 4.598.

indiget: a plain word in a plain line; common in prose, found in Plautus, Terence, and Lucilius, but elsewhere in poetry only here and in Sen. *Tro.* 55.

72. uiminibus mollique . . . detexere iunco: Corydon (like Polyphemus) would be better occupied weaving a crate or *fiscella* from pliable twigs and soft rushes in which to make his cheese; so Tibullus, who apparently took *uiminibus* and *iunco* as a hendiadys, 2.3.15–16 'tunc fiscella leui detexta est uimine iunci / raraque per nexus est uia facta sero' (*TLL* s.v. *detexo* 811.36). Cf. 10.71 and see Mynors on *G.* 3.402.

73. inuenies . . .: cf. Theocr. 11.76 εὑρησεῖς Γαλάτειαν ἴσως καὶ καλλίον' ἄλλαν, 'You will find another, perhaps, and a prettier Galatea'.

ECLOGUE 3

Introduction

The herdsmen of Theocritus are much given to singing; all the pastoral *Idylls* except the Fourth contain songs[1]—singing-matches (5, 6, [8]), quasi-competitive songs (7, [9], 10), and solos (1, 3, 11)—and even in the Fourth one of the speakers, an amateur piper who knows the tunes of two professional musicians, sings a snatch of a song (30–2). Virgil's herdsmen are similarly occupied, with singing-matches (3, 7), quasi-competitive songs (5, 8), and solos (2, 6, 10). Of the remaining *Eclogues*, the First begins with an evocation of song, and in the Ninth singers are heard and fragments of song; the absence of song in the Fourth is an indication of its original character.[2]

The range of Theocritean pastoral is ample and various, extending from the scurrilous conversation of the herdsmen in the Fifth *Idyll* to the elevation of Thyrsis' lament for Daphnis in the First. The character of the Fifth *Idyll*—its 'realistic' character[3]—should not, however, be attributed to any failure of art or imagination: the Fifth *Idyll* is no less artificial, no more real, than the First. No doubt there was an old tradition in Sicily,[4] older than literature, of rustic singing-matches, with improvised capping of verses, umpires, and prizes. And Theocritus' desire to represent such an occasion in literary form accounts for the character, or the limitations, as they may appear to be, of the Fifth *Idyll*.

The Fifth *Idyll* consists of two main parts: a conversation between two herdsmen, Comatas and Lacon, which sets the scene and prepares for their singing-match (1–79); and the singing-

[1] See Gow, ii. 76.

[2] See *E.* 4, Introduction, p. 126.

[3] A confusion, which few commentators on Theocr. 5 avoid, between the subject and the poetry of the subject. Cf. V. Nabokov, *Lectures on Literature* (New York, 1980), 138: 'The subject may be crude and repulsive. Its expression is artistically modulated and balanced. This is style. This is art.' G. Serrao, *Problemi di poesia alessandrina*, i: *Studi su Teocrito* (Rome, 1971), 72, emphasizes the art of Theocr. 5.

[4] In Sicily and elsewhere; see R. Merkelbach, '*ΒΟΥΚΟΛΙΑΣΤΑΙ* (Der Wettgesang der Hirten)', *RhM* 99 (1956), 97–133. Merkelbach 115 n. 54 remarks that shepherds can never have sung in hexameters.

match itself of twenty-nine couplets (80–137), which is won by Comatas (138–50).

Comatas accuses Lacon of stealing his goatskin: Lacon accuses Comatas of stealing his pipe.

> *Αἶγες ἐμαί, τῆνον τὸν ποιμένα, τὸν Συβαρίταν,*
> *φεύγετε, τὸν Λάκωνα· τό μευ νάκος ἐχθὲς ἔκλεψεν.* (1–2)
>
> Away from that shepherd, my goats, the Sybarite,
> Lacon, the one who stole my goatskin yesterday.
>
> *οὐκ ἀπὸ τᾶς κράνας; σίττ', ἀμνίδες· οὐκ ἐσορῆτε*
> *τόν μευ τὰν σύριγγα πρόαν κλέψαντα Κομάταν;* (3–4)
>
> Back from the spring, hoy, lambs; don't you see Comatas,
> The one who stole my pipe day before yesterday?

The assimilation[5] of Lacon's unpremeditated response (as dramatically it must be) to Comatas' accusation renders both the accusation and the response, or rather, the combination of the two, elegant. Couplet follows couplet as in a singing-match, as if Lacon were capping—supposing a pipe more valuable than a goatskin—Comatas' couplet. The first two couplets of the singing-match proper, with Comatas leading off, may be compared.

> *ταὶ Μοῖσαί με φιλεῦντι πολὺ πλέον ἢ τὸν ἀοιδόν*
> *Δάφνιν· ἐγὼ δ' αὐταῖς χιμάρως δύο πρᾶν ποκ' ἔθυσα.* (80–1)

The Muses love me much better than the minstrel Daphnis; and two goats I sacrificed to them the other day.

> *καὶ γὰρ ἔμ' Ὡπόλλων φιλέει μέγα, καὶ καλὸν αὐτῷ*
> *κριὸν ἐγὼ βόσκω· τὰ δὲ Κάρνεα καὶ δὴ ἐφέρπει.* (82–3)

Aye, and me Apollo dearly loves. And a fine ram I feed for him, and already the Carnea are coming on.[6]

[5] As indicated by the underlining.
[6] Gow's translations.

Theocritus was concerned for the coherence of his poem and took pains to relate part to part. Certain moments in the conversation anticipate the singing-match, or, conversely, certain moments in the singing-match recall the conversation.[7] In general, however, the singing-match is more elegant, as the reader will feel especially at the end of the poem (141–50), when a jubilant and uninhibited Comatas abruptly reverts to the tone and manner of his conversation with Lacon.

The Third *Eclogue* is modelled on the Fifth *Idyll*, with incidental reference to the Fourth,[8] First,[9] Eighth,[10] and Third.[11] That Virgil's imitations should be 'ambitiously frequent'[12] in an early work[13] is hardly surprising. Like the Fifth *Idyll*, the Third *Eclogue* consists of two main parts: a bickering conversation between two herdsmen, Menalcas and Damoetas, which sets the scene and prepares for their singing-match (1–54); and the singing-match itself of twenty-four couplets (60–107), which ends in a draw (108–11). The umpire Palaemon intervenes with a brief, practical speech of five lines (55–9), which however contains a lyrical moment, apparently unmotivated:[14]

> et nunc omnis ager, nunc omnis parturit arbos,
> nunc frondent siluae, nunc formosissimus annus

—two lines, suggestive of a couplet, which supply a liquid transition to the singing-match.

[7] e.g. 1–4 ~ 100–4, 116–19 ~ 41–4.

[8] Imitation of Theocr. 4 is confined almost exclusively to the opening lines; see notes on 1–2, 3, 5, 100. Apparently Virgil could make nothing of Theocr. 5.1 *τὸν Συβαρίταν* (where see Gow) or 5 *δῶλε Σιβύρτα*. He did, however, imitate Theocr. 5.5–7 later in the *Eclogue*; see 25–7n.

[9] See notes on ll. 29–30, 38–9, 43.

[10] See notes on ll. 32–4, 80–1.

[11] See notes on ll. 20, 71.

[12] Samuel Johnson of Pope's *Pastorals*, also an early work.

[13] See *E.* 8, Introduction, p. 238.

[14] In the sense that the lines would not be missed were they absent.

The conversation of the herdsmen in the Fifth *Idyll* proceeds for the first twenty-two lines with strict formality, a series of equal exchanges that may be schematized as follows: Comatas two lines (1–2), Lacon two (3–4); C. three (5–7), L. three (8–10); C. three (11–13), L. three (14–16); C. three (17–19), L. three (20–2). Here Lacon proposes a singing-match. And from this point onward their conversation, while remaining symmetrical, becomes more involved. For the most part, Lacon ceases to speak the same number of lines as Comatas, but Comatas invariably reciprocates by speaking the same number of lines as Lacon, correcting him, as it were, and righting the imbalance: C. two (23–4), L. three (25–7); C. three (28–30), L. four (31–4); C. four (35–8), L. two (39–40); C. two (41–2), L. two (43–4);[15] C. five (45–9), L. five (50–4); C. five (55–9), L. three (60–2); C. three (63–5), L. four (66–9);[16] C. four (70–3),[17] L. two (74–5); C. two (76–7), L. two (78–9).

Virgil's herdsmen, by contrast, seem determined to be informal. Except for an initial exchange of single lines (1–2) and a final exchange of three lines each (49–54), which unobtrusively frame their conversation,[18] the conversation of Menalcas and Damoetas is unequal throughout[19]–deliberately so, for Virgil was writing with an eye to Theocritus. Unlike Theocritus, yet no less concerned for

[15] Two lines following two lines, a couplet-like effect with which Theocritus defines the herdsmen's conversation, at the beginning (1–4), at the end (76–9) and here, where the 'couplets' are followed by descriptions of *loci amoeni* (45–59: 5 + 5 + 5) and the selection of an umpire.

[16] Disregarding the slight irregularity caused by Comatas' interjection in l. 66.

[17] For the difficulty of this passage see Gow ad loc. In any case, the deletion of l. 73 (Wilamowitz) would spoil the symmetry.

[18] Virgil avoids an exchange of two lines, a couplet-like effect that would anticipate the singing-match; see above, n. 15.

[19] Menalcas one line (1), Damoetas one (2); M. four (3–6), D. three (7–9); M. two (10–11), D. four (12–15); M. five (16–20), D. four (21–4); M. three (25–7), D. four (28–31); M. twelve (32–43), D. five (44–8); M. three (49–51), D. three (52–4). In so far as Virgil had a model, it was [Theocr.] 8. 11–27: Menalcas one (11), Daphnis one (12); M. one (13), D. one (14); M. two (15–16), D. one (17); M. three (18–20), D. four (21–4); M. one (25), D. two (26–7), in which the umpire is chosen; and then five lines (28–32), which introduce the singing-match.

the coherence of his poem, Virgil did not relate part to part (as he might easily have done) by means of verbal or stylistic similarities; rather, he made the two parts of his poem as dissimilar as possible, opposing, and thereby relating, the informality of the conversation to the formality of the singing-match.[20]

Why does Comatas win? Was the unheard melody with which he accompanied his words indeed sweeter?[21] Is it because he is more truthful than Lacon?[22] Or more skilled?[23] Or less coarse?[24] Or simply older and more experienced?[25] All these answers—all such answers—have this in common, that they are, more or less, unsatisfactory and unconvincing; which suggests that the wrong question has been asked. The question to be asked is not 'Why does Comatas win?' but 'What was the literary problem?' On a real occasion, one of the

[20] See La Penna 157.

[21] Gow, ii. 94: 'The poet tells us the words of the songs but not the music; the decision must turn on a combination of the two, and if we complain that we cannot understand the verdict, the poet may reasonably reply that we have heard only half the evidence and that we must take on trust the judgment of those who have heard the whole.' Dover 127 has no comment on the victory of Comatas.

[22] Lacon calls Comatas a windbag for telling the umpire that Lacon's flock is not his own, and Comatas protests that he is merely telling the truth, 76–7 *ἐγὼ μὲν ἀλαθέα πάντ' ἀγορεύω | κοὐδὲν καυχέομαι*, 'I am speaking the whole truth and not boasting'. See J. Van Sickle, 'The Unity of the *Eclogues*: Arcadian Forest, Theocritean Trees', *TAPA* 98 (1967), 495–6, E. A. Schmidt, 'Der göttliche Ziegenhirt', *Hermes*, 102 (1974), 218–19, G. Serrao, 'L'idillio V di Teocrito', *QUCC* 19 (1975), 88–9, G. Giangrande, 'Victory and Defeat in Theocritus' Idyll V', *Mnemosyne*, 4th ser., 29 (1976), 143–54, C. Segal, 'Thematic Coherence and Levels of Style in Theocritus' Bucolic Idylls', *WS* 90 (1977), 59. It is not permissible, however, to abstract a statement from its dramatic context and make it a principle of interpretation.

[23] As indicated by Lacon's failure to answer his last couplet (136–7). See Merkelbach (above, n.4), 110–13, A. Köhnken, 'Komatas' Sieg über Lacon', *Hermes*, 108 (1980), 122–5.

[24] Such ethical considerations tend to be partial. Gow has this comment on Lacon's use of the verb *μολύνει* in l. 87: 'perhaps intended as an indication of coarseness or brutality', and resorts to Latin in his translation. Twice in the poem Comatas is responsible for an obscenity, in the conversation (41–2) and in the singing-match (116–17); Gow resorts to Latin in his translation but makes no comment on the character of Comatas. In the opinion of G. Lawall, *Theocritus' Coan Pastorals* (Washington, DC, 1967), 64, Comatas, whom he finds less clever and witty than Lacon, dominates the *Idyll* with 'his erotic impetuosity'.

[25] See T. E. Rinkevich, 'Theokritos' Fifth *Idyll*: The Education of Lacon', *Arethusa*, 10 (1977), 295–303.

singers, usually the second, as may be assumed,[26] would show himself incompetent by faltering or failing altogether; but in a poem neither singer can be permitted to falter, at least not very noticeably, or to fail, for then the poem would be ruined. Yet verisimilitude, the poet's desire to construct a plausible fiction, requires that the singing-match end with a decision. Hence the literary problem, the poet's embarrassment. How is he to manage? Since neither singer can be exposed as incompetent, the poet's decision must be arbitrary,[27] though he may attempt, discreetly, to justify it. Perhaps the reader will have noticed the responsiveness of Comatas to Lacon's deviations in their conversation and thus be prepared to accept Comatas as the more adroit and accomplished performer. Or will recall, with distaste, Lacon's gratuitous sneer, 51–2 *ταὶ δὲ τραγεῖαι | ταὶ παρὰ τὶν ὄσδοντι κακώτερον ἢ τύ περ ὄσδεις*, 'Your goatskins there stink worse than even you stink', which disrupts a *locus amoenus*. Or (unlike the critic) accept the decision unquestioningly.

Why does the singing-match of Damoetas and Menalcas end in a draw?[28] Because they are, except for sexual orientation, virtually indistinguishable; and Virgil employed the umpire Palaemon to arrange a neutral decision. In the Fifth *Idyll*, Comatas is both the first to speak and the first to sing; the umpire Morson, who has been chosen with some little trouble,[29] remains silent, as though it were understood that Comatas should begin.[30] Likewise, in the

[26] Gow's argument (ii. 93), anticipated by Servius on 3.59, that the second singer labours under 'a prohibitive handicap' may be true in life but is not true in poetry; see L. E. Rossi, 'Vittoria e sconfitta nell'agone bucolico letterario', *GIF* 23 (1971), 13–18. There are intimations of reality in Theocr. 5.22 *ἀλλά γέ τοι διαείσομαι ἔστε κ' ἀπείπῃς*, 'why, I'll sing a match with you until you've had enough' (Gow), and in [Theocr.] 8.7 *φαμί τυ νικασεῖν, ὅσσον θέλω αὐτὸς ἀείδων*, 'I swear I shall defeat you, if I sing as long as I please', 10 *οὔποκα νικασεῖς μ', οὐδ' εἴ τι πάθοις τύγ' ἀείδων*, 'you will never defeat me, not even if you sing your heart out'.

[27] Cf. Gow, loc. cit.: 'it is legitimate to observe that there is little or no visible difference between the performance of victor and vanquished'; to which he adds in a footnote: 'Lacon's responses seem a little lame on perhaps two occasions (110, 118), but in view of his handicap he does remarkably well ... In *Id.* 8 it might be maintained that the loser's final octet is better rounded than the winner's'.

[28] O. Skutsch, *BICS* 18 (1971), 26–9, relates the question to 'the poet's concern for the balance and contrasting symmetry of his book' (29).

[29] The selection of an umpire (61–73) resembles the selection of an arbitrator in Menander, *Epitrep.* 218–39.

[30] Lacon urges Comatas to begin (78), though, as Gow observes (loc. cit.), he had previously been invited to begin himself (30).

Eighth *Idyll*, Menalcas is both the first to speak and the first to sing; and the unnamed umpire, like Morson, remains silent. Why, with these models before him, did Virgil decide to alter the pattern? Why is Menalcas, who is the first to speak, not also the first to sing? Because Palaemon intervenes and determines otherwise, 58 'incipe, Damoeta; tu deinde sequere, Menalca'.[31] Menalcas and Damoetas, Damoetas and Menalcas—an effect of parity is achieved, and their singing-match appropriately ends in a draw. 'Non nostrum inter uos . . .', says Palaemon (108), without embarrassment.

Bibliography

Klingner 50–9.

C. P. Segal, 'Vergil's 'Caelatum Opus': An Interpretation of the Third *Eclogue*', *AJP* 88 (1967), 279–308.

A. La Penna, 'Lettura della terza bucolica', in Gigante 131–69.

1–2. Cf. Theocr. 4. 1–2 *Εἰπέ μοι, ὦ Κορύδων, τίνος αἱ βόες; ἦ ῥα Φιλώνδα; | οὔκ, ἀλλ' Αἴγωνος· βόσκειν δέ μοι αὐτὰς ἔδωκεν*, 'Tell me, Corydon, whose cows are these? Philondas's?' 'No, Aegon's; he gave me them to graze' (Gow).

1. Dic mihi, Damoeta, cuium pecus?: a remarkably close translation, including the bucolic diaeresis, of Theocritus. But a Roman reader would immediately be reminded, as V. no doubt intended he should be, of Roman comedy; cf. Donatus on Ter. *Andr.* 667 'semper *τὸ dic mihi* iniuriosum est, ut ille "dic mihi, Damoeta, cuium pecus?"' The phrase *dic mihi* frequently introduces a question in Plautus and Terence, e.g. *Bacch.* 600–1 'dic mihi, / quis tu est?', *Capt.* 987 'dic mihi, isne istic fuit?', *Andr.* 931–2 'CH. . . . eho dic mihi, / quid eam tum? suamne esse aibat? CR. non. CH. cuiam igitur? CR. fratris filiam'; see McGlynn, *Lexicon Terentianum* s.v. *dico* XI. (I am indebted for this observation to J. Wills.)

cuium pecus: '*cuium* autem antique ait uitans homoeoteleuton, ne diceret *cuius pecus*' (Serv.). V. did take pains to avoid homoeote-

[31] Cf. [Theocr.] 9. 1–2 *Βουκολιάζεο, Δάφνι· τὺ δ' ᾠδᾶς ἄρχεο πρᾶτος, | ᾠδᾶς ἄρχεο, Δάφνι, ἐφεψάσθω δὲ Μενάλκας*, 'Make rustic music, Daphnis, and begin the song first, begin the song, Daphnis, and let Menalcas follow'. The singing-match, such as it is, ends in a draw.

leuton (see Norden 405–7), but such a concern hardly explains so extreme an archaism so prominently placed. Nor is 'cuium pecus' rustic speech, as implied by the malevolent Numitorius in his *Antibucolica*: 'Dic mihi, Damoeta: "cuium pecus" anne Latinum? / non, uerum Aegonis nostri sic rure loquuntur' (*Vita Donati* 43); see Wackernagel ii. 81. Although the adjective *cuius-a-um* had long since become obsolete, it is common in Plautus and Terence. Here two slaves are wrangling, and the obvious model for such a scene was Plautus; see H. MacL. Currie, 'The Third *Eclogue* and the Roman Comic Spirit', *Mnemosyne*, 4th ser., 29 (1976), 411–20. V., the youthful poet, establishes this initial reference to Plautus with an audacity that has startled generations of readers. The fact that Menalcas and Damoetas repeatedly avail themselves of Plautine language or language reminiscent of comedy (3, 10, 13, 15, 17, 21, 26, 27, 32, 35–6, 51, 52, 53) confirms this interpretation of 'cuium pecus'. For reminiscence of Plautus elsewhere in the *E.* cf. 7.7, 8.

Damoeta: 2.37n.

Meliboei: 1.6n.

3. infelix o semper, oues, pecus: for the word-order see 1.57n. Cf. Theocr. 4.13 δείλαιαί γ' αὖται, τὸν βουκόλον ὡς κακὸν εὗρον, 'poor beasts, it's a sorry herdsman they found' (Gow). Note how *pecus* is repeated before the bucolic diaeresis for emphasis; Menalcas is surprised and disgusted.

ipse: Aegon is first named, then referred to as *ipse* 'the master', a sense *ipse* (*ipsus*) commonly has in colloquial Latin. Cf. Plaut. *Aul.* 353 'Megadorus iussit Euclioni haec mittere', 356 'si a foro ipsus redierit'.

Neaeram: ever since [Dem.] 59 a name associated with easy virtue: in Parthenius, Ἐρωτικὰ παθήματα 18, the wife who seduces her husband's friend; a beautiful Nymph in Ov. *Am.* 3.6.27–8 'nondum Troia fuit lustris obsessa duobus, / cum rapuit uultus, Xanthe, Neaera tuos'; (cf. Hom. *Od.* 12.131–3); and in Hor. *Epod.* 15, as in the elegies of Lygdamus, the poet's mistress. See also A. J. P. Woodhouse and D. Bush, *A Variorum Commentary on the Poems of John Milton* (New York, 1972), ii. 2 on *Lycidas*, 68–9 'To sport with Amaryllis in the shade, / Or with the tangles of Neaera's hair'.

4. fouet: 'verbum sumptum ex penu amatorio', La Cerda, citing *A.* 8.387–8 'diua lacertis / cunctantem amplexu molli fouet', Ov.

Am 1.4.5 'alteriusque sinus apte subiecta fouebis?' See *TLL* s.v. 1219.32.

ac: instead of *et* to avoid the cacophony *fouet et*, as also in 4.9 *desinet ac*; *ac* does not occur elsewhere in the *E*. So Axelson 82; cf., however, 9.42.

5. alienus . . . custos: a stranger to the flock, a *mercennarius*, who abuses the ewes for his own gain, 'because he is an hireling, and careth not for the sheep' (John 10:13). Cf. Theocr. 4.3 ἦ πᾳ ψε κρύβδαν τὰ ποθέσπερα πάσας ἀμέλγες;, 'And you, maybe, milk them all on the sly in the evening?' (Gow).

bis . . . in hora: the older construction with *in* when the ablative is unmodified: thus Cato, *De agr.* 26 'bis in die', Plaut. *Bacch.* 1127 'ter in anno' (Hofmann–Szantyr 148). See 2.42n. Twice an hour, as Servius explains, instead of twice a day: a physical impossibility; see below, 34n.

6. sucus: the ewes are drained of their 'juice'—that is, of their milk and strength—and the lambs starved. One of the speakers in Varro, *RR* 2.2.17, says that it is better not to milk ewes at all while they are giving suck.

pecori et: hiatus occurs six times at this point in the line in the *E.*: below, 63, 7.53, 8.41, 44, 10.13; see Wagner, *Quaest. Virg.* 11.2, Fordyce on *A.* 7.178.

subducitur: with a suggestion of stealth, as in *A.* 6.524 'fidum capiti subduxerat ensem'.

7. uiris: Damoetas may be dishonest but at least he is a man, and might say with Vespasian (Suet. *Vesp.* 13) 'ego tamen uir sum' (La Cerda).

8–9. The goats leered, and the easy Nymphs laughed. Not the usual reaction in such circumstances; for engaging in sexual intercourse in a woodland shrine sacred to her Cybele transformed Atalanta and Hippomenes into lions (Ov. *Met.* 10.686–704).

8. nouimus et qui te: Quintil. 9.3.60 'hanc quidam aposiopesin putant, frustra: nam illa quid taceat incertum est aut certe longiore sermone explicandum, hic unum uerbum et manifestum quidem desideratum'. Servius supplies a verb, though hardly the desired verb: 'subaudis *corruperint*, quod suppressit uerecunde, licet Theocritus aperte ipsam turpitudinem ponat et exprimat'; but the reference seems to be to a single occasion and there is no reason to suppose Menalcas had more than one lover. Servius, or a rather more learned predecessor, had in mind Theocr. 4.58–61, 5.41–3, 87, 116–17, where description of sexual congress is explicit and brutal. In 1.105, however, Theocritus suppresses the offensive verb: *οὐ λέγεται τὰν Κύπριν ὁ βουκόλος;* 'Is it not told how the herdsman with Cypris—?', as do other poets both Greek and Latin, e.g. Meleager 72.5 G.–P. (= *AP* 5.184) *οὐχ ὁ περίβλεπτός σε Κλέων;*, 'Did not the gallant Cleon and you—?', Philodemus, *AP* 5.4.5 *καὶ σύ, φίλη Ξανθώ, με*, 'And you, dear Xantho, with me—', Ter. *Eun.* 479 'ego illum eunuchum, si opus siet, uel sobrius', *Heaut.* 913 'qui se uidente amicam patiatur suam?' See Gow on Theocr. 1.105, J. N. Adams, 'A Type of Sexual Euphemism in Latin', *Phoenix*, 35 (1981), 120–8.

transuersa tuentibus hircis: 'the Goats observ'd with leering Eyes' (Dryden). As in Theocr. 5.41–2 the goats are stimulated by observing human sexuality. For suggestive sidelong glances cf. *Priapea* 73.1 'obliquis quid me, pathicae, spectatis ocellis?', Quintil. 11.3.76 (*oculi*) 'lasciui et mobiles et natantes et quadam uoluptate suffusi aut limi et, ut sic dicam, uenerii', and see A. Hudson-Williams, *CQ*, NS 30 (1980), 124–5.

transuersa tuentibus: cf. *A.* 6.467 'torua tuentem', with Norden's note. *Tueor* in this sense is archaic and poetic.

9. sacello: a rustic shrine; cf. Prop. 2.19.13 (Cynthia in the country) 'atque ibi rara feres inculto tura sacello'. Here a *Nymphaeum*, a grotto or cave, which would offer some concealment: Hom. *Od.* 13.103–4 *ἄντρον ἐπήρατον ἠεροειδές, | ἱρὸν Νυμφάων*, 'a pleasant, shadowy cave, sacred to the Nymphs', Theocr. 7.137 *Νυμφᾶν ἐξ ἄντροιο*, 'from the cave of the Nymphs', Longus 1.4.1 *Νυμφῶν ἄντρον*, 'a cave of the Nymphs', *A.* 1.166–8 'antrum; / intus aquae dulces uiuoque sedilia saxo, / Nympharum domus'.

10. credo: 'no doubt'; colloquial Latin from Plautus onward (*TLL* s.v. 1137.19). Here ironic, for Menalcas of course means not *me* but *te*; 'The day, no doubt, they saw *me* hacking Mico's vines' (Lee).

arbustum . . . Miconis: cf. Theocr. 5.112 *τὰ Μίκωνος*, 'Micon's vineyard'. The name occurs again in 7.30.

11. mala . . . falce: cf. the legal phrase *dolo malo*. Wrongfully felling another man's trees was punishable by a fine in the Twelve Tables (Pliny, *NH* 17.7).

nouellas: almost a technical term; cf. Cato, *De agr.* 33.4 'uineas nouellas', Varro, *RR* 1.31.1 'uineas nouellas', Pliny, *NH* 17.195 'uitem nouellam resecari tum erit tempus ubi ualebit'. Being too tender for a pruning-hook, the leaves of young vines should be plucked off (*G.* 2.363–6). Columella 4.11.1 disagrees with V.

13. calamos: properly a reed to which an arrow-head was fitted (Pliny, *NH* 16.159 'calamis spicula addunt'); like *harundo* (Pease on *A.* 4.73), or *reed* in English (*OED* s.v. 2.6.a), a poetic synecdoche, which first occurs here (*TLL* s.v. 123.46), again in *A.* 10.140, and a few times in later poetry.

peruerse: found in Plautus as an adjective and especially as an adverb; not in Terence. Evidently Menalcas had been defeated in a singing-match with Daphnis and then had destroyed the prize, the bow and arrows, that Daphnis had won.

Menalca 2.15n.

13–15. Cf. Theocr. 5.11–13 *τὸ Κροκύλος μοι ἔδωκε, τὸ ποικίλον, ἀνίκ' ἔθυσε | ταῖς Νύμφαις τὰν αἶγα· τὺ δ', ὦ κακέ, καὶ τότ' ἐτάκευ | βασκαίνων, καὶ νῦν με τὰ λοίσθια γυμνὸν ἔθηκας*, 'It was the skin Crocylus gave me, the dappled one, when he sacrificed the goat to the Nymphs. And you, you rascal, were consumed with envy then, and now, at last, you've stripped me bare'—'which accounts for the repetition of *et*' (Conington).

14. puero: Daphnis, much younger than Menalcas; hence, in part, his envy and anger.

15. aliqua: occurs in Plautus and Terence, once in Lucilius, then here and only here in V. (*TLL* s.v.).

mortuus esses: 'you'd have died'; the hyperbole is colloquial and Plautine, e.g. *Cas.* 622 'cor metu mortuomst'.

16. quid domini faciant, audent cum talia fures?: V. seems to have been thinking of Catull. 66.47 'quid facient crines, cum ferro talia cedant?', although here *talia* looks forward; see below, 40n. Damon, though a slave, is nevertheless the owner, in effect the *dominus*, of his goats; see 1.32n., 2.21n.

17. pessime: a Plautine vocative of objurgation (*TLL* s.v. *malus* 219.50 'in conviciis'); in non-dramatic poetry first here, and only here in V., then in Hor. *Serm.* 2.7.22, Pers. 2.46. Cf. Theocr. 5.12 *ὦ κακέ* (above, 13–15n.), 75 *κάκιστε*.

18. excipere: so Orestes catches Pyrrhus by surprise, *A.* 3.332 'excipit incautum' (Serv.). Cf. Prop. 2.19.23–4 'haec igitur mihi sit lepores audacia molles / excipere', Hor. *Carm.* 3.12.11–12 'arto latitantem / fruticeto excipere aprum', and 2.15.14–16 'nulla . . . / . . . opacam / porticus excipiebat Arcton', 'no portico caught the shady (cool) north (wind)', where L. P. Wilkinson, *CQ*, NS 9 (1959), 188, detects a witty allusion to the northern constellation, the Bear (*Arcton*); see Nisbet–Hubbard ad loc. Cf. 1.52 'frigus . . . opacum'.

Lycisca: the name of the twenty-seventh in Ovid's catalogue of Actaeon's hunting-dogs (*Met.* 3.220 'uelox . . . Lycisce'). Properly a mongrel bitch; cf. Serv. 'lycisci sunt, ut etiam Plinius dicit, canes nati ex lupis et canibus, cum inter se forte misceantur'. According to Aristotle, *HA* 607a 2–3, whom La Cerda cites here, such cross-breeding occurs in Cyrene.

20. Tityre: here a fellow-slave to whom Damon entrusted his goats, as the anonymous goatherd in Theocr. 3.1–5 entrusts his goats to Tityrus; cf. 9.23–5.

carecta: sedge or sheargrass (Abbe 44), a very rare word not found elsewhere in poetry. Cf. *G.* 3.231 'carice . . . acuta' and see 5.17n.

21–2. cantando . . . / . . . fistula: 5.48n.

21. non redderet: cf. Plaut. *Trin.* 133 'CAL. non ego illi argentum redderem? MEG. non redderes' (Conington).

23. si nescis: colloquial and rather rude, 'I'd have you know'; cf. Prop. 2.15.12, Ov. *Am.* 3.8.13, *Trist.* 4.9.11 'omnia, si nescis, Caesar mihi iura reliquit', *Ex Pont.* 3.3.28, Juv. 5.159.

24. posse: the reflexive pronoun is sometimes omitted where it can easily be understood, especially in comedy; see Kühner–Stegmann i.701.

25–7. Cf. Theocr. 5.5–7 *τὰν ποίαν σύριγγα; τὺ γάρ ποκα, δῶλε Σιβύρτα, | ἐκτάσω σύριγγα; τί δ' οὐκέτι σὺν Κορύδωνι | ἀρκεῖ τοι καλάμας αὐλὸν ποππύσδεν ἔχοντι;*, 'What pipe was that? Have you, Sibyrtas's slave, ever come by a pipe? And why aren't you still content to toot upon an oaten whistle with Corydon?' (Gow).

26. indocte: found in Plautus as an adjective and as an adverb; not in Terence nor elsewhere in V.

27. stridenti: Milton, *Lycidas*, 123-4 'their lean and flashy songs / Grate on their scrannel pipes of wretched straw'. A similar effect had been achieved by Catullus, 64.264 'barbaraque horribili stridebat tibia cantu'.

stipula: contrasted with *fistula* (25), the usual term for the pan-pipe (2.37n.), as *καλάμας* is contrasted with *σύριγγα* in Theocr. 5.5–7 (above, 25–7n.). Like *καλάμα*, *stipula* is unique in this sense.

disperdere: an old verb (Plaut. *Cas.* 248 'disperde rem'), employed here for the harsh sound it makes. La Cerda compares Prop. 2.33.10 (Io) 'et pecoris duro perdere uerba sono'; see O. Zwierlein, *Hermes*, 119 (1991), 120n.2.

28–51. The structure of this passage is based on [Theocr.] 8.11–27.

28. uicissim: in turn, in amoebean song; cf. 5.50.

29. experiamur: a full stop after the second trochee occurs only here in the *E.*, and only once in the *G.*, where it is powerfully expressive, 1.501 'ne prohibete'.

29–30. Cf. Theocr. 1.25–6 *αἶγά τέ τοι δωσῶ διδυματόκον ἐς τρὶς ἀμέλξαι, | ἃ δύ' ἔχοισ' ἐρίφως ποταμέλγεται ἐς δύο πέλλας*, 'I'll give you a goat, the mother of twins, to milk three times, and even though she has two kids her milk yet fills two pails'.

29. uitulam: evidently a heifer that has just calved. Some ancient readers objected: 'male enim quidam quaestionem mouent, dicentes uitulam paruam esse nec congruere ut eam iam enixam esse dicamus' (Serv.). V. was thinking of Theocritus' she-goat.

30. bis uenit ad mulctram: cf. Hor. *Epod.* 16.49 'iniussae ueniunt ad mulctra capellae'. Cows' milk, mentioned also in *G.* 2.524–5, 3.177, was not ordinarily drunk but consumed as cheese; cf. Varro, *RR* 2.11.1 'lac est omnium rerum, quas cibi causa capimus, liquentium maxime alibile, et id ouillum, dein caprinum'.

binos . . . fetus: according to Aristotle, *HA* 575a 30, whence Pliny, *NH* 8.177, cows seldom give birth to twins; V.'s reminiscence of Theocritus has become a recommendation of Damoetas' cow.

31. depono: 32 *deponere*, 36 *ponam*—the compound form of the verb followed by the simple in the same sense, of which there are many instances in Latin poetry. But this is unusual for two reasons: (i) it is the only instance in the *E.*, and (ii) it is an imitation of [Theocr.] 8.11 *καταθεῖναι*, 12 *καταθεῖναι*, 13 *θησεύμεσθ'*, 14 *θησῶ*, *θές*, 15 *θησῶ*, 17 *θησεῖς*. See Clausen, *AJP* 76 (1955), 49–51 and 86 (1965), 97–8, C. Watkins, *HSCP* 71 (1966), 115–19, R. Renehan, *CP* 72 (1977), 243–8.

certes: V.'s word for amoebean singing: 4.58, 5.8, 9, 15, 7.16 (*certamen*), 8.3.

32–4. Cf. [Theocr.] 8.15–16 *οὐ θησῶ ποκα ἀμνόν, ἐπεὶ χαλεπὸς ὁ πατήρ μευ | χἀ μάτηρ, τὰ δὲ μῆλα ποθέσπερα πάντ' ἀριθμεῦντι*, 'Never will I stake a lamb, for my father is stern, my mother too; and they count all the sheep in the evening'.

32. non ausim: cf. Plaut. *Bacch.* 1056 'ego cum illo pignus haud ausim dare.'

33. iniusta nouerca: father and mother in [Theocr.] 8 have become father and stepmother in V.; cf. Catull. 23.3 'et pater et nouerca'. The stepmother always has a bad character; see Otto, no. 1239.

34. bisque die: 'et cum uadit ad pascua et cum inde reuertitur' (Serv.). Once a day, in the evening, seems to have been normal; cf. 6.85, Calp. Sic. 3.64.

alter et haedos: one or other of them counts the kids, sometimes the father sometimes the stepmother. Servius is concerned to explain *alter*: 'nec nos moueat quod *alter* dixit de femina . . . scimus, quia quotiens haec duo genera iunguntur, femininum non praeponderat'. Cf. Livy 39.10.1 (a prostitute and her young lover) 'alter ab altero', Ter. *Eun.* 840 'uterque, mater et pater', Ov. *Am.* 1.3.10 'parcus uterque parens'.

35–6. Menalcas' presentation of his cups is artfully delayed to emphasize their value: 'multo . . . maius', linked by alliteration and reinforced by 'tute ipse', which is colloquial and Plautine; a parenthesis following; then, with notable assonance, 'pocula ponam'—a pair of cups, as the reader learns in l. 44.

36–42. These lines constitute an ecphrasis, the description, that is, of an imagined work of art (the poet's intention being to display his own work of art). Ecphrasis, which is as old as Homer's Shield of Achilles (*Il.* 18.478–608), was developed and refined by the Hellenistic poets with much ingenuity; see Clausen 17, 77–82. As R. F. Thomas has shown, *HSCP* 87 (1983), 175–84, almost every post-Homeric ecphrasis—V.'s model here, the elaborately carved cup in Theocr. 1.27–56, is an exception—contains a reference to its centre, p. 178: 'But there is a further development with Virgil. The medial reference itself appears in the center of the passage: Menalcas has the phrase *in medio* midway through his description of the cup (40), but he is capped by Damoetas, for whom *in medio* occurs at the medial caesura in the central line of his five-line response (46)'. In Theocr. 1, the figures of the woman, the old fisherman, and the little boy are described as adorning the interior of the cup (32 ἔντοσθεν δὲ γυνά . . .). This, the apparent meaning, is rejected by Dover 79: 'an odd bit of work, extraordinarily difficult to execute'—but not, surely, for a poet. In any case, *in medio*, given V.'s use of the phrase in ecphrasis, must mean 'in the centre', in the interior of the cup, and not 'in the fields, the spaces enclosed by the vine and ivy' (Conington). V. had in mind silver cups with a figured medallion occupying the centre of the bowl; see D. E. Strong, *Greek and Roman Gold and Silver Plate* (London, 1966), 151; also *The Search for Alexander* (New York Graphic Society, 1980), col. pl. 33. (I am indebted for these references to M. J. Rein.) But Alcimedon's cups are made, in keeping with pastoral decorum, not of silver but of beechwood; see below, 37n. (*caelatum*).

37. fagina: the first occurrence of this form ('sed *faginea* rectius', Philargyrius); *fagineus* is as old as Cato, *De agr.* 21.5 (*TLL* s.v.). After V., and perhaps because of V., beechwood cups were associated with primitive simplicity; cf. Tib. 1.10.7–8 'diuitis hoc uitium est auri, nec bella fuerunt / faginus astabat cum scyphus ante dapes', Ov. *Met.* 8.669–70 (next note), *Fast.* 5.522 'pocula fagus erant'.

caelatum: implies metal-working; cf. Ov. *Met.* 8.668–70 (Philemon and Baucis) 'omnia fictilibus; post haec caelatus eodem / sistitur argento crater fabricataque fago / pocula'.

diuini opus: a harsh elision, unparalleled in the *E.*; but cf. *A.* 1.332 'ignari hominumque', 2.741 'respexi animumue', and see A. Siedow, *De elisionis aphaeresis hiatus usu in hexametris Latinis ab Ennii usque ad Ovidii tempora* (diss. inaug., Greifswald, 1911), 72.

Alcimedontis: a suitably pretentious name. The work of art is identified by attribution to an artist who has, however, no separate identity, as in Hom. *Od.* 8.373 (*Πόλυβος*), 19.57 (*Ἰκμάλιος*), *A.* 5.359 ('Didymaonis artes'), 9.304–5 ('Lycaon / Cnosius'), 10.499 ('Clonus Eurytides').

38. lenta . . . uitis: Catull. 61.102–3.

torno: apparently a wood-carver's chisel; see Mynors on *G.* 2.449.

facili: 'expresses the ease of perfect mastery' (T. E. Page); cf. Tib. 1.1.7–8 'ipse seram teneras maturo tempore uites / rusticus et facili grandia poma manu'. For the text see S. Timpanaro, *Per la storia della filologia virgiliana antica* (Rome, 1986), 151–3.

superaddita: this compound does not occur before V. Cf. *superaddite* (*E.* 5.42), *superuenit* (*E.* 6.20, *A.* 12.356), *superinice* (*G.* 4.46), *supereminet* (*A.* 1.501, 6.856, where see Norden, 10.765), *superuolat* (*A.* 10.522), *superiacit* (*A.* 11.625), *superimminet* (*A.* 12.306)—all, it would seem, V.'s inventions. See 2.70n., 6.81n.

38–9. lenta quibus torno facili superaddita uitis / diffusos hedera uestit pallente corymbos: 'On which a lithe vine, added by easy lathe, / Is clothing berry-clusters scattered on pale ivy' (Lee). The lines V. had in mind—Theocr. 1.29–31 *τῶ ποτὶ μὲν χείλη μαρύεται ὑψόθι κισσός, | κισσὸς ἑλιχρύσῳ κεκονιμένος· ἃ δὲ κατ' αὐτόν | καρπῷ ἕλιξ εἱλεῖται ἀγαλλομένα κροκόεντι*, 'Along the lips above trails ivy, ivy dotted with its golden clusters, and along it winds the tendril glorying in its yellow fruit' (Gow)—are difficult and possibly corrupt. Nor are V.'s lines without difficulty, on a first reading. For the reader will naturally take *hedera pallente* not as a local ablative with *diffusos*, but as instrumental with *uestit* ('Virgil cannot be acquitted of obscurity, as the ablative at first sight seems clearly to belong to "vestit"', Conington). For *diffundo* (*diffusus*) with a local ablative cf. *A.* 11.464–5 'equitem . . . / . . . latis diffundite campis', Cic. *De nat. deor.* 2.95 'toto caelo luce diffusa', Suet. *Calig.* 42 'aureorum aceruos patentissimo diffusos loco'. The vine is carved in relief (*superaddita*) so that it 'clothes' the ivy and only allows its clusters to show through here and there (T. E. Page). Finally, it should not be supposed that V. was trying to explicate Theocritus; Wilamowitz, *Die Textgeschichte der griechischen Bukoliker* (Philol. Untersuch. 18; Berlin, 1906), 224: 'Er hatte die Aufgabe etwas Vorstellbares zu geben, nicht den Theokrit zu erklären'.

39. hedera . . . pallente: *G.* 4.124 'pallentis . . . hederas', pale ivy. Cf. 7.38 'hedera . . . alba' and see 6.54n.

40. Conon: a Samian mathematician and astronomer resident in Alexandria. But Conon's prominence here is not owing to his scientific discoveries: in about 245 BC, he observed a dishevelled brightness in the heavens between Leo and Boötes and identified it as Queen Berenice's missing lock of hair. And Callimachus commemorated the event in an elegant poem, which was translated by Catullus, 66.7–8 (the lock of hair speaking) 'idem me ille Conon caelesti in lumine uidit / e Bereniceo uertice caesariem'. See above, 16n.

quis fuit alter?: ancient (and modern) answers vary. The Verona scholiast offers seven names: Eudoxus, whose *Phaenomena* was versified by Aratus; Conon's friend Archimedes, who devised a marvellous celestial sphere; Hipparchus; Euctemon; Hesiod; and Euclid. See D. E. W. Wormell, *CQ*, NS 10 (1960), 29–32, E. Brown, *Numeri Vergiliani* (Collection Latomus 63; Brussels, 1963), 88–92, R. S. Fisher, *Latomus*, 41 (1982), 803–14, C. Springer, *CJ* 79 (1983–4), 131–4. Menalcas' lapse of memory, while serving to characterize him ('errantem rusticum induxit', [Probus]), may also be a metrical expedient. Wormell 32 makes the interesting point that, of the names suggested by the ancient commentators, only the name of Archimedes could not be fitted into a dactylic hexameter.

41. descripsit radio totum qui gentibus orbem: see A. C. Cassio, *RIF* 101 (1973), 329–32, who discerns here an allusion to Callim. *Aet.* fr. 110.1 Pf. (*Coma Berenices*), and cf. *A.* 6.849–50 'caelique meatus / describent radio et surgentia sidera dicent'.

radio: 'for geometrical and other figures drawn by the finger or the rod (*radius*) either in the dust of the ground or on *abaci* or tables covered with a thin layer of sand cf. *Fin.* 5, 50 *quem enim ardorem studii censetis fuisse in Archimede, qui dum in pulvere quaedam describit attentius, ne patriam captam esse senserit*'—so Pease, introducing an erudite note on Cic. *De nat. deor.* 2.48 (*eruditum pulverem*).

totum . . . orbem: the whole circle of the heavens, with the risings and settings of the constellations.

gentibus: for all people, but especially for the reaper and the ploughman (42). Cf. Hes. *Works and Days* 383–4 *Πληιάδων*

Ἀτλαγενέων ἐπιτελλομενάων | ἄρχεσθ' ἀμήτου, ἀρότοιο δὲ δυσομενάων, 'When the Pleiads, the daughters of Atlas, rise, begin your reaping; your ploughing, when they set'.

42. **curuus:** a standing epithet of the plough (*TLL* s.v. 1550.33), here transferred to the ploughman. Cf. *G.* 1.494 'agricola incuruo terram molitus aratro', Pliny, *NH* 18.179 'arator nisi incuruus praeuaricatur' (Voss).

43. Cf. Theocr. 1.59–60 *οὐδέ τί πω ποτὶ χεῖλος ἐμὸν θίγεν, ἀλλ' ἔτι κεῖται | ἄχραντον*, 'but never yet has it touched my lips; it lies unsullied still' (Gow).

44. Damoetas, too, owns a pair of cups made by Alcimedon, which he considers less valuable than his cow (48). Cf. 5.67, *A.* 9.263–4 'bina ... / pocula', Hor. *Serm.* 1.6.117 'pocula ... duo', where Palmer comments: 'A pair of cups are put on the table because such articles were generally in pairs', citing, after Orelli, Cic. *Verr.* 2.2.47 'scyphorum paria complura'.

45. **molli ... acantho:** a Mediterranean native, *Acanthus mollis* L., whose foliage was much imitated in ancient art; see Abbe 179, Gow on Theocr. 1.55 *παντᾷ δ' ἀμφὶ δέπας περιπέπταται ὑγρὸς ἄκανθος*, 'And all round the cup is spread the flowing acanthus leaf'. Cf. Pliny, *Epist.* 5.6.16 'acanthus ... mollis et paene dixerim liquidus' (La Cerda).

46. **Orphea ... siluasque sequentis:** the first appearance of Orpheus in Latin poetry, as P. E. Knox points out, *Ovid's Metamorphoses and the Traditions of Augustan Poetry* (Cambridge Philol. Soc., Suppl. 11; Cambridge, 1986), 48.

47. The repetition of 1.43 may be read as mocking (Voss); better perhaps as suggesting the artlessness of rustic speech; cf. [Theocr.] 8.18–19, 21–2.

48. **si ad:** for this rare elision see Soubiran 412: '(au début du vers: double infraction, dans un dialogue d'allure rustique et familière)'.

si ad uitulam spectas: the tone seems to be colloquial; cf. Varro, *RR* 3.6.1 (keeping peafowl) 'ii aliquanto pauciores esse debent mares quam feminae, si ad fructum spectes; si ad delectationem, contra; formosior enim mas' (Forbiger).

49. **numquam hodie effugies:** cf. Naevius, *Equus Troianus* 13 R.[3] 'numquam hódie effugies quín mea moriarís manu'. It was

presumably a reminiscence of Naevius that allowed V. to write in *A.* 2.670 'numquam omnes hodie moriemur inulti', for *numquam hodie* had become, as it is here, emphatically colloquial (*TLL* s.v. *hodie* 2851.14). See Austin on *A.* 2. 670, Tarrant on Sen. *Ag.* 971. The translation of a phrase or passage from one metre into another is well attested in antiquity; see R. Kassel, 'Dichterspiele', *ZPE* 42 (1981), 11–20, E. W. Handley, 'Hidden Verses', *BICS*, *Suppl.* 51 (1988), 166–74. Here the reader will hear, or rather overhear, the beginning of a senarius, 'numquam hódie effugies'. Similarly, Horace translates the beginning of a senarius in Ter. *Eun.* 49 'exclúsit; reuocat: rédeam?' into the beginning of a hexameter, *Serm.* 2.3.264 'exclusit; reuocat: redeam?'

ueniam quocumque uocaris: so confident is Menalcas of winning that he now agrees to meet Damoetas on his own ground, i.e. to stake a cow (cf. 109). Possibly suggested by, but quite unlike, the rudeness of Lacon, Theocr. 5.44 *ἀλλὰ γὰρ ἕρφ', ὧδ' ἕρπε, καὶ ὕστατα βουκολιαξῇ*, 'But come, come here, and you shall sing for the last time' (but Comatas refuses to budge, 45 *οὐχ ἑρψῶ τηνεί*, 'I'll not come there'). In [Theocr.] 8.14–24, Daphnis offers to stake a calf if Menalcas will stake a lamb; Menalcas declines, for the lamb is his father's, and offers his pan-pipe instead, which Daphnis, although he has a pan-pipe of his own, accepts (84).

50. 'Here Menalcas begins as if he wished for someone in particular, but corrects himself, and offers to take the chance of a man just then approaching, whom he identifies at the end of the verse as Palaemon' (Conington). Cf. Theocr. 5.61–5, [Theocr.] 8.25–7, *E.* 7, Introduction, p. 212.

uenit ecce Palaemon: cf. Phaedr. 3.5.6 'uenit ecce diues et potens' (*TLL* s.v. *ecce* 26.55).

Palaemon: a very strange name—the name of a sea-deity, *A.* 5.823 'Inousque Palaemon'—to find in this context. Remmius Palaemon, the most eminent and no doubt the most immoral *grammaticus* of his time (he was denounced as unfit to teach the young by both Tiberius and Claudius), discovered a prophetic allusion to himself here; cf. Suet. *De gramm.* 23.4 'adrogantia fuit tanta ut . . . secum et natas et morituras litteras iactaret, nomen suum in Bucolicis non temere positum sed praesagante Vergilio fore quandoque omnium poetarum ac Palaemonem iudicem'.

51. posthac: common in Plautus and Terence but infrequent in later poetry (*TLL* s.v. 223.21); see 1.75n.

uoce lacessas: cf. *A*. 10.644 'uoce lacessit'; the phrase seems to be V.'s own (*TLL* s.v. *lacesso* 832.31).

52. quin age, si quid habes: the tone of *quin age* is challenging, 'come on, then'; found elsewhere in V. only in *G*. 4.329 'quin age et ipsa manu felicis erue siluas' and *A*. 5.635 'quin agite et mecum infaustas exurite puppis'. The construction here may have been suggested by Theocr. 5.78 (an impatient Lacon urges Comatas to begin) εἶα λέγ', εἴ τι λέγεις, 'come, say it, if you've anything to say'; cf. Plaut. *Epid*. 196 'age, si quid agis' (and five times elsewhere), a colloquial idiom illustrated by Headlam on Herodas 7.47 φέρ', εἰ φέρεις τι, 'give, if you've anything to give'. Cf. 9.32.

in me mora non erit ulla: cf. Plaut. *Curc*. 461 'caue in te sit mora mihi', *Stich*. 710 'non mora erit apud me', Ter. *Andr*. 420 'neque istic neque alibi tibi erit usquam in me mora', *A*. 12.11 'nulla mora in Turno'.

53. nec quemquam fugio: cf. Plaut. *Vid*. fr. 6 Leo 'haud fugio sequestrum'.

uicine Palaemon: 'beniuolum reddit ex uicinitatis commemoratione' (Serv.).

55. dicite: first attested in this sense in Catullus, 61.39, 62.4 'iam dicetur hymenaeus', 18 (*TLL* s.v. 977.65). Cf. 5.2, 10.3.

quandoquidem: see Fordyce on *A*. 7.547, Lyne on *Ciris* 323.

consedimus: the pastoral musician is usually seated while performing: Theocr. 1.12, 21 ὑπὸ τὰν πτελέαν ἑσδώμεθα, 'let us sit beneath the elm', 5.31–2, 6.3–4, 8, 11.17, [Moschus], *Epitaph. Bion*. 21, *E*. 1.1, 5.3, 7.1, 10.70–1, Longus 1.13.4.

56. et nunc omnis ager, nunc omnis parturit arbos: in this metaphorical sense *parturio* is found in classical poetry only here, in *G*. 2.330 'parturit almus ager', and Columella 10.413 'parturit arbos' (Columella in his prose books uses the form *arbor*); see *TLL* s.v. 533.15. Heyne compares Bion, fr. 2.17 Gow εἴαρι πάντα κύει, πάντ' εἴαρος αδέα βλαστεῖ, 'in the spring all things are fruitful, all sweet things blossom in the spring'.

arbos: 'forma *arbos* poetis usitatissima, inprimis Vergilio, qui *arbor* omnino sprevit' (*TLL* s.v. *arbor* 419.57).

59. alternis dicetis: amant alterna Camenae: cf. Catull. 62.16 'iure igitur uincemur: amat uictoria curam'; E. Fraenkel, *JRS* 45

(1955), 3–4 = *Kl. Beiträge* (Rome, 1964), ii. 91: 'The rhythm as well as the syntactical structure and a stylistic device (*vincemur* . . . *victoria*) of this line recur in an early poem of Catullus' great admirer Virgil, *Ecl.* 3, 59 . . .; cf. also ibid. 10, 3'. See 7. 19n.

Camenae: the old native goddesses of song in Livius Andronicus and Naevius, here recalled by V. (who never mentions them again), then, in an elegant compliment to V., by Horace, *Serm.* 1. 10. 44–5 'molle atque facetum / Vergilio adnuerunt gaudentes rure Camenae'; for their subsequent history see *TLL* Onomast. s.v. The *Camenae* and the Saturnian metre had been expelled from Latin poetry by Ennius, who introduced the Greek Muses and the hexameter, *Ann.* 1. 1 'Musae, quae pedibus magnum pulsatis Olympum', where see Skutsch. See also J. H. Waszink, '*Camena*', *C&M* 17 (1956), 139–48 = *Opuscula Selecta* (Leiden, 1979), 89–98.

60. ab Ioue principium Musae: 'uel Musae meae ab Ioue est principium, uel, o Musae, sumamus ab Ioue principium' (Serv.); 'the question is as nearly balanced as possible' (Conington). In Theocr. 5, the singing-match begins with Comatas stating that the Muses love him (80) and Lacon replying that Apollo loves him (82). But an invocation to the Muses following on 'amant alterna Camenae' would be jarring and inept; for Damoetas would seem to be 'correcting' Palaemon, whose good will he had sought to conciliate in advance (above, 53n.). Furthermore, had Menalcas heard *Musae* as a vocative he would have answered with a vocative; cf. below, 74 ~ 76, 85 ~ 86 (implied), 88 ~ 90, 93 ~ 94, 96 ~ 98; Theocr. 5. 100 ~ 102, 108 ~ 110, 120 ~ 122. *Musae* must therefore be a genitive, 'From Jove begins my song'; see 1. 2n. and cf. *A.* 7. 219 'ab Ioue principium generis, Ioue Dardana pubes'. Damoetas' beginning is from Aratus, *Phaen.* 1 Ἐκ Διὸς ἀρχώμεσθα, 'From Zeus let us begin', imitated by Theocritus 17. 1 (see *E.* 8. 11n.) and translated by Cicero, *Arat.* fr. 1 Soubiran 'A Ioue Musarum primordia'. Ovid may have heard *Musae* as a vocative, *Met.* 10. 148–9 'Ab Ioue, Musa parens (cedunt Iouis omnia regno), / carmina nostra moue'.

Iouis omnia plena: cf. Arat. *Phaen.* 2–4 μεσταὶ δὲ Διὸς πᾶσαι μὲν ἀγυιαί, | πᾶσαι δ' ἀνθρώπων ἀγοραί, μεστὴ δὲ θάλασσα | καὶ λιμένες, 'full of Zeus are all the streets and all the market places of men, full is the sea and its harbours'.

62. et me Phoebus amat: a translation of Theocr. 5.82 *καὶ γὰρ ἔμ' Ὡπόλλων φιλέει.*

apud me: 'at home', 'chez moi', a colloquial expression (*TLL* s.v. *apud* 339.58) found only here in V.

63. lauri et suaue rubens hyacinthus: Comatas and Lacon base their claims to divine favour on the gifts they offer: to the Muses two goats, to Apollo a ram. Menalcas' gifts to Apollo are much less grand: bay and hyacinth. (Is there an allusion here, as Servius suggests, to Apollo the lover of Daphne and Hyacinthus?)

suaue rubens: 2.49n.

hyacinthus: a name that seems to have been borne by several flowers; see Gow on Theocr. 10.28, Abbe 53–63; also below, 106n.

64–7. Heterosexual and homosexual passion, as in Theocr. 5.88–91.

64. malo me Galatea petit: since the apple was the fruit of love, throwing an apple at somebody was obviously provocative; see Gow on Theocr. 5.88 *βάλλει καὶ μάλοισι τὸν αἰπόλον ἁ Κλεαρίστα*, 'With apples too Clearista pelts the goatherd' (Gow), and A. R. Littlewood, 'The Symbolism of the Apple', *HSCP* 72 (1967), 154–5. Cf. Theocr. 6.6–7 *βάλλει τοι, Πολύφαμε, τὸ ποίμνιον ἁ Γαλάτεια | μάλοισιν*, 'Galatea pelts your flock with apples, Polyphemus'.

Galatea: 1.30n.

65. fugit ad salices: 'She runs, but hopes she does not run unseen' (Pope, *Spring* 58). The willow-grove would provide a cover for dalliance; cf. 10.40. The willow is a fairly prominent feature of V.'s pastoral landscape but does not appear in that of either the genuine or the spurious Theocritus; cf. 1.54, 78, 3.83, 5.16, 10.40. Gellius 9.9.5 has this comment on V.'s adaptation of Theocritus: 'quod substituit pro eo quod omiserat non abest quin iucundius lepidiusque sit'.

66. Amyntas, by contrast, comes to his lover unasked; cf. Theocr. 5.90–1 *κὴμὲ γὰρ ὁ Κρατίδας τὸν ποιμένα λεῖος ὑπαντῶν | ἐκμαίνει*, 'Yes, and when in gentle temper Cratidas runs to meet the shepherd, he maddens me' (Gow).

meus ignis: first here, but modelled on such phrases as *mea uita, uoluptas mea, meum suauium, mel meum*, etc.; imitated by Ovid

(*TLL* s.v. *ignis* 295.77). The metaphor itself—the burning intensity of love, love as a flame—while very old, is especially frequent in Hellenistic poetry, e.g. Ap. Rhod. 3.286–7, Theocr. 2.26, 40, 131, 3.17, 7.56, 102; see Pease on *A*. 4.2. Cf. Hor. *Epod*. 14.13–15 'quodsi non pulchrior ignis / accendit obsessam Ilion, / gaude sorte tua'.

Amyntas: like Antigenes, the name of a real person in Theocr. 7.2 *Ἀμύντας*, 4 *κἀντιγένης*; V. makes no distinction, however, and treats both as fictitious pastoral names. But whereas Antigenes occurs only once in the *E.* (5.89), Amyntas occurs eleven times and thus became established in the pastoral tradition: cf. Calp. Sic. 4.17, 78, 81, Nemes. 3.1, 4.62, Tasso's *Aminta*; also a pastoral fragment (iii–iv AD) in Gow, *Bucolici Graeci* (Oxford, 1952), p. 168, l. 15 *ἢ Λυκίδας ἢ Θύρσις, Ἀμύντιχος ἠὲ Μενάλκας*. The homosexual connotation of the name is traceable to the affectionate diminutive with *καλός* in Theocr. 7.132 *χὠ καλὸς Ἀμύντιχος*, 'and the pretty boy Amyntas'. See K. J. Dover, *Greek Homosexuality* (Cambridge, Mass., 1978), 120–1.

67. Delia: presumably his mistress or *contubernalis*; a slave-name like *Lesbia* (the mid-wife in Terence's *Andria*), *Samia*, *Phrygia*, etc. 'To understand it of Diana (as 7.29) and say that she accompanied the shepherd hunting is absurd' (T. E. Page). Still, ancient readers were of two minds—'Deliam alii amicam priorem uolunt, alii Dianam' (Serv.)—and the collocation with 'meae Veneri' in the next line is curious. Diana and Venus are sometimes associated in amatory contexts; cf. Prop. 2.19.17–18 'iam nunc me sacra Dianae / suscipere et Veneris ponere uota iuuat' and see *TLL* Onomast. s.v. *Diana* 135.35. See also 7.29n.

68–71. In Theocr. 5.94–5, Lacon answers Comatas by praising wild apples for their sweetness, whereupon Comatas boasts that he will give his girl the ring-dove he plans to catch in a juniper tree (96–7). Here Damoetas, imitating Comatas, says that he will give his 'Venus' the eggs from a wood-pigeon's nest he has observed high in a tree, and Menalcas, imitating the anonymous goatherd in Theocr. 3.10–11, replies that he has sent his boyfriend ten golden apples picked from a tree in the woods and will send him ten more tomorrow.

68. parta: 'peculiarly used of things that are virtually, though not actually realized' (Conington on *A*. 2.784).

meae Veneri: cf. Plaut. *Curc.* 192 'tun meam Venerem uituperas?', Lucr. 4.1185 'nec Veneres nostras hoc fallit', Hor. *Carm.* 1.33.13 'ipsum me melior cum peteret Venus'.

69. ipse: 'emphasizes the fact that he has taken personal trouble' (T. E. Page).

aeriae: not flying high, as in Lucr. 1.12, *G.* 1.375, but nesting high, like the turtle-doves in 1.58. Raiding the nest, he implies, is a risky business; cf. 2.40.

congessere: 'nidificauere' (Serv.); without an expressed object only here and in Gellius 2.29.5 (*TLL* s.v. 279.1).

palumbes: 1.57n.

70. quod potui: he would have liked to pick more; for the idiom see Kroll on Catull. 68.149 'hoc tibi, quod potui, confectum carmine munus', Palmer on Ov. *Her.* 8.5 'quod potui, renui'.

71. aurea mala decem: cf. Theocr. 3.10–11 *ἠνίδε τοι δέκα μᾶλα φέρω · τηνῶθε καθεῖλον | ὦ μ' ἐκέλευ καθελεῖν τύ· καὶ αὔριον ἄλλα τοι οἰσῶ*, 'Look, ten apples I bring you, picked from the very place where you bade me pick them, and others will I bring you tomorrow.' Did *αὔριον ἄλλα (μᾶλα)* suggest 'aurea mala'? Cf. 8.52–3.

aurea mala: bright perfect fruit; ten, and ten more tomorrow—an absurd boast, since wildings tend to be blemished and misshapen.

72–3. 'He prays that the gods may hear some of her vows of affection and so compel her to keep her word' (T. E. Page). It was an ancient commonplace that lovers' vows were dispersed by the winds and never reached the ears of the gods; cf. Tib. 1.4.21–2 'Veneris periuria uenti / irrita per terras et freta summa ferunt', with K. F. Smith's note. See 8.19n.

75. ego retia seruo: minding or carrying the nets—a menial task undertaken by the willing lover (but Menalcas complains a little)—is an erotic motif; cf. Tib. 1.4.49–50 'nec, uelit insidiis altas si claudere ualles, / dum placeas, umeri retia ferre negent', with K. F. Smith's note.

76–7. Send me Phyllis for my birthday, Iollas; when I purify my fields, come yourself. The insulting sexual innuendo is plain enough; a Roman's birthday was celebrated in great style, with feasting and merrymaking (Plaut. *Persa* 768–9 'hunc diem suauem / meum natalem agitemus amoenum. date aquam manibus, apponite mensam', 773–4 'bene mihi, bene uobis, bene meae amicae, optatus hic mihi dies datus hodiest / ab dis, quia te licet liberam med amplecti'), whereas strict chastity was enjoined during a religious ceremony (Tib. 2.1.11–12 'uos quoque abesse procul iubeo, discedat ab aris / cui tulit hesterna gaudia nocte Venus'); Phyllis is evidently the sweetheart or *contubernalis* of Iollas. The reference here is to a modest, private *ambarualia* of the sort described in *G.* 1.338–50, Tib. 1.1.21–4, 2.1.1–26 and alluded to in *E.* 5.75. The victim, followed by a group of worshippers, was led around the boundaries of the land three times and then sacrificed. See Wissowa 143.

76. Phyllida: cf. 5.10, 7.14n., 59, 63, 10.37, 41. For the structure of the line cf. 6.24, 7.9, 8.30, 9.50.

Iolla: 2.57n.

77. cum faciam uitula: pretentious for someone in Damoetas' position; the poor man's victim was a lamb: Tib. 1.1.21–2 'tunc uitula innumeros lustrabat caesa iuuencos, / nunc agna exigui est hostia parua soli'. *Vitula* is instrumental; cf. Plaut. *Stich.* 251 'quot agnis fecerat?', Varro, *LL* 6.16 (*flamen Dialis*) 'agna Ioui facit', and see *TLL* s.v. *facio* 97.37, Hofmann–Szantyr 121.

78–9. Menalcas replies in the person of Iollas, protesting that he loves Phyllis most of all and that she was sad to see him go; see 7.29–32n.

78. fleuit: here first with the accusative and infinitive; see *TLL* s.v. 900.52, Hofmann–Szantyr 358.

79. longum: rightly construed with *uale* by Servius; so also by La Cerda, who compares Claudian, *De rapt. Pros.* 2.234 'longumque uale', and by Norden on *A.* 6.401 'aeternum latrans', who traces the development of the construction. Cf. Hosidius Geta, *Medea* 460–1 'et longum, formose, uale, et quisquis amores / aut metuet dulces aut experietur amaros' (cf. below, 109–10).

'uale, uale', inquit, 'Iolla': cf. Ov. *Met.* 3.501 'dictoque "uale", "uale" inquit et Echo', *Am.* 2.13.21 'precibusque meis faue, Ilithyia'; also 6.44 'ut litus "Hyla, Hyla" omne sonaret'. In the case of *uale*, the final vowel retains its original quantity under the ictus (*ualĕ* was the normal pronunciation) and is then shortened in hiatus before *inquit*; *breuis breuians* did not, of course, affect the pronunciation of *Hyla*. Cf. Callim. *Hymn* 4.204 πέρα, πέρα εἰς ἐμέ, Λητοῖ, 'cross, cross over, Leto, to me', where the α of πέρα is shortened in hiatus before εἰς ('for which there seems to be no parallel in Greek', Mineur ad loc.). Similar rhythmical effects may be observed in Callim. *Hymn* 2.2, *Aet.* fr. 75.4 Pf., and Theocr. 6.8, where see Gow. For vowels shortened in hiatus in V. see R. D. Williams (Oxford, 1960) on *A.* 5.261 'Ilio alto'.

80–1. For the form, though not the content, of this elaborate comparison V. is indebted to [Theocr.] 8.57–9 δένδρεσι μὲν χειμὼν φοβερὸν κακόν, ὕδασι δ' αὐχμός, | ὄρνισιν δ' ὕσπλαγξ, ἀγροτέροις δὲ λίνα, | ἀνδρὶ δὲ παρθενικᾶς ἁπαλᾶς πόθος, 'Dread plague to trees is tempest, drought to the waters, the springe to birds, and nets to game; and to a man desire for a tender maiden' (Gow).

81. Amaryllidis irae: cf. 2.14.

82. depulsis . . . haedis: 1.21n., 7.15n.

arbutus: the strawberry tree, *Arbutus unedo* L., the κόμαρος of Theocr. 5.129 and Longus 2.16.3; goats were especially fond of its foliage (*G.* 3.300–1). See Abbe 155–6.

84. Pollio: V. everywhere elides the final *o* (3.86, 88, 4.12), which Horace scans as short (*Serm.* 1.10.42, 85, *Carm.* 2.1.14). See Soubiran 210–11.

quamuis est: the indicative is found in V. only here and in *A.* 5.542, and here, according to W. M. Lindsay, *AJP* 37 (1916), 33, because V. wished to avoid a *cacemphaton* or 'phrase of malodorous suggestion' (*uissit*); cf. Cic. *Ad fam.* 9.22.4 'non honestum uerbum est *diuisio*? at inest obscenum, cui respondet *intercapedo*'.

85. Pierides: V. confines the Pierian Muses to the *E.*: 6.13, 8.63, 9.33, 10.72. Cf. Theocr. 10.24 Μοῖσαι Πιερίδες, 11.3 Πιερίδες. See K. F. Smith on [Tib.] 4.2.21 (=3.8).

lectori: E. Fraenkel, *JRS* 56 (1966), 146, on *Dirae* 26: '*nostris cantata libellis* may sound odd to a modern ear, but that is how the poet's model, and so he himself, looks at this literary genre; their bucolic Muse produces literature'.

86. facit noua carmina: cf. the New Gallus 6 'fecerunt carmina Musae' and see D. E. Keefe, 'Gallus and Euphorion', *CQ*, NS 32 (1982), 237–8; cf. also Prop. 2.8.11 'qualia carmina feci!', 34.79 'tale facis carmen'.

noua carmina: being a poet himself, a writer of tragedies, Pollio will be a discriminating judge of poetry. Although a darling of the New Poets in his youth (Catull. 12.8–9 'est enim leporum / differtus puer ac facetiarum'), Pollio did not become a New Poet; and the style of his tragedies, of which nothing, not even a title, survives, was, in the judgement of Tacitus, old-fashioned, harsh, and dry, *Dial.* 21.7 'Pacuuium certe et Accium non solum tragoediis sed etiam orationibus suis expressit, adeo durus et siccus est'. For *carmina* referring to tragedy cf. 8.10 'sola Sophocleo tua carmina digna coturno', Tac. *Ann.* 11.13.1 'Pomponium consularem (is carmina scaenae dabat)', *TLL* s.v. 466.64.

pascite taurum: in contrast to 85 'uitulam ... pascite'—the more important animal for the more important genre.

87. This line is reused in *A.* 9.629.

88. quo te quoque gaudet: 'subaudis *uenisse*' (Serv.). The felicity attained by Pollio involves, as ll. 90–1 indicate, his poetry.

89. mella fluant illi: cf. Theocr. 5.126 *ῥείτω χἀ Συβαρῖτις ἐμὶν μέλι*, 'And for me let Sybaris flow with honey' (Gow). V. mentions honey only twice in the *E.*, here and in 4.30, and in both places as symbolic of the Golden Age; but for Theocritus it is a delicious food (1.146–7, 3.54, 5.59, 15.117). Cf. A. Cartault, *Étude sur les Bucoliques de Virgile* (Paris, 1897), 463: 'Théocrite, dans ses Idylles, mentionne volontiers les choses qui se mangent; Virgile est plus réservé sur ce point, excepté en ce qui concerne les fruits, qui lui apparaissent comme quelque chose de poétique; les expressions de gourmandise naïve n'ont pas trouvé place chez lui'. Cf. J. Griffin, *Latin Poets and Roman Life* (London, 1985), 81–3.

amomum: an unidentifiable aromatic plant imported from the East; see P. Wagler, *RE* i. 1873–4, and cf. Plaut. *Truc.* 539–40 'ex Arabia tibi / attuli tus, Ponto amomum', Pliny, *NH* 12.49 'nascitur et in Armeniae parte quae uocatur Otene et in Media et in Ponto'. See 4.25 n.; also above, p. xxii.

90–1. Menalcas converts Damoetas' benediction into a witty curse, a thrust at two contemporary poetasters.

90. Bauium: recorded by St. Jerome as having died in Cappadocia in 35 BC; see F. Marx, *RE* iii. 152.

tua carmina, Meui: cf. *A.* 3.371 'tua limina, Phoebe', 9.271 'tua praemia, Nise'. V. might have written 'mala carmina Meui' or the like; cf. Hor. *Epist.* 1.1.67 'lacrimosa poemata Pupi'. But the unexpected vocative adds immediacy to the insult and balances *Pollio* in l. 88. Quite possibly Mevius is the unsavory character attacked by Horace in his Tenth *Epode*, 1–2 'Mala soluta nauis exit alite, / ferens olentem Meuium'; see Fraenkel, *Horace*, 25–7, W. Kroll, *RE* xv. 1508. *Meuius* is the correct spelling (Fraenkel, ibid. 26n.2).

91. iungat uulpes et mulgeat hircos: 'milking he-goats' (*τράγον ἀμέλγειν*) was a proverbial expression for futile effort, and 'yoking foxes', though not attested, is evidently a comparable expression; see Otto, nos. 812 and 1939.4. 'Here, however, "iungere vulpes" and "mulgere hircos" appears to be a sort of comic purgatory, opposed to the paradise of v. 89' (Conington).

92. humi nascentia fraga: growing close to the ground; for *nascor* in this sense cf. below, 107, Catull. 62.39 'ut flos in saeptis secretus nascitur hortis', Cic. *De diu.* 2.135 (*radicula*) 'quo illa loci nasceretur'. Ovid's Polyphemus entices Galatea with a promise of wild strawberries for her to pick, *Met.* 13.815–16 'ipsa tuis manibus siluestri nata sub umbra / mollia fraga leges', and men of the Golden Age, as Ovid imagines it, were content with strawberries and other wild fruit (Ov. *Met.* 1.103–5); but the Romans seem never to have acquired a taste for strawberries, which grow abundantly in the hilly districts of Italy (Sargeaunt 48). See J. K. Anderson, *CW* 77 (1983–4), 303, and J. D. Morgan, *CW* 78 (1984–5), 579–80.

93. The disrupted word-order suggests excitement and confusion (Voss).

frigidus . . . anguis: 8.71; cf. Theocr. 15.58 *τὸν ψυχρὸν ὄφιν*, 'the cold snake', with Gow's note. Not a special kind of snake, for all snakes are cold-blooded.

94–7. Cf. Theocr. 5.100–4, where Comatas interrupts the singing-match to order his goats away from the wild olive (100) and Lacon replies with a similar couplet. V. imagines a more dramatic scene with sheep and goats grazing by an Italian river in spate. Cf. Hor. *Serm.* 1.1.54–8 'ut tibi si sit opus liquidi non amplius urna / uel cyatho, et dicas "magno de flumine mallem / quam ex hoc fonticulo tantundem sumere". eo fit, / plenior ut siquos delectet copia iusto,/ cum ripa simul auulsos ferat Aufidus acer'.

94. parcite . . . procedere: for the construction cf. *A.* 3.42 and see Kühner–Stegmann i. 206.

94–5. non bene ripae / creditur: the impersonal construction, as in Hor. *Serm.* 2.4.21 (mushrooms) 'aliis male creditur'.

95. ipse aries: cf. 4.43, 7.7. As in *G.* 3.445–7, the ram is singled out for description.

96–7. Only here does Damoetas take his cue from Menalcas.

96. reice: for the synizesis cf. Ter. *Phorm.* 18 *reicere*, 717 *reiciat*, Hor. *Serm.* 1.6.39 *deicere*, and see Leumann, *Lat. Laut- und Formenlehre*[2], 128–9.

97. omnis in fonte lauabo: cf. Theocr. 5.145–6 (Comatas to his goats) *αὔριον ὔμμε | πάσας ἐγὼ λουσῶ Συβαρίτιδος ἔνδοθι λίμνας*, 'Tomorrow I'll wash you all in Sybaris lake', with a variant *κράνας*, 'spring', which V. may have read.

erit: for the prosody see 1.38n.

98. cogite ouis: into the shade, to avoid the noonday heat; cf. *G.* 3.331 (*iubebo*) 'aestibus at mediis umbrosam exquirere uallem'.

lac praeceperit aestus: cf. Lucr. 6.1049–50 'aestus ubi aeris / praecepit ferrique uias possedit apertas', Columella, *De arb.* 2.2 'uuae, quae integrae et incorruptae ad maturitatem perueniunt, longe melioris saporis uinum faciunt quam quae praecipientur aestu'.

99. pressabimus ubera palmis: cf. *G.* 3.310 'pressis manabunt flumina mammis', *A.* 3.642 'ubera pressat', Ov. *Met.* 15.472 'ubera dent saturae manibus pressanda capellae'.

100. heu heu: 2.58n.

pingui macer . . . taurus in eruo: since his bull remains lean in the rich vetch, Damoetas assumes that he too must be suffering from love. Not so with Aegon's scrawny bull, Theocr. 4.20 *λεπτὸς μὰν χὼ ταῦρος ὁ πυρρίχος*, 'The bull's lean too, the red one'; missing their master, who has gone off to the Olympic games, his poor animals refuse to feed. Lovesick animals are post-Theocritean: Longus 2.7.4 *ἔγνων δὲ ἐγὼ καὶ ταῦρον ἐρασθέντα καὶ ὡς οἴστρῳ* πληγεὶς ἐμυκᾶτο, καὶ τράγον φιλήσαντα αἶγα καὶ ἠκολούθει πανταχοῦ, 'I've even known a bull that fell in love, and used to bellow as if he'd been stung by a gadfly. And I've known a he-goat that loved a she-goat and followed her everywhere' (Turner), *E.* 8.85–8, Ov. *Met.* 13.871 (the Cyclops in love) 'ut taurus uacca furibundus adempta'. See G. Jachmann, *N. Jahrb.* 49 (1922), 102–3 = *Ausgewählte Schriften* (Königstein im Taurus, 1981), 304–5. For *macies* as a symptom of lovesickness cf. Ov. *Her.* 11.27–8 'macies adduxerat artus, / sumebant minimos ora coacta cibos', *Met.* 11.793 'fecit amor maciem'.

eruo: *Ervum ervilia* L., on which see F. Olck, *RE* vi. 556–61.

101. idem amor exitium pecori pecorisque magistro: cf. [Theocr.] 8.43, 48 (the fair Nais) αἰ δ' ἂν ἀφέρπῃ, | χὠ τὰς βῶς βόσκων χαἰ βόες αὐότεραι, 'But if she depart, then withered is the cowherd and withered his cows'. See Mynors on *G.* 4.283.

102. his certe neque amor causa est; uix ossibus haerent: so Mynors, with most editors and commentators, apparently taking *neque* as = *ne . . . quidem*; cf. Hor. *Serm.* 2.3.262–3, 'nec nunc, cum me uocet ultro, / accedam?' and see Hofmann–Szantyr 450—an interpretation rejected by Madvig, Cic. *De fin.*[3] (Copenhagen, 1876), 808–9, Excursus 3.12: 'Vergilium in Ecl. III, 102 ita locutum non esse, ineptissima sententia (quid enim ibi agit *ne amor quidem*?), certissimum est; quid poeta scripsisset, pridem a Stephano et Heinsio intellectum'. *Neque* cannot be taken as = *non*, a usage confined to archaic formulae and phrases; cf. 9.6 and see Hofmann–Szantyr 448–9, Kühner–Stegmann i.817. Nor can *his* be taken as = *hi*, for although *hisce* = *hice* is found in Plautus and Terence, *his* = *hi* is unattested in literature; see *TLL* s.v. *hic* 2699.84, Leumann, *Lat. Laut- und Formenlehre*[2], 427. And why should V. so confuse his reader here? On the whole, it seems best to accept Stephanus' emendation, with T. E. Page, and read: 'hi certe (neque amor causa est) uix ossibus haerent'. For *neque (nec)* introducing a parenthesis cf. 10.46, *A.* 3.173, 484, 9.813, 11.568. *His* will then be a very early corruption, of which there are several in V.'s text, e.g. 4.62, *A.* 9.514, 10.705, 710; see S. Timpanaro, *Per la storia della filologia virgiliana antica* (Rome, 1986), 179–80. V. may have had in mind Theocr. 4.15–16 τήνας μὲν δή τοι τᾶς πόρτιος αὐτὰ λέλειπται | τὠστία, 'Of that heifer now only the bones are left'.

103. nescio quis: colloquial in tone; see 8.43n.

teneros oculus mihi fascinat agnos: cf. Catull. 7.12 'mala fascinare lingua', where the verb first occurs; also Hor. *Epist.* 1.14.37–8 'non istic obliquo oculo mea commoda quisquam / limat'. Not love but the evil eye is causing Menalcas' lambs to waste away. Cf. F. T. Elworthy, *Hastings' Encyclopaedia of Religion and Ethics*, v. 609: 'Young animals of all kinds are now, as ever, thought to be specially liable to injury' (quoting this line); La Cerda: 'Ex natura fascini, cuius vis maior in tenera, atque imbecillia'. Servius makes a nice point: 'et per transitum, pulchrum se pecus habere significat, quod meruit fascinari'. Cf. 7.28.

104–7. dic quibus in terris . . .: the singing-match ends surprisingly, with two riddles; nothing of the sort is to be found in Theocritus. Riddles are, however, a feature of comedy, and of Plautine comedy in particular; see Currie (above, 1n.), 31, who refers to Marx on Plaut. *Rud.* 150–1, 514–19; also E. Fraenkel, *Elementi plautini in Plauto* (Florence, 1960), 38–44. V.'s riddles are not Plautine in form ('a short and baffling remark, after which comes a farcical explanation of it', Currie) but are designed to recall the beginning of the *Eclogue*, 'Dic mihi, Damoeta, cuium pecus?', and, in general, the Plautine colouring of the herdsmen's conversation.

104–5. Servius records two solutions to this riddle: (i) the tomb of Caelius, a bankrupt who in selling up his property reserved only enough ground for a tomb. Philargyrius 1 preserves two confused notes relating to Caelius: Cornutus (born about AD 20) states that he heard the story of Caelius, now become a Mantuan, from Virgil himself; and Asconius Pedianus (9 BC–AD 76) that he heard Virgil say he had set a trap here for interpreters ('hoc loco se grammaticis crucem fixisse'). For the Virgilian studies of Cornutus see Ribbeck, *Prolegomena critica ad P. Vergili Maronis opera maiora* (Leipzig, 1856), 124–8, Timpanaro (above, 102n.), 87–9; for those of Asconius, Wissowa, *RE* ii. 1525. And (ii) the well at Syene (Aswan) used by Eratosthenes, who is not, however, mentioned, in calculating the earth's circumference. Servius rejects both of these solutions as being unsuited to a rustic, but thinks a well is meant, 'cuiuslibet loci puteus, in quem cum quis descenderit, tantum caeli conspicit spatium quantum putei latitudo permiserit'.

Of various other solutions, that hesitantly advanced by Martyn ad loc. seems the least improbable: 'Might not the shepherd mean a celestial globe or sphere?' More precisely, the famous sphere of Archimedes, which Marcellus sequestered on taking Syracuse and brought to Rome. Wormell (above, 40n.) accepts Martyn's solution but points out that all solutions ignore the repeated phrase 'quibus in terris', 'in what lands?', 'where?'. Both riddles, he argues, are designedly ambiguous and admit of two answers. The first riddle may refer to the wonderful sphere of Archimedes or to the equally wonderful sphere of Cicero's friend Posidonius, described by Cicero in *De nat. deor.* 2.88 (where see Pease) along with that of Archimedes. Thus, to the question 'where?' the answer could be either 'at Rome' or, presumably, 'at Rhodes'. Similarly, the answer to the second riddle could be either 'at Troy' or 'at Sparta' (see below, 106n.). If Wormell is right, the two riddles are closely related, and that is as it should be in a singing-match.

For older solutions see Voss ad loc., with ironic comment; for recent solutions, W. W. Briggs, Jr., *ANRW* II 31.2 (1981), 1310–11, adding K. Schöpsdau, *Hermes*, 102 (1974), 286n.56, E. L. Brown, *Vergilius*, 24 (1978), 25–31, H. Hofmann, *Hermes*, 113 (1985), 468–80, E. Malaspina, *RCCM* 28 (1986), 7–15.

104. eris mihi magnus Apollo: an unparalleled use of the god's name (*TLL* s.v. *Apollo* 245.18). The epithet is found elsewhere only in Hor. *Serm.* 2.5.60 'magnus . . . Apollo'.

105. tris . . . non amplius ulnas: about five feet; for the construction cf. *A.* 1.683, and see Kühner–Stegmann ii. 471, Hofmann–Szantyr 110. For the contrast between the dimensions of a sphere and the immensity of the sky cf. Ov. *Fast.* 6.277–8 'arte Syracosia suspensus in aere clauso / stat globus, immensi parua figura poli'.

caeli spatium: a Lucretian phrase, 4.202, 6.452, 820.

106. inscripti nomina regum: for the construction see 1.54n. and Austin on *A.* 2.273; T. E. Page compares Soph. *Trach.* 157–8 δέλτον ἐγγεγραμμένην | ξυνθήματ', 'a tablet inscribed with signs'. The flower inscribed with the names of princes ('*reges* duces Graecorum voce Homerica', Heyne) is the hyacinth, Theocr. 10.28 ἁ γραπτὰ ὑάκινθος, where see Gow; also Abbe 53–63. The markings on its petals were thought to resemble the letters AI, of which there were two competing explanations or *αἴτια*: the flower sprang from the blood of Ajax, who killed himself at Troy, and the markings stand for the first two letters of his name (*Αἴας*); or from the blood of Hyacinthus, who was accidentally killed at Sparta by his lover Apollo, and the markings denote the sound of Apollo's grief. Cf. Ov. *Met.* 10.215–16 (Apollo) 'ipse suos gemitus foliis inscribit, et AI AI / flos habet inscriptum', 13.397–8 'littera communis mediis pueroque uiroque / inscripta est foliis, haec nominis, illa querellae'. Menalcas probably means Hyacinthus (cf. above, 63 n.), although 'nomina regum' is vague, perhaps intentionally so.

108. Servius proposes and defends a perverse interpretation of this line, and implies that most readers agree with him: '*non* hic distinguendum, ut sit sensus: officii iudicis est ferre sententiam, non eorum qui inter se contendunt. unde male quidam totum iungunt "non nostrum inter uos tantas componere lites"'.

109–10. quisquis amores / aut metuet dulcis: so Dido fears the perfect security of her love, *A.* 4.298 'omnia tuta timens'. Graser's conjecture (see app. crit.) is impossible for stylistic reasons: *haud* belongs to the high style of poetry and, while fairly common in the *G.* and *A.*, is not found in the *E.*; see Axelson 91–2, Tränkle 45–6.

amores / . . . amaros: traditional word-play, 'bittersweet love'; cf. Plaut. *Cist.* 68 'an amare occipere amarum est, opsecro?', *Trin.* 259 'Amor amara dat tamen', *Rhet. ad Her.* 4.21 'nam amari iucundum sit, si curetur ne quid insit amari', with Quintilian's disapproving comment (9.3.69–70). See R. D. Brown, *Lucretius on Love and Sex* (Leiden, 1987), 265–6.

111. riuos: irrigation ditches, as in *G.* 1.106, 269.

sat prata biberunt: the meadows are saturated, and the singing-match ends. The completion of song has an analogy in nature; elsewhere, in *E.* 2, 6, and 10, it is the close of day, of which there may be a suggestion here.

ECLOGUE 4

Introduction

> Vltima Cumaei uenit iam carminis aetas;
> magnus ab integro saeclorum nascitur ordo. (4. 4–5)

The last age of the Cumaean prophecy has now come, and a great temporal order is born anew. The year is 40 BC, a time of crisis in the Roman world, of wars and rumours of wars,[1] of oracles and direst prophecy.[2] Why Virgil took notice of the Sibyl's annunciation, which would otherwise be unknown, and made it the pretext for a poem—a beautiful, mysterious poem—we can only guess. At any rate, he unhesitatingly identifies the last age (whatever the Sibyl may have meant) with reference to Hesiod and Aratus as the mythical Golden Age.

In the *Works and Days* 106–201, Hesiod introduces, after the story of Prometheus and Pandora, another story—a λόγος that men tell (he says) to account for the human condition. It is a story of progressive degeneration from a golden race[3] of men who lived like gods in the reign of Kronos, through the intermediate races of silver and bronze, to the iron race, the present debased and corrupt race of which Hesiod despairs. To this symbolic, traditional scheme Hesiod adds, between the bronze and iron races, the race of heroes; theirs being, he observes, the generation before our own. Although necessary—as Hesiod must have felt—to a more comprehensive view of the legendary Greek past, his interpolation evidently disrupts an inherited pattern, for the heroes are associated with no metal and are out of place, besides, in this descending scale of value. They were the mighty men of old, demigods, who fought and died at Thebes and Troy, passing thereafter to the Isles of the Blest, where they enjoy Nature's spontaneous and unfailing

[1] See R. Syme, *The Roman Revolution* (Oxford, 1939), 205–13.

[2] Cf. Luc. 1. 564–5 'diraque per populum Cumanae carmina uatis/uolgantur'. According to Thucydides 2. 8. 2, many prophecies were spoken or sung by the oracle-mongers on the eve of the Peloponnesian War.

[3] Greek writers refer to a golden race, Latin poets sometimes to a golden age; see H. C. Baldry, *CQ*, NS 2 (1952), 88–90.

bounty for ever—a condition indistinguishable, save for its duration, from that of the golden race.[4]

iam redit et Virgo, redeunt Saturnia regna. (6)

Virgil assumes a knowledge of the story which Aratus tells in describing the constellation of the Virgin (*Παρθένος*, *Virgo*).[5] The Virgin once lived on earth and mingled freely with the men of the golden race, who called her Justice (*Δίκη*).[6] For she would gather the elders of the people together in the market place or in the streets and exhort them to be kinder and more just. There was no strife then, no hatred, no tumult of battle; the cruel sea and its commerce were unknown to them; oxen and the plough, and Justice herself, abundantly supplied their every want. With the silver race Justice mingled less freely, longing as she did for the golden race. She would come alone, in the late afternoon, from the echoing hills to rebuke them for their wickedness, then leave them gazing after her as she returned to the hills. They too died and were succeeded by the bronze race. Loathing them—the first men to forge the highwayman's dagger, to eat the flesh of the plough-ox—she fled to the heavens; yet still in the night she appears to men, to remind them of their ancient sin, near the far-seen constellation of Boötes.

The Virgin is synonymous with the golden race, and by implication therefore—an implication which Virgil exploits—with the reign of Kronos. Hence 'Saturnia regna,' for the Latins had long since identified Kronos with Saturn,[7] who reigned in that golden

[4] Cf. M. L. West, *CQ*, NS 11 (1961), 113, commenting on the Hesiodic *Catalogue* fr. 204. 100–4 M.-W.[3]: 'Thus the heroic age is not distinguished from the golden race of the *Erga*'.

[5] *Phaen.* 96–136.

[6] The identification of *Παρθένος* with *Δίκη* was suggested by Hesiod, *Works and Days* 256 ἡ δέ τε παρθένος ἐστὶ Δίκη; see Wilamowitz, *Hellenistische Dichtung* (Berlin, 1924), ii. 265. In Hesiod, Aidos and Nemesis personified—'their lovely bodies wrapped in white robes'—leave the earth in the iron age, the last. In Aratus, the Virgin leaves the earth in the bronze age, the last in Aratus; 'drei Weltalter (mehr konnte er nicht brauchen)' (Wilamowitz, ibid.). Cf. Hor. *Epod.* 16. 63–6 'Iuppiter illa piae secreuit litora genti, / ut inquinauit aere tempus aureum, / aere, dehinc ferro durauit saecula, quorum / piis secunda uate me datur fuga'.

[7] Cf. Livius Andronicus, *FPL*, fr. 2 Büchner 'pater noster, Saturni filie', translating *Od.* 1. 45 ὦ πάτερ ἡμέτερε Κρονίδη.

springtime (as Virgil imagines it)[8] so soon faded and vanished of primeval Latium.

Two surprising innovations are involved in Virgil's vision of the Golden Age: the Golden Age is about to be—indeed, is now being[9]—restored to mankind; and the restoration coincides with the birth of a child. Ever so slightly Virgil labours the coincidence: with the birth of the child (8 'nascenti') is born (5 'nascitur') a new order of time. The Ancients conceived of no such prodigious birth or rebirth; for them the Golden Age was a mythical paradise irretrievably lost.

> tu modo nascenti puero, quo ferrea primum
> desinet ac toto surget gens aurea mundo,
> casta faue Lucina: tuus iam regnat Apollo. (8–10)

The tender, almost homely appeal to Lucina, the old Roman goddess of childbirth,[10] is immediately qualified by the statement that her brother is now reigning, which re-establishes the elevation of the opening line.[11]

This blessed event will be initiated—we now approach contemporary reality—when Pollio is consul; that is, in the year 40 BC. Virgil is alluding to the Pact of Brundisium, a political settlement between Antony and Octavian, which was soon disregarded or violated, but which, for the moment, offered hope of lasting peace, or at least freedom from lasting fear (14 'perpetua . . . formidine'), to a world burdened with guilt and despair. The Pact was negotiated in September, with Pollio acting as Antony's lieutenant.[12]

The Pact of Brundisium was solemnized, in the high Roman fashion, with a dynastic marriage as Antony took to wife Octavian's sister, the blameless Octavia. To contemporary readers the vexed question 'Who is the boy?' would not have occurred. They

[8] Virgil is the first to describe Saturn's reign and Saturn himself as golden, *G*. 2. 538 'aureus hanc uitam in terris Saturnus agebat', *A*. 6. 792–4 'Augustus Caesar, diui genus, aurea condet / saecula qui rursus Latio regnata per arua / Saturno quondam'.

[9] See on 4 'iam'.

[10] See ad loc.

[11] Line 4, that is, the opening line of the original poem; see F. Jacoby, *RhM* 65 (1910), 77 n. 1, Jachmann 49. The style of reference, however, 'tuus . . . Apollo', is suitably affectionate and familiar; cf. Plaut. *Capt*. 157 (of a son) 'Philopolemus tuus', *Most*. 182 (of a lover) 'Philolaches tuus', *A*. 5. 804 (Neptune to Venus) 'Aeneae . . . tui'.

[12] See Syme (above, n. 1), 217–20.

knew well enough who was meant: the expected son of Antony and Octavia and heir to Antony's greatness—the son that never was; a daughter was born instead. Antony claimed descent from Hercules as proudly as Julius Caesar (and Octavian, his adopted son) claimed descent from Venus;[13] thus the boy would have been descended on his father's side from Hercules, on his mother's from Venus: a symbol incarnate of unity and peace. Like the deified Hercules (Virgil implies) he will be exalted to heaven and there see gods mingling with heroes:

> ille deum uitam accipiet diuisque uidebit
> permixtos heroas et ipse uidebitur illis,
> pacatumque reget patriis uirtutibus orbem. (15–17)

Scholars have failed to recognize that the allusion to Hercules in ll. 15–16 is continued in l. 17 'pacatumque reget patriis uirtutibus orbem', 'and he will rule a world pacified by his father's valour'. The verb *paco* is ordinarily used of military conquest, e.g. Caes. *BG* 2. 1. 2 'omni pacata Gallia', but it had also a well-attested, special sense or reference: to Hercules and the labours by which he pacified the world.[14] Virgil was to use this verb only once again, years later, in a similar if far more elaborate context, the encomium of Augustus in Book 6 of the *Aeneid*; Augustus is extolled for the magnitude of his achievements and compared favourably with 'the god Hercules, whom Antony lov'd' (Shakespeare), and Dionysus.[15]

> nec uero Alcides tantum telluris obiuit,
> fixerit aeripedem ceruam licet, aut Erymanthi
> pacarit nemora et Lernam tremefecerit arcu. (801–3)

The larger reference in l. 17 and at the end of the *Eclogue*, where again there is an allusion to Hercules, is to Hellenistic ruler-worship. For a hundred and fifty years now Roman proconsuls in

[13] See Appian, *BC* 3. 16, 19, Plutarch, *Antony* 4, 36, P. Zanker, *The Power of Images in the Age of Augustus*, trans. A. Shapiro (Ann Arbor, Mich., 1988), 45–6.

[14] Cf. Cic. *Tusc. disp.* 2. 22 (ll. 19–20) 'haec dextra Lernam taetra mactata excetra / pacauit?', Ov. *Her.* 9. 13 (Deianira to Hercules) 'respice uindicibus pacatum uiribus orbem', Prop. 3. 11. 19 'qui pacato statuisset in orbe columnas'. Note the reminiscence of Virgil in Ovid and Propertius, especially in Ovid. See *TLL* s.v. 21. 58.

[15] Augustus did not claim descent from Hercules or Dionysus, but the comparison, Hellenistic in origin, had become a feature of such encomia; see Austin ad loc.

the East had been paid divine honours, a subservient people merely transferring the language and gestures of such worship from Greek potentates—Alexander's successors and their successors—to Roman magistrates, men like Pompey, Julius Caesar, Antony.[16] But in the West, in Rome, there was no such tradition nor, as yet, practice; hence the conciseness and indirection of Virgil's reference.

The *Eclogue* closes on a note of tenderness and intimacy as the poet addresses the new-born child, urging him to smile at his mother:

> Incipe, parue puer, risu cognoscere matrem
> (matri longa decem tulerunt fastidia menses)
> incipe, parue puer . . . (60–2)

Much in the tone and manner of these lines is Hellenistic—is owing, that is to say, to Hellenistic poetry and especially to two poets Virgil had much in mind at the time, Callimachus and Theocritus. In Callim. *Hymn* 4, a travel-worn and impatient Leto appeals to the child in her womb, the nascent god Apollo:

> τί μητέρα, κοῦρε, βαρύνεις; . . .
> γείνεο, γείνεο, κοῦρε, καὶ ἤπιος ἔξιθι κόλπου. (212, 214)
>
> Why, child, do you burden your mother? . . .
> Be born, be born, child, and come gentle
> from the womb.

Closer in tone, certainly, is Theocr. 24, *The Infant Heracles*; after bathing and nursing Heracles and his little brother Iphicles, Alcmena puts them to bed in a great bronze shield, then speaks as she strokes their heads and rocks the shield, like a cradle (an exquisite homeliness, that curious blending of the domestic and the heroic so typically Hellenistic):

> εὕδετ', ἐμὰ βρέφεα, γλυκερὸν καὶ ἐγέρσιμον ὕπνον·
> εὕδετ', ἐμὰ ψυχά. (7–8)
>
> Sleep, my babes, a light and delicious sleep.
> Sleep, soul of my soul.[17]

[16] See Syme (above, n. 1), 263; also *E.* 1. 43 n.

[17] Note the endearment of the repetition εὕδετ', ἐμὰ βρέφεα . . . εὕδετ', ἐμὰ ψυχά and cf. 'incipe, parue puer . . . incipe, parue puer'.

The tenderness of the scene at the end of the *Eclogue* is qualified by the last line with its grander reference:

nec deus hunc mensa, dea nec dignata cubili est.

No god invites him to table, no goddess to bed—the sort of child, that is, who refuses his poor mother a smile. It was the deified Heracles who enjoyed such transcendent gratification; and Virgil may have been thinking of Theocr. 17, which begins with an encomium of the reigning Ptolemy's father, Ptolemy I Soter, now dead and deified, who had been one of Alexander's favourite commanders:

Ἐκ πατέρων οἷος μὲν ἔην τελέσαι μέγα ἔργον
Λαγεΐδας Πτολεμαῖος . . .
τῆνον καὶ μακάρεσσι πατὴρ ὁμότιμον ἔθηκεν
ἀθανάτοις, καί οἱ χρύσεος θρόνος ἐν Διὸς οἴκῳ
δέδμηται· παρὰ δ' αὐτὸν Ἀλέξανδρος φίλα εἰδὼς
ἑδριάει, Πέρσαισι βαρὺς θεὸς αἰολομίτρας.
ἀντία δ' Ἡρακλῆος ἕδρα κενταυροφόνοιο
ἵδρυται στερεοῖο τετυγμένα ἐξ ἀδάμαντος·
ἔνθα σὺν ἄλλοισιν θαλίας ἔχει Οὐρανίδῃσι,
χαίρων υἱωνῶν περιώσιον υἱωνοῖσιν . . .
τῷ καὶ ἐπεὶ δαίτηθεν ἴοι κεκορημένος ἤδη
νέκταρος εὐόδμοιο φίλας ἐς δῶμ' ἀλόχοιο,
τῷ μὲν τόξον ἔδωκεν ὑπωλένιόν τε φαρέτραν,
τῷ δὲ σιδάρειον σκύταλον κεχαραγμένον ὄζοις·
οἳ δ' εἰς ἀμβρόσιον θάλαμον λευκοσφύρου Ἥβας
ὅπλα καὶ αὐτὸν ἄγουσι γενειήταν Διὸς υἱόν.
(17. 13–14, 16–23, 28–33)

What a man was Ptolemy the high-born son of Lagus to accomplish some great deed . . . Him Zeus made equal in honour even with the blessed immortals, for him a golden throne is erected in Zeus' hall; and beside him in friendship sits Alexander, god of the glittering turban, the Persians' bane. Opposite them, and wrought of solid adamant, is placed the chair of Heracles, slayer of the Centaurs, and there with the other celestials he holds revel, rejoicing exceedingly in the sons of his sons . . . And when he has drunk his fill of fragrant nectar and leaves the feast for his dear wife's house, he hands the bow and quiver from beneath his arm to one, and to the other his knotted club of iron, and to the ambrosial chamber of white-ankled Hebe they escort, with his weapons, Zeus' bearded son.[18]

[18] Translation after Gow.

In the year 40 BC—on earth—Antony, not Octavian, was 'the greatest prince o' the world', and of this their contemporaries, spectators of the mighty drama, could be in no doubt. In the year 40 BC, Octavian was a sickly if determined and ruthless young man; the future Augustus unimaginable. Failure of historical perspective vitiates much that has been written about the Fourth *Eclogue*.[19]

The Golden Age coincides, as already remarked, with the birth of the child. While he is still in his cradle—his miraculously flowering cradle—Earth will pour forth her gifts of plants and flowers; she-goats will return home of their own volition with udders full; and the snake will perish and the plant that conceals its poison. But the Golden Age is not yet completely realized; it will attain perfection as the child develops. (The perfectability of the Golden Age is another of Virgil's innovations.) When the boy becomes old enough to read of the glories of heroes and of his father's deeds, and to understand what true manhood (*uirtus*) is, the fields will gradually turn golden with grain, the grape hang reddening on the wild briar, and stubborn oaks exude dewdrops of honey. Yet some few vestiges of ancient wrong will remain. Men will sail the sea, build walled cities, and inflict deep furrows on the Earth. A second Argo will transport the chosen heroes; there will be new wars, and again great Achilles will be sent to Troy. During this period of military expansion in the eastern Mediterranean and beyond, the Golden Age will insensibly be merged, as the boy grows to manhood, with the age of heroes; like the golden, in Hesiod, an age of preternatural felicity, but, unlike the golden, immune to deterioration. Now the merchantman will quit the sea and intercourse of trade will cease, for Earth everywhere will bear everything. Agricultural labour too will cease; there will be no dyeing of wool—a sign of decadence—for the ram in the meadow will alter his own fleece, now to glowing sea-purple, now to saffron yellow.

ipse sed in pratis aries iam suaue rubenti
murice, iam croceo mutabit uellera luto. (43–4)

(The portentous Sibylline tone must seem very faint and far away as the reader contemplates Virgil's polychromatic ram.)

[19] Cf. Heyne, *E.* 4, Argumentum, p. 127: 'de Augusto rerum domino nondum illo anno quisquam cogitare poterat . . . est enim res ex eius anni, quo pax facta est, actis dispicienda'.

Since the Fourth *Eclogue* was not originally conceived as an *Eclogue*, it has, or rather had, no distinctively pastoral features, not even—a significant technical detail noticed by Walter Savage Landor[20]—the bucolic diaeresis. When Virgil published it in his Book of *Eclogues*, he prefixed a brief pastoral apology[21]—

> Sicelides Musae, paulo maiora canamus!
> non omnis arbusta iuuant humilesque myricae;
> si canimus siluas, siluae sint consule dignae. (1–3)

—and added, perhaps, the emphatic—the rather too emphatic—reference to Pan and Arcadia near the end:

> Pan etiam, Arcadia mecum si iudice certet,
> Pan etiam Arcadia dicat se iudice uictum. (58–9)

In the absence of these lines a slight emphasis falls on 'Apollo' at the end of l. 57, 'Lino formosus Apollo', which recalls the earlier mention of Apollo with Lucina—'tu modo nascenti puero . . . casta faue Lucina: tuus iam regnat Apollo' (8–10)—and thus prepares for the intimacy of the closing scene: 'Incipe, parue puer . . .'.

Much had happened in the years after the Pact of Brundisium while Virgil was writing and rewriting his *Eclogues* and meditating his book. His consulate over, Pollio departed to govern Macedonia for Antony; Virgil was drawn into the circle of Maecenas and became acquainted with Octavian. And so, when Virgil decided to represent his epithalamium (so to call it) as an *Eclogue* and publish it, some five years later, as the fourth in his book, it is likely that he made certain changes in it, whether of addition or subtraction.[22] Hence, it may be, something of the mystery, or mystification, of the Fourth *Eclogue*. Time obliterates: the political circumstances of the year 40 BC, the hopes then entertained, were soon forgotten, so that in the following generation Pollio's son, the rash and ambitious Asinius Gallus, who was to die an old man in prison, could claim to be the marvellous child.[23]

The Fourth *Eclogue* was accessible and attractive to Christian apologists for several reasons: it was beautifully, conveniently

[20] 'The Idyls of Theocritus', *The Complete Works of Walter Savage Landor*, ed. T. E. Welby, xii (London, 1931), 5–6.

[21] See above, n. 11.

[22] As suggested by A. D. Nock; see W. W. Tarn, *JRS* 22 (1932), 156 n. 4.

[23] DServ. on l. 11: 'Asconius Pedianus a Gallo audisse se refert hanc eclogam in honorem eius factam'.

mysterious, for they, like their pagan contemporaries,[24] were unable to appreciate its poetic or political context; it referred to the Sibyl, whose oracles, mostly Jewish and Christian forgeries, the Christians triumphantly adduced in confirmation of their faith;[25] and it proposed to imagination a new birth of time with a virgin, a child, and a perfected Edenic world.

Here are a few examples of Christian interpretation from the fourth and fifth centuries:[26]

6 'iam redit et Virgo': the Virgin Mary, who returns after Eve.

7 'noua progenies': baptized Christians as distinct from Jews; or Christ; or, more precisely, for Prudentius, the new Adam.

22 'nec magnos metuent armenta leones': Christian congregations will no longer fear persecution from pagan emperors, those great lions.

24 'occidet et serpens': the Old Serpent, of course.

25 'Assyrium uulgo nascetur amomum': this prized exotic, springing up everywhere, signified for the Emperor Constantine—if he did indeed expound the Fourth *Eclogue* to the Assembly of Saints on Good Friday, AD 313—the propagation of the Christian faith.

Saint Jerome and Saint Augustine knew better, for they had been educated in the tradition of pagan letters. Jerome characterizes such interpretations as puerile; Augustine identifies the Fourth *Eclogue* as an adulatory poem to a Roman noble, but supposes that the Sibyl may have divined the advent of Christ.

The Christian, or Messianic, interpretation prevailed unchallenged for centuries, supported by, and supporting, Virgil's reputation as a seer, a Christian before Christ. Virgil's first great

[24] Thus Virgil's pagan commentator Servius, writing towards the end of the 4th c., has only a vague and confused notion of the historical circumstances, and, apart from a note on *Sicelides*, which reveals his essential ignorance, makes no reference to Hellenistic poetry.

[25] *Oracula Sibyllina* in fourteen books; see *OCD* s.v. *Sibylla*. Lactantius, *Diu. inst*. 7. 24. 11–14, quotes ll. 38–41, 28–30, 42–5, and 21–2 of the Fourth *Eclogue* with the comment, 'quae poeta secundum Cumaeae Sibyllae carmina prolocutus est', then quotes *Oracula Sibyllina* 3. 787–91, 619–23, and 5. 281–3.

[26] See P. Courcelle, 'Les Exégèses chrétiennes de la quatrième églogue', *REA* 59 (1957), 298–315.

modern commentator, the Spanish Jesuit Juan Luis de la Cerda (1617), accepted the Christian interpretation in principle, in his *Explicatio*, but largely ignored it in his notes. Virgil's second great modern commentator, the German Protestant Christian Gottlob Heyne[27] (1767), rejected the Christian interpretation in a characteristically vigorous, plain statement, noting, however, that most learned men still accepted it. It would be difficult, if not impossible, to determine exactly when the Christian interpretation was given up. Samuel Johnson, Heyne's contemporary, firmly believed it; in fact, his first published work (1728) was a translation into Latin hexameters of Pope's *Messiah, A Sacred Eclogue, in Imitation of Virgil's Pollio*, in which he tacitly corrected, as they appeared to him, the famous older poet's theological errors. But in 1858, Conington, the most important nineteenth-century commentator on Virgil, and himself a devout Christian, could dismiss the Christian interpretation of the Fourth *Eclogue* with this brief statement: 'The coincidence between Virgil's language and that of the Old Testament prophets is sufficiently striking: but it may be doubted whether Virgil uses any image to which a classical parallel cannot be found' (p. 46). In England, however, a sense of the Fourth *Eclogue* as inspired utterance seems to have persisted through most of the nineteenth century; perhaps even longer.

In 1907, a small book was published in London with the title *Virgil's Messianic Eclogue*; it contained three essays by three English classical scholars, J. B. Mayor, W. W. Fowler, and R. S. Conway. The author of the first essay, Conway, begins by expressing regret at 'the complete decay of the reverence with which Virgil's Fourth Eclogue was once regarded' (p. 11). Conway knows that as a scholar he cannot accept the Christian interpretation, and yet can hardly bring himself to reject it. The occasional involution of his prose may betray some distress of spirit: 'Understood in the only way possible to the mind of the early centuries, that Eclogue made him a direct prophet, and therefore an interpreter of Christ; and it is not the deepest students of Virgil who have thought him unworthy of that divine ministry' (his concluding sentence, p. 48).

Seventeen years later, however, in 1924, a book was published in Leipzig that made something like the discredited Christian interpretation acceptable: Eduard Norden's *Die Geburt des Kindes*.

[27] See 'The Life of Heyne', Thomas Carlyle, *Critical and Miscellaneous Essays* (New York, 1900), i. 319–54.

Norden connected the Fourth *Eclogue* with eastern theology and ritual, especially with two religious festivals celebrated annually in Alexandria—that of Helios on 24–5 December (Christmas Eve) and that of Aion on 6 January (Epiphany). He sought to explain the Fourth *Eclogue* by relating it not to Christianity as such, but to a theosophy originating in Egypt and diffused throughout the Middle East, a gnosis immemorially old of which Christianity was but a particular manifestation. *Die Geburt des Kindes*, though not a large book, is impressive—obscurely learned, occasionally enlightening: the consistent aberration of a great scholar. It is not necessary to reconstruct Norden's elaborate argument, for it seems never to have been accepted in detail,[28] and in any case was demolished by Jachmann in 1952. Yet the general effect of *Die Geburt des Kindes* is still felt, and to this extent Norden was successful; he succeeded, that is, in making a religious or mystical interpretation of the Fourth *Eclogue* seem intellectually respectable. He might, however, be accused with some justice of readmitting the Babylonian darkness into the interpretation of this brilliant little poem: brilliant and playful, with overtones of grandeur, delicate, intentionally vague perhaps—at once urgent and elusive.

Bibliography

D. A. Slater, 'Was the Fourth Eclogue Written to Celebrate the Marriage of Octavia to Mark Antony?—A Literary Parallel', *CR* 26 (1912), 114–19.

E. Norden, *Die Geburt des Kindes. Geschichte einer religiösen Idee* (Leipzig, 1924).

G. Jachmann, 'Die vierte Ekloge Vergils', *ASNP*, 2nd ser., 21 (1952), 13–62.

H. C. Gotoff, 'On the Fourth Eclogue of Virgil', *Philologus*, 111 (1967), 66–79.

G. Williams, *Tradition and Originality in Roman Poetry* (Oxford, 1968), 274–85.

—— 'A Version of Pastoral', in T. Woodman and D. West (eds.), *Quality and Pleasure in Latin Poetry* (Cambridge, 1974), 31–46.

R. G. M. Nisbet, 'Virgil's Fourth Eclogue: Easterners and Westerners', *BICS* 25 (1978), 59–78.

[28] For a critical, though not unsympathetic, review see H. J. Rose, *CR* 38 (1924), 200–1.

W. Kraus, 'Vergils vierte Ekloge: Ein kritisches Hypomnema', *ANRW* II 31. 1 (1980), 604–45.

1–3. See Introduction, n. 11.

1. Sicelides Musae: a Greek poet would have put the noun first, e.g. Theocr. 10. 24 *Μοῖσαι Πιερίδες*, 7. 148 *Νύμφαι Κασταλίδες*, Callim. *Hymn* 4. 109 *Νύμφαι Θεσσαλίδες*, but V. tends to admit a spondaic word in the first foot of his hexameter only under certain circumstances; see Norden 435–6. For the artificial lengthening of the first syllable of *Σικελίδες*, for which V. had Hellenistic precedent, see Leumann, *Lat. Laut- und Formenlehre*², 115, and cf. [Theocr.] 8. 56 *Σικελικάν*, Archimelus, *Suppl. Hell.* 202. 17 *Σικελίας*, and especially the refrain in [Moschus], *Epitaph. Bion.* 8 *ἄρχετε Σικελικαί, τῶ πένθεος ἄρχετε, Μοῖσαι*, 'Begin, Sicilian Muses, begin the dirge'. *Σικελίδες* is evidently V.'s invention.

paulo maiora: the construction—*paulo* with a comparative—seems to be colloquial and occurs in comedy, satire, and elegy (though not in Propertius or Tibullus), e.g. Plaut. *Men.* 681 'paulo prius', Ter. *Ad.* 831 'omissiores paulo', Lucil. 833 M. 'meliore paulo facie', Hor. *Serm.* 1. 9. 71 'sum paulo infirmior', Ov. *Trist.* 2. 577 'tutius exilium pauloque quietius oro'; see Gotoff 67 n. 4, Axelson 95, *TLL* s.v. *paulus* 832. 4. *Paulo* is not found elsewhere in V.

canamus: 'verbum heroicum' (La Cerda).

2. arbusta: 1. 39 n.

humiles . . . myricae: V.'s tamarisks are borrowed from Theocritus; see above, p. xxix. Tamarisks grow by the seashore and in other arid and barren places; quite possibly, as a young poet from the Po valley, V. had never seen one. In any case, the tamarisk, a bush or small tree, is not remarkably low, except, perhaps, as compared with taller trees, *arbusta* (cf. 1. 24–5). But such a comparison would have no point here, and V. was thinking rather of his pastoral poetry (see 6. 1–2 n.), of which he makes tamarisks a symbol; cf. 6. 6–12, where V. apologizes to a consul for not celebrating his deeds in epic song.

3. siluae sint consule dignae: 9. 35–6 n.

4. ultima . . . aetas: cf. Hor. *Epod.* 16. 1 'Altera iam teritur bellis ciuilibus aetas.' Lines framed by an adjective and a noun are especially frequent in Cicero's *Aratea* and Catullus 64, the latter a poem V. had in mind as he wrote *E.* 4; see below, 15–16 n., 28–30 n., 46–7 n., 63 n. It is no accident, therefore, that such lines are more frequent in *E.* 4 than in any other *E.*: 1. 83, 2. 1, 3. 38, 39, 4. 4, 5, 17, 21, 28, 41, 47, 5. 17, 20, 84, 6. 8, 7. 15, 32, 8. 14, 71, 9. 13, 10. 20.

Cumaei: 'poeticae dictionis epitheton cum Sibylla et Sibyllinis rebus coniungitur' (*TLL* Onomast. s.v. *Cumae* 744. 68). This, the Greek form of the adjective (*Κυμαῖος*), is first attested here; cf. *A.* 6. 98 'Cumaea Sibylla', Prop. 2. 2. 16 'Cumaeae saecula uatis'. The Latin form is *Cumanus*; cf. Luc. 1. 564 'Cumanae carmina uatis'.

uenit: strictly speaking, the last age has not yet come, but the poet eagerly anticipates it.

iam: 6 *iam*, 7 *iam*, 10 *iam*—importing a note of urgency; the time is at hand.

5. magnus . . . saeclorum . . . ordo: not a reference, as has commonly been supposed, to the Stoic concept of the Great Year (*μέγας ἐνιαυτός*), a period variously calculated as lasting thousands of years, at the end of which the heavenly bodies return to their original places, the earth is dissolved in a great fire or flood, and time begins anew; see Pease on Cic. *De nat. deor.* 2. 51, B. L. van der Waerden, 'Das Große Jahr und die ewige Wiederkehr', *Hermes*, 80 (1952), 129–55. V. imagines no such cyclical process, no vast regression of time, but a state of constant felicity, a world endlessly redeemed. The adjective *magnus* occurs six times in this *E.* (but only seven times elsewhere in the *E.*, 1. 23, 47, 3. 104, 6. 31, 55, 7. 16, 8. 6) and modifies not so much the individual noun to which it is attached as the whole poem: 5 'magnus . . . ordo', 12 'magni . . . menses', 22 'magnos . . . leones', 36 'magnus . . . Achilles', 48 'magnos . . . honores', 49 'magnum . . . incrementum'.

ab integro: a phrase unusual enough to be noticed by Servius and DServius, who quotes Cato (see *TLL* s.v. *integer* 2080. 66); the usual phrase is *de* or *ex integro* (ibid. 44, 69).

6. redit et Virgo, redeunt Saturnia regna: instead of a second *et*—that is, *et Virgo et Saturnia regna redeunt*—the verb is repeated, a construction briefly discussed by Wagner ad loc. Cf. *A.* 7. 327 'odit et ipse pater Pluton, odere sorores', where see Fordyce, 516 'audiit et Triuiae longe lacus, audiit amnis', 8. 91–2 'mirantur et undae, / miratur nemus', Ov. *Met.* 2. 248 'arsit et Euphrates Babylonius, arsit Orontes'.

Virgo: see Introduction, p. 120.

Saturnia regna: see Introduction, p. 120.

7. noua progenies: a new race of men, the 'gens aurea' of line 9.

caelo demittitur alto: perhaps suggested by Lucr. 2. 1153–4 'haud, ut opinor, enim mortalia saecla superne / aurea de caelo demisit funis in arua', where, however, *aurea* modifies *funis.*

8. tu modo: the sense seems to be 'only bless the boy and all will be well'; cf. *A.* 4. 50 (Anna to Dido) 'tu modo posce deos ueniam'. The phrase *tu modo*, placed as here at the beginning of the line and followed by an imperative or the like, is found also in *G.* 3. 73, *A.* 2. 160, Prop. 2. 15. 49, Manil. 1. 458.

8–10. nascenti puero . . . / . . . / . . . faue: cf. Cic. *Ad fam.* 1. 7. 8 'iam olim nascenti prope nostrae laudi dignitatique fauisti'.

8. quo: the boy is both cause and agent.

9. desinet ac: 3. 4n.

10. Lucina: in the Republican period an epithet of Juno, Juno Lucina, whose aid was invoked by women in labour; cf. Ter. *Ad.* 487 'Iuno Lucina, fer opem! serua me, obsecro!' and see Wissowa 183–4. But from the Augustan period onward the epithet is applied, as here, to Diana; cf. Hor. *Carm. saec.* 1 'Phoebe siluarumque potens Diana', 13–15 'rite maturos aperire partus / lenis, Ilithyia, tuere matres, / siue tu Lucina probas uocari', and see Pease on Cic. *De nat. deor.* 2. 68.

V. may have had in mind Theocritus' description of Ptolemy's birth on the island of Cos, 17. 60–7: the daughter of Antigone called on Ilithyia to ease her pains, and in the dear likeness of his father a child was born. Cos saw and cried aloud, embracing the child and blessing him, 'Blessed be thou, child, and mayest thou honour me as Phoebus Apollo honoured Delos of the azure crown'.

tuus iam regnat Apollo: a child has been conceived, that the words of the Sibyl might be fulfilled, and now your Apollo reigns. Apollo figures here in a dual capacity, as the brother of Diana Lucina and as the god who inspired the Sibyl. A closer definition of *regnat* is hardly to be looked for, a certain vagueness, an indefinable resonance, being appropriate to such prophetic intoning.

11. teque adeo: as usual in V., *adeo* emphasizes the word which it follows and stands second in the line, e.g. 9. 59, *G.* 1. 24 'tuque adeo', *A*. 6. 498, 11. 275; see Mynors on *G*. 4. 197, Fordyce on *A*. 7. 427.

decus hoc aeui: 'this glorious age' (Conington). The verb *inibit* shows that a period of time and not the child is meant.

inibit: 'inchoabit, exordium accipiet, aureum scilicet saeculum' (Serv.). Cf. Lucr. 2. 743 'ex ineunte aeuo'.

12. Pollio: 3. 84 n.

13. sceleris: the guilt of civil war; cf. *G*. 1. 506 'tam multae scelerum facies', Hor. *Epod*. 7. 1–2 'Quo, quo scelesti ruitis? aut cur dexteris / aptantur enses conditi?', 17–18 'acerba fata Romanos agunt / scelusque fraternae necis', *Carm*. 1.2.29–30 'cui dabit partis scelus expiandi / Iuppiter?'

14. inrita: nullified, of no effect.

15–16. 'The son shall lead the life of gods, and be / By gods and heroes seen, and gods and heroes see' (Dryden). Cf. Catull. 64. 384–6 'praesentes namque ante domos inuisere castas / heroum et sese mortali ostendere coetu / caelicolae nondum spreta pietate solebant', Hom. *Od*. 7. 201–3 (Alcinous to Odysseus) *αἰεὶ γὰρ τὸ πάρος γε θεοὶ φαίνονται ἐναργεῖς | ἡμῖν, εὖτ' ἔρδωμεν ἀγακλειτὰς ἑκατόμβας, | δαίνυνταί τε παρ' ἄμμι καθήμενοι ἔνθα περ ἡμεῖς*, 'For always in the past the gods have appeared to us in bodily shape, whenever we perform our splendid hecatombs, and dined with us, sitting where we sit'. See Introduction, p. 122.

16. heroas: three times in this poem (see ll. 26, 35), but nowhere else in the *E*.

uidebitur illis: he will see them and be seen by them; *illis* is therefore dative of agent, as in *A*. 1. 440 'miscetque uiris neque cernitur ulli'.

17. See Introduction, p. 122.

18–30. The Golden Age was a time of ease and abundance, for the earth produced everything spontaneously; but human life changed for the worse when Jupiter became king of the gods, *G*. 1. 127–32 'ipsaque tellus / omnia liberius nullo poscente ferebat. / ille malum uirus serpentibus addidit atris / praedarique lupos iussit pontumque moueri, / mellaque decussit foliis ignemque remouit / et passim riuis currentia uina repressit'. V. imagines a restoration of that golden time coincident with the birth and early manhood of the boy.

18. nullo . . . cultu: may be taken with *fundet* (20) or with *munuscula*, or with both *munuscula* and *fundet*. The spontaneous bounty of earth is a feature of the Golden Age (Hes. *Works and Days* 117–18).

munuscula: 'a charming diminutive, the gifts being for a child' (T. E. Page). Cf. Hor. *Epist.* 1.7.17 'non inuisa feres pueris munuscula paruis'.

19. errantis hederas: cf. Catull. 61. 34–5 'ut tenax hedera huc et huc / arborem implicat errans'. *Hedera* is singular in V. except where, as here and in *G.* 4. 124 'pallentisque hederas', the plural is metrically necessitated; so Maas 528 = 568 and Löfstedt i. 45; but see M. P. Cunningham, *CP* 44 (1949), 9–10.

baccare: 7. 27 n.

20. ridenti: for the metaphor ('Pulcerrima Metaphora', La Cerda) see 7. 55 n. 'The flowers are described as putting on their sweetest smiles to lure the child's approval, cf. line 23 *blandos flores*' (T. E. Page).

colocasia: Scholfield on Nicander, *Georg.* fr. 81–2: 'The plant is the Indian lotus, *Nelumbium speciosum*, on which see Theophr. *H.P.* 4. 8. 7, Diosc. 2. 106, *RE* 13. 1518. The "Egyptian bean" is its seed . . . It has a large pink flower, and an edible root (*κολοκάσιον*)'.

fundet: cf. 9. 40–1 'uarios hic flumina circum / fundit humus flores', *G.* 2. 460 'fundit humo facilem uictum iustissima tellus'.

acantho: another Egyptian plant, 'Aegyptia spina' (Pliny, *NH* 13. 66), or rather, a substantial tree; see Theophrastus, *HP* 4. 2. 8, who says that garlands are made of its flower, which is very beautiful. See also P. Wagler, *RE* i. 1159–62, Abbe 129. This acanthus is to be distinguished from the acanthus in 3. 45, where see note.

21–2. See Appendix, p. 149.

22–3. There is, after the caesura in the second foot, an exact correspondence between these lines: adjective modifying object, future plural verb, neuter plural subject, object. Moreover, these four words rhyme: *magnos / blandos, metuent / fundent, armenta / cunabula, leones / flores*; although V. ordinarily avoids—indeed takes pains to avoid—rhyming effects in his verse. Somewhat similar are 5.25–6 (rhyme before the main caesura) and *G.* 1.134–5, where see Thomas. (I owe this observation to D. Christianson.)

23. The child's spontaneously flowering cradle, though not a traditional feature of the Golden Age, is not without analogy in accounts of supernatural births; thus the infant Dionysus is swaddled with sudden tendrils of ivy (Eur. *Phoen.* 649–54) and the Delian olive-tree puts forth golden foliage at the birth of Apollo (Callim. *Hymn* 4. 262). See F. Pfister, *RE Suppl.* iv. 319.

Standing here, between mollified lions and the perished snake, V.'s cradle seems oddly placed, and W. Klouček, *Progr. Obergymn. zu Leitmeritz* (1870–3) suggested that l. 23 should be placed after l. 20; a suggestion made independently by J. F. Mountford, *CR* 52 (1938), 54–5, and thereafter modified by A. Y. Campbell, ibid. 55–6, who, finding the conception of a 'self-efflorescent' cradle 'almost grotesque', conjectured *fundet*, with *ipsa (tellus)* as its subject. Klouček's transposition was accepted by B. Snell, *Hermes*, 73 (1938), 239–40, and has been accepted more recently, with Campbell's *fundet*, by Geymonat and by Thomas on *G.* 2. 459–60. But *ipsae* (21) extends the idea of spontaneity from inanimate nature (18 'nullo . . . cultu') to animals and ought not to be separated from ll. 18–20; further, *ipsa* followed so closely and emphatically by *ipsae* in a different sense would surely be very awkward. As for Campbell's ingenious conjecture, it would oblige the reader to hear *ipsa* apart from *cunabula* and *cunabula* in apposition to *blandos . . . flores*. It might also be objected to Klouček's transposition that it spoils the pattern—a syntactical and phonetic pattern too precise not to have been designed—of ll. 22 and 23 as described above. See also Appendix, n. 9.

24. occidet et serpens: cf. Hor. *Epod.* 16. 52 'neque intumescit alta uiperis humus'.

fallax herba ueneni: as explained by Housman, *CR* 14 (1900), 258–9 = *Class. Papers*, ii. 521–2, *ueneni* is an objective genitive with *fallax*, that is, *quae fallit uenenum*, the plant that disguises or conceals its poison. For *fallo* in this sense Housman compares Prop. 4. 5. 14 'sua nocturno fallere terga lupo' (she disguises her own shape under that of a werewolf), Ov. *Met.* 8. 578 'spatium discrimina fallit' (distance conceals the divisions), *Fast.* 3. 22, Manil. 1. 240 (adding in his edition Prop. 3. 14. 5 and Ov. *Am.* 2. 5. 5), Tac. *Hist.* 3. 23. 3. See 1. 65 n.

25. Assyrium: used again by V. in *G.* 2. 465 'alba neque Assyrio fucatur lana ueneno', and both there and here meaning 'Syrian',

Syrius being a less poetic adjective than *Assyrius*. Perfumes were called Syrian because spices from the East were brought overland by caravan and shipped from Syrian ports; see 3. 89 n. Cf. Catull. 68. 144 'Assyrio . . . odore', Tib. 1. 3. 7 'Assyrios . . . odores', Hor. *Carm*. 2. 11. 16 'Assyriaque nardo', where see Nisbet–Hubbard.

26–7. V. seems to have adapted the language of Hellenistic aretalogy to a Roman context: *laudes* (ἔπαινοι), *facta* (πράξεις), *uirtus* (ἀρετή); see A. Henrichs, *HSCP* 82 (1978), 209–10 and 88 (1984), 148–58. Cf. *A*. 8. 287–8 'hic iuuenum chorus, ille senum, qui carmine laudes / Herculeas et facta ferunt'.

28–30. These three lines recall, with a certain abruptness, the style of Catullus 64 and thus prepare for the reference to that poem below in ll. 46–7. All three lines are end-stopped, and all three have a molossus, a verb, after the main caesura, *flauescet, pendebit, sudabunt*; cf. Catull. 64. 39–41 'non humilis curuis purgatur uinea rastris, / non glebam prono conuellit uomere taurus, / non falx attenuat frondatorum arboris umbram'; cf. also 63–5. This archaic, rather ponderous rhythm, which occurs some 145 times in Catullus 64 (408 lines), is used sparingly by V. and later poets; see L. P. Wilkinson, *CQ* 34 (1940), 33. The Catullan effect of these lines is mitigated by the varied disposition of nouns and adjectives, so concerned was V. for the integrity of his style.

28. There would be nothing marvellous about a field gradually turning yellow with grain; presumably V. means that the field has not been cultivated, like the briars in the next line; cf. Hor. *Epod*. 16. 43–4 (the Isles of the Blest) 'reddit ubi Cererem tellus inarata quotannis / et imputata floret usque uinea', Ov. *Met*. 1. 109–10 (the Golden Age) 'mox etiam fruges tellus inarata ferebat, / nec renouatus ager grauidis canebat aristis'.

molli: with no particular emphasis; 'the soft *strokeable* look of a field of waving corn?' (A. W. Pickard-Cambridge, *JRS* 8 (1918), 204). V. likes to apply this adjective to growing things, 2. 50 'mollia . . . uaccinia', 72 'molli . . . iunco', 3. 45 'molli . . . acantho', 55 'molli . . . herba', 5. 31 'foliis . . . mollibus', 38 'molli uiola' (*A*. 11. 69), 6. 53 'molli . . . hyacintho' (*G*. 4. 137), *G*. 2. 12 'molle siler', *A*. 1. 693 'mollis amaracus', 4. 147–8 'molli . . . / fronde'.

paulatim: so the passing of the Golden Age is described in *A*. 8. 326–7 'deterior donec paulatim ac decolor aetas / et belli rabies et amor successit habendi'.

flauescet: cf. Cato, *De agr.* 151. 2 'hordeum flauescit'; only here in V., and, apart from Ovid (five times), quite a rare verb in poetry of the classical period; see *TLL* s.v.

29. pendebit: 'Signatissime ex more Italiae, ubi vites arbustivae, et ideo uvae semper pendentes' (La Cerda).

30. durae quercus sudabunt roscida mella: cf. Hor. *Epod.* 16. 47 'mella caua manant ex ilice'.

sudabunt: 'Pro, stillabunt. Phrasis digna Virgilio' (La Cerda). Cf. 8. 54 (an *adynaton*) 'pinguia corticibus sudent electra myricae', Ov. *Met.* 2. 364–5 'stillataque sole rigescunt / de ramis electra nouis', *G.* 2. 118–19 'odorato . . . sudantia ligno / balsama', where, however, *sudantia* is intransitive. V. does not use the verb *stillo*.

roscida mella: 'honey-dew, a glutinous saccharine substance which in sultry weather is found covering the leaves of many trees, especially oaks, elms, and limes, and is eagerly consumed by bees. It is generally regarded as an exudation of sap . . . but much of it is also secreted by various species of aphides which live upon the leaves' (T. E. Page on *G.* 4. 1). Cf. 8. 37 'roscida mala'.

mella: the poetic plural is first attested in Varro Atac. *FPL* fr. 18. 3 Büchner 'dulcia cui nequeant suco contendere mella'.

31. priscae uestigia fraudis: cf. above, 13 'sceleris uestigia nostri'.

32–3. The vestiges of ancient wrong—note the tricolon crescendo with anaphora—are three: trespassing on the sea, surrounding cities with walls, and wounding the earth with ploughshares.

32. temptare Thetin ratibus: an old and persistent theme: Hes. *Works and Days* 236–7. *οὐδ' ἐπὶ νηῶν | νίσονται, καρπὸν δὲ φέρει ζείδωρος ἄρουρα*, 'nor do they travel on the sea, but the grain-giving earth bears them fruit', Arat. *Phaen.* 110–11 *χαλεπὴ δ' ἀπέκειτο θάλασσα, | καὶ βίον οὔπω νῆες ἀπόπροθεν ἠγίνεσκον*, 'The cruel sea lay apart, and ships did not yet bring their livelihood from afar', Tib. 1. 3. 37–8 'nondum caeruleas pinus contempserat undas, / effusum uentis praebueratque sinum', where see K. F. Smith, Hor. *Carm.* 1. 3. 21–4 'nequiquam deus abscidit / prudens Oceano dissociabili / terras, si tamen impiae / non tangenda rates transiliunt uada'.

Thetin: a metonymy found before V. only in Lycophron 22; see 10. 5 n. and below, 46–7 n.

33. infindere: a rare verb, found before V. only in Accius, *FPL* fr. 4 Büchner 'fraxinus fixa ferox infensa infinditur ossis'. And a strong verb, here reinforced by a harsh elision: not ploughing merely, but ploughing described as a violent act, an act of violation. *Infindere sulcos* appears to be V.'s phrase, which he reuses in *A*. 5. 142–3 'infindunt pariter sulcos, totumque dehiscit / conuulsum remis rostrisque tridentibus aequor'; see *TLL* s.v. and cf. Cic. *De diu*. 2. 50 'sulcus altius . . . impressus', Palladius 2. 10. 2 'sulcus imprimitur'. The scribe of the Romanus, or a predecessor, evidently tried to lessen the harshness of V.'s phrase.

34–6. The structure of these lines approximates that of a tricolon crescendo, *alter* . . . *altera* . . . / . . . *altera*.

34–5. alter erit tum Tiphys et altera quae uehat Argo / delectos heroas: cf. Hor. *Epod*. 16. 57 'non huc Argoo contendit remige pinus'.

34. Tiphys: the helmsman of the Argo.

35. delectos heroas: the Argonauts were traditionally referred to as heroes: Herod. 4. 145. 3 *τῶν ἐν τῇ Ἀργοῖ πλεόντων ἡρώων*, 'the heroes who sailed in the Argo', and Ap. Rhod. repeatedly, e.g. 4. 831 *λεκτοὺς ἡρώων*, 'the choicest of the heroes'. Cf. Enn. *Medea exul* 250 V.^{2} = 212 J. 'Argo, quia Argiui in ea delecti uiri', Catull. 64.4 'cum lecti iuuenes, Argiuae robora pubis', whence *A*. 8. 179 'tum lecti iuuenes', 518–19 'robora pubis / lecta'.

36. Achilles: opportunely introduced here, for his father, Peleus, was an Argonaut and ll. 46–7 below allude to Catullus 64, that is, to the song foretelling the birth of Achilles which the Parcae sang at the wedding of Peleus and Thetis, 338 'nascetur uobis expers terroris Achilles'.

38. cedet et ipse mari uector: and even the merchant will quit the sea; for *ipse* used of behaviour that is surprising or unexpected cf. 1. 38–9, 5. 62–4, 7. 7, and see Mynors on *G*. 3. 255.

uector: properly a passenger, e.g. Livy 24.8.12 'quilibet nautarum uectorumque tranquillo mari gubernare potest'; but DServius here explains *uector* as 'tam is qui uehitur quam qui uehit' and T. E. Page as the merchant 'who will no longer charter a ship to convey himself and his goods to foreign lands for purposes of barter or sale'. In any case, *mercator* was evidently too low a word for V., though not for Horace, *Serm.* 1. 1. 4, 6, *Carm.* 1. 1. 16, 31. 11, 3. 24. 40, *Epist.* 1. 1. 45, 16. 71, *Ars* 117, to use; apart from Plautus and Terence (Plautus especially), the only other poets who use the word are Manilius 5. 535 and Juvenal 13. 154; see *TLL* s.v.

nautica: an adjective found in Greek tragedy, e.g. Aesch. *Pers.* 383, Soph. *Phil.* 561, Eur. *Hec.* 607, which V. seems to have brought into Latin poetry; cf. *A.* 3. 128 'nauticus ... clamor', 5. 140–1 'clamor / nauticus', Hor. *Serm.* 2. 3. 106 'nautica uela', and see *OLD* s.v.

pinus: for the synecdoche cf. Hor. *Epod.* 16. 57 (above, 34–5 n.), *A.* 10. 206 'infesta ... pinu', 8. 91 'labitur uncta uadis abies'. Cf. also Catull. 64. 10 (of the Argo) 'pinea ... texta'.

39. mutabit merces: cf. Hor. *Serm.* 1.4. 29–30 'hic mutat merces surgente a sole ad eum quo / uespertina tepet regio', Tib. 1. 3. 39–40 'nec uagus ignotis repetens compendia terris / presserat externa nauita merce ratem', Manil. 1. 87–8 'et uagus in caecum penetrauit nauita pontum, / fecit et ignotas inter commercia terras' (so P. E. Knox, *CQ*, NS 39 (1989), 564–5), Ov. *Trist.* 1. 2. 75–6 'non ego diuitias auidus sine fine parandi / latum mutandis mercibus aequor aro'.

omnis feret omnia tellus: cf. Lucr. 1. 166 'ferre omnes omnia possent', *G.* 2. 109 'nec uero terrae ferre omnes omnia possunt'.

40–1. non rastros patietur humus, non uinea falcem; / robustus quoque iam tauris iuga soluet arator: cf. Lucr. 5. 933–6 (the condition of primitive man) 'nec robustus erat curui moderator aratri / quisquam, nec scibat ferro molirier arua / nec noua defodere in terram uirgulta neque altis / arboribus ueteres decidere falcibus ramos'; also Catull. 64. 39–41 (above, 28–30 n.), Hor. *Epod.* 16. 43–4 (above, 28n.).

40. patietur: the latent metaphor was developed by Ovid, *Met.* 1. 101–2 'ipsa quoque immunis rastroque intacta nec ullis / saucia uomeribus per se dabat omnia tellus', 2.286–7 'quod adunci uulnera aratri / rastrorumque fero' (La Cerda).

41. tauris: 1. 45 n.

42. nec uarios discet mentiri lana colores: cf. *G.* 2. 465 'alba neque Assyrio fucatur lana ueneno', 4. 334–5 'uellera ... / ... hyali saturo fucata colore', Hor. *Epist.* 2. 1. 207 'lana Tarentino uiolas imitata ueneno'.

mentiri: first attested here in this sense (*TLL* s.v. 780. 27).

43–5. The practice of dyeing wool will be abolished with the restoration of the Golden Age, because the wool of sheep in the meadow will spontaneously assume the brilliant colours formerly produced by dyes. For various judgements on these fanciful, yet rhetorically perfect, lines see B. Thornton, *AJP* 109 (1988), 226.

43–4. iam . . . / . . . iam: in the sense of *modo* . . . *modo* ('significat *modo*', DServ.) or *nunc* . . . *nunc* first attested here; see *TLL* s.v. 118. 60, Hofmann–Szantyr 520.

43. suaue rubenti: 2. 49 n.

44. murice: the famous Tyrian purple, a dark deep dye obtained from the juice of a shellfish; see Pease on *A*. 4. 134 and cf. *A*. 4. 262 'Tyrioque ardebat murice laena', 9. 614 'fulgenti murice uestis'.

croceo: first attested here (*TLL* s.v. 1212. 53).

luto: 'the common dyer's weed or weld (*Reseda luteola*) is to be found in many parts both of Italy and of England . . . It yields a yellow dye, which is obtained by boiling the whole plant when in flower' (Sargeaunt 73).

45. sandyx: a red dye derived from oxides of lead and iron; cf. Prop. 2. 25. 45 'illaque plebeio uel sit sandycis amictu'.

46–7. An allusion to the song of the Parcae in Catull. 64. 320–1, 'haec tum clarisona uellentes uellera uoce / talia diuino fuderunt carmine fata', and especially to the refrain of their song, 327 'currite ducentes subtegmina, currite, fusi'. But V. may also have had in mind a two-line passage in Lycophron's *Alexandra*, a poem he apparently knew (above, 32n.), 584–5 καὶ ταῦτα μὲν μίτοισι χαλκέων πάλαι | στρόμβων ἐπιρροιζοῦσι γηραιαὶ κόραι, 'These dooms the Ancient Maidens whirl on with whistling threads of brazen spindles'. V.'s imitations, even his earlier imitations, are rarely simple.

46. talia: retrospective and summarizing, as in Catull. 64. 265 'talibus amplifice uestis decorata figuris', *A*. 8. 729–30 'talia per clipeum Volcani . . . / miratur'; see W. Bühler, *Die Europa des Moschos* (Hermes, Einzelschr. 13; Wiesbaden, 1960), 108 n. 3.

talia saecla: accusative or vocative? The question appears to be ancient, for DServius glosses *currite* with *euoluite*, as if a reader might be in doubt. Symmachus, however, was not: *Or*. 3. 9 (to the Emperor Gratian) 'si mihi nunc altius euagari poetico liceret eloquio, totum de nouo saeculo Maronis excursum uati similis in tuum nomen exscriberem . . . et uere . . . iamdudum aureum saeculum currunt fusa Parcarum'. La Cerda and, of more recent scholars, J. B. Hofmann (*TLL* s.v. *curro* 1515. 39), Norden, *Die Geburt des Kindes*, 10, and Klingner 73 construe *talia saecla* as accusative. Heyne, however, construes it as vocative ('O talia saecla, currite!'), Conington finds the accusative 'exceedingly harsh', and T. E. Page considers the vocative 'simpler'. But V. uses *curro* twice elsewhere with the accusative (*A*. 3. 191, 5. 862), and Lycophron, if V. was indeed thinking of Lycophron, indicates that the accusative is preferable here.

47. concordes: an adjective as old as Naevius, *Bell. Poen.* 1 Strz. 'Nouem Iouis concordes filiae sorores'.

concordes stabili fatorum numine: 'unanimous by fate's steadfast decree'. Cf. Cic. *Tusc. disp.* 1. 115, where Cicero translates *Εὐθύνοος κεῖται μοιριδίῳ θανάτῳ*, a verse by the Academic philosopher Crantor, as 'Euthynous potitur fatorum numine leto' (the phrase 'fatorum numine' does not occur elsewhere, *TLL* s.v. *fatum* 367. 55); also Hor. *Carm. saec.* 25–8 'uosque ueraces cecinisse, Parcae, / quod semel dictum est stabilisque rerum / terminus seruet, bona iam peractis / iungite fata'.

Parcae: originally goddesses of childbirth, but owing to a false etymology identified with the Greek Moirai (Wissowa 264–5)—the fatal sisters, Clotho, who spins, Lachesis, who twists, and Atropos, who severs, the thread of human life. Cf. Hes. *Theog.* 217–19, Hor. *Carm.* 2. 3. 15–16 'dum res et aetas et sororum / fila trium patiuntur atra', and see S. Eitrem, *RE* xv. 2479–83.

48–52. 'Dicere haec ἐνθουσιῶν de puero, iam ad virilem aetatem adulto, existimandus est poeta, impatienter ferens moram, et iam animo providens praesagia ac prodigia, quae magnam rerum mutationem portendunt' (Heyne).

48. o: 'Optantis particula, et huc mire inserta' (La Cerda). *O* intensifies the imperative *adgredere*, as in *G.* 2. 35, where see Mynors, *A*. 6. 194–6, 12. 314. See Fraenkel, *Horace*, 242 n. 1.

magnos . . . honores: possibly a reference to the *cursus honorum*, more likely to unspecified honours and glory.

49. One of the three spondaic hexameters in the *E.*: 5. 38 'purpureo narcisso', 7. 53 'castaneae hirsutae'.

incrementum: a common word, as Servius recognized, 'et est uulgare', adding—and betraying a misapprehension of Virgilian pastoral—'quod bucolico congruit carmini'; a common word, moreover, that draws attention to itself by its weight and place in the line. The line consists of two cola similar in shape (adjective-genitive-noun), the second 'magnum Iouis incrementum' being in effect a definition of the first 'cara deum suboles'; *incrementum* ought not, therefore, to differ appreciably from *suboles* in meaning, 'Dear scion of the gods, great aftergrowth of Jupiter' (Lee); see Norden, *Die Geburt des Kindes*, 129–30. Although *suboles* has some of the rustic associations of *incrementum*, it is mentioned as a poetic word by Cicero, *De or*. 3. 153, and occurs in Lucr. 4. 1232. That V. dared to introduce such a word as *incrementum* into such a context may perhaps be taken as evidence of his youthful *audacia* (*G*. 4. 565–6).

50–2. Resonant, artfully composed lines: *aspice conuexo / aspice uenturo, nutantem / laetantur* (R: *laetentur* P, but *laetantur* is the *lectio difficilior* and conforms to V.'s practice; cf. 5. 6–7, *A*. 6. 855–6, 8. 191–2), *mundum / caelumque profundum* (for assonant verse-endings in V. see Pease on *A*. 4. 54). See A. Traina, '*Conuexo nutantem pondere mundum* (Verg. *Ecl*. 4, 50). Cosmologia e poesia', *Poeti latini (e neolatini): Note e saggi filologici*[2] (Bologna, 1986), i. 197–218 = *Studi in onore di C. Diano* (Bologna, 1975), 435–47.

50. aspice ... nutantem ... mundum, 52. aspice ... laetantur ut omnia: for the varied construction cf. *A*. 8. 190–2 'iam primum saxis *suspensam* hanc *aspice rupem*, / disiectae procul *ut* moles desertaque montis / *stat domus*'.

50. conuexo: refers to the curvature or sphericity of the universe (*mundus*), defined by Cicero as *rotundus* (*De nat. deor*. 1. 18) and *globosus* (ibid. 2. 116, where see Pease).

nutantem: a sign of cosmic joy (*laetantur*); see Traina 210–15.

mundum: the universe, the constituent parts of which—earth, sea, and sky—are enumerated in the next line and summed up in *omnia* (52). Cf. Cic. *Tusc. disp*. 5. 105 'in hoc ipso mundo caelum, terras, maria cognoscimus'.

51. Reused in *G*. 4. 222, where see Thomas. The language of this line goes back to a time before philosophy, when no single term for the universe (κόσμος) yet existed; cf. Hom. *Il*. 15. 187–93, Hes.

Theog. 736–9, with West's note. But the triple division persisted, especially in poetry; see Munro on Lucr. 1. 2–3, Traina 200 n. 2.

terrasque tractusque: the first *que* is lengthened under Greek influence, e.g. Hom. *Il.* 18.43 *Δωτώ τε Πρωτώ τε*. There are sixteen instances of this lengthening in V.: here, *G.* 1.153, 164, 352, 371, 3. 385, 4. 222, 336, *A.* 3. 91, 4. 146, 7. 186, 8. 425, 12. 89, 181, 363, 443; it is probably significant that it occurs only here in the *E.*, in this scarcely pastoral poem. In every case, the first word fills the whole foot and the following word, to which a second *que* is attached, is a spondee or an anapaest. See Mynors on *G.* 1.153, Pease on *A.* 4.146.

terrasque tractusque maris caelumque profundum: cf. Lucr. 5. 417 'terram et caelum pontique profunda', Enn. *Ann.* 559 Skutsch 'caelus profundus', *G.* 4.222, *A.* 1.58. For the plural *terras* see 6. 32 n.

53. o mihi tum ... maneat: cf. 2. 28.

54. tua dicere facta: 8. 8.

55–6. See 5. 25–6 n.

56. huic ... huic: instead of *huic ... illi*; a construction found in comedy and introduced into higher poetry by V. (Hofmann–Szantyr 181).

57. formosus Apollo: cf. Tib. 2.3.11 'pauit et Admeti tauros formosus Apollo', where, however, the adjective has an erotic colouring. The beauty of Apollo is traditional (see F. Williams on Callim. *Hymn* 2. 36), but only here in V. is he called *formosus*; in *A.* 3. 119 he is 'pulcher Apollo'. See 1. 5 n. (*formosam*).

58–9. Such symmetrical repetition is an occasional feature of Hellenistic poetry; see Dover xlvii f. and cf. e.g. Theocr. 11. 22–3 *φοιτῇς δ' αὖθ' οὕτως ὅκκα γλυκὺς ὕπνος ἔχῃ με, | οἴχῃ δ' εὐθὺς ἰοῖσ' ὅκκα γλυκὺς ὕπνος ἀνῇ με*, 'Why thus do you approach when sweet sleep holds me, and straightway are gone when sweet sleep releases me?', Callim. *Hymn* 4. 84–5 *Νύμφαι μὲν χαίρουσιν, ὅτε δρύας ὄμβρος ἀέξει, | Νύμφαι δ' αὖ κλαίουσιν, ὅτε δρυσὶ μηκέτι φύλλα*, 'The Nymphs rejoice when the rain gives growth to oaks, the Nymphs again weep when the leaves are no longer on the oaks', Nonnus, *Dionys.* 36.55–6 *σῶν ἐλάφων ἀλέγιζε καὶ εὐκεράου σέο δίφρου, | σῶν ἐλάφων ἀλέγιζε· τί σοὶ Διὸς υἷα γεραίρειν*, 'Attend to your stags and your horned team, attend to your stags: why should you honour the son of Zeus?' Such repetition becomes a mannerism with Ovid, e.g. *Met.* 1. 325–6 'et superesse uirum de tot modo milibus unum / et superesse uidet de tot modo milibus unam', 361–2, 481–2, 2. 82–3, 3. 611–12, 4. 575–6, 5. 578–9, 9. 488–9, 12. 148–9.

58. Arcadia ... iudice: for the legal metaphor cf. Livy 27. 11. 11 'Q. Fabium Maximum quem tum principem Romanae ciuitatis esse uel Hannibale iudice uicturus esset', where Weissenborn–Mueller cite Hor. *Serm.* 1. 2. 134 'Fabio uel iudice uincam'. Cf. also Hor. *Serm.* 2. 1. 49 'grande malum Turius, si quid se iudice certes'.

61. decem ... menses: a normal pregnancy of ten (sidereal) months, e.g. Plaut. *Stich.* 159 'nam illa me in aluo menses gestauit decem'; see Pease on Cic. *De nat. deor.* 2. 69, O. Neugebauer, *AJP* 84 (1963), 64–5, C. Watkins, 'Language Typology 1988', in W. P. Lehmann and H.-J. Jakusz Hewitt (eds.), *Current Issues in Linguistic Theory,* 81 (1991), 139.

62. qui non risere parenti: the MSS and Servius have 'cui non risere parentes', which gives the wrong sense; so far from being wonderful, it is natural for parents to smile at a new-born child. Quintilian 9. 3. 8 evidently read 'qui non risere parentes', but this again gives the wrong sense; *rideo* with the accusative can only mean 'laugh at' or 'mock', as in Hor. *Epist.* 1. 14. 39 'rident uicini glaebas et saxa mouentem'. J. Schrader saw that *parenti* was wanted; cf. Catull. 61. 209–12 'Torquatus uolo paruulus / . . . / . . . / dulce rideat ad patrem' (*ad patrem* being equivalent to *patri*). The marvellous child is urged to greet his mother with a smile ('risu cognoscere matrem')—a recognition of which a new-born child is incapable, except in the fond imagination of his mother—for no god invites to table those who have not smiled at their mother, no goddess to bed. The transition from a generalizing plural to the singular is Greek; P. Maas, *Textkritik*[4] (Leipzig, 1960), 23, compares Eur. *Herc.* 195–7; for other examples see Fraenkel on Aesch. *Ag.* 1521 ff. (p. 717 n. 3). Schrader also conjectures *hos* for *hunc*, but the singular, as Maas remarks, will be intelligible to anyone who thinks of the goddess's bed. See Norden, *Die Geburt des Kindes*, 62 n. 2.

parenti: for the feminine see *TLL* s.v. 354. 31, Hofmann–Szantyr 7.

63. nec deus hunc mensa, dea nec dignata cubili est: the allusion is to Hercules; cf. Hom. *Od.* 11.602–4 *αὐτὸς δὲ μετ' ἀθανάτοισι θεοῖσι | τέρπεται ἐν θαλίῃς καὶ ἔχει καλλίσφυρον Ἥβην | παῖδα Διὸς μεγάλοιο καὶ Ἥρης χρυσοπεδίλου*, 'but he himself rejoices in the feasts of the immortal gods and has to wife Hebe of the lovely ankles, daughter of great Zeus and Hera of the golden sandals', Hor. *Carm.* 4.8.29–30 'sic Iouis interest / optatis epulis impiger Hercules', and see Introduction, p. 122. The table appears to be a Hellenistic detail; cf. Nonnus, *Dionys.* 8.416–18 (Semele translated to heaven) *σύνθρονον Ἄρτεμιν εὗρε καὶ ὡμίλησεν Ἀθήνῃ | καὶ πόλον ἕδνον ἔδεκτο, μιῆς ψαύουσα τραπέζης | Ζηνὶ καὶ Ἑρμάωνι καὶ Ἄρεϊ καὶ Κυθερείῃ*, 'she found Artemis seated beside her and communed with Athena and received heaven as a wedding-gift, sharing the same table with Zeus and Hermaon and Ares and Cytherea'. (I am indebted for the reference to R. Kassel.) There may also be a reminiscence in l. 63 of the closing lines of Catull. 64.407–8 'quare *nec* talis *dignantur* uisere coetus, / *nec* se contingi patiuntur lumine claro'.

APPENDIX

The Fourth *Eclogue* and Horace's Sixteenth *Epode*

Altera iam teritur bellis ciuilibus aetas,
  suis et ipsa Roma uiribus ruit.
quam neque finitimi ualuerunt perdere Marsi
  minacis aut Etrusca Porsenae manus,
aemula nec uirtus Capuae nec Spartacus acer
  nouisque rebus infidelis Allobrox,
nec fera caerulea domuit Germania pube
  parentibusque abominatus Hannibal,
impia perdemus deuoti sanguinis aetas,
  ferisque rursus occupabitur solum.
barbarus heu cineres insistet uictor et Vrbem
  eques sonante uerberabit ungula,
quaeque carent uentis et solibus ossa Quirini,
  nefas uidere! dissipabit insolens.
forte quid expediat communiter aut melior pars

malis carere quaeritis laboribus.
nulla sit hac potior sententia, Phocaeorum
uelut profugit exsecrata ciuitas
agros atque Lares patrios, habitandaque fana
apris reliquit et rapacibus lupis,
ire pedes quocumque ferent, quocumque per undas
Notus uocabit aut proteruus Africus.
sic placet? an melius quis habet suadere? secunda
ratem occupare quid moramur alite?
sed iuremus in haec: simul imis saxa renarint
uadis leuata, ne redire sit nefas;
neu conuersa domum pigeat dare lintea, quando
Padus Matina lauerit cacumina,
in mare seu celsus procurrerit Appenninus,
nouaque monstra iunxerit libidine
mirus amor, iuuet ut tigris subsidere ceruis,
adulteretur et columba miluo,
credula nec rauos timeant armenta leones,
ametque salsa leuis hircus aequora.
haec et quae poterunt reditus abscindere dulcis
eamus omnis exsecrata ciuitas,
aut pars indocili melior grege; mollis et exspes
inominata perprimat cubilia!
uos quibus est uirtus, muliebrem tollite luctum,
Etrusca praeter et uolate litora.
nos manet Oceanus circumuagus: arua beata
petamus, arua diuites et insulas,
reddit ubi Cererem tellus inarata quotannis
et imputata floret usque uinea,
germinat et numquam fallentis termes oliuae,
suamque pulla ficus ornat arborem,
mella caua manant ex ilice, montibus altis
leuis crepante lympha desilit pede.
illic iniussae ueniunt ad mulctra capellae,
refertque tenta grex amicus ubera,
nec uespertinus circumgemit ursus ouile,
neque intumescit alta uiperis humus;
nulla nocent pecori contagia, nullius astri 61
gregem aestuosa torret impotentia. 62
pluraque felices mirabimur, ut neque largis 53
aquosus Eurus arua radat imbribus, 54
pinguia nec siccis urantur semina glaebis, 55
utrumque rege temperante caelitum.

non huc Argoo contendit remige pinus,
  neque impudica Colchis intulit pedem;
non huc Sidonii torserunt cornua nautae,
  laboriosa nec cohors Vlixei. 60
Iuppiter illa piae secreuit litora genti, 63
  ut inquinauit aere tempus aureum, 64
aere, dehinc ferro durauit saecula, quorum 65
  piis secunda uate me datur fuga. (*Epode* 16)[1]

The similarities of *E.* 4 and *Epode* 16 are too many and too close to be owing to chance.[2] Both poems describe a Golden Age, Horace's a Golden Age that can be realized only in the Isles of the Blest in the Western Sea, Virgil's a Golden Age that will be inaugurated in Italy with the birth of a marvellous child. Hence the much-debated question, 'Which poet is the imitator?'

In resolving this question, the verbal coincidence of the following passages is, as F. Skutsch perceived,[3] crucially important.

credula nec rauos timeant armenta leones. (*Epode* 16. 33)

illic iniussae ueniunt ad mulctra capellae,
  refertque tenta grex amicus ubera. (*Epode* 16. 49–50)

ipsae lacte domum referent distenta capellae
ubera, nec magnos metuent armenta leones. (*E.* 4. 21–2)

Line 33 in Horace is not, like l. 22 in Virgil, a feature of the Golden Age but one in a series of *adynata*: we shall return (Horace writes) when the Po washes the mountain-tops of Apulia, when the Apennines thrust forward into the sea, and a monstrous passion joins tiger to deer, dove to kite, 'nouaque monstra iunxerit libidine / mirus amor, iuuet ut tigris subsidere ceruis, / adulteretur et columba miluo, / credula nec rauos timeant armenta leones' (30–3). In literature before Virgil, as Skutsch points out, the harmonious

[1] For an excellent discussion of Horace's poem see Fraenkel, *Horace*, 42–55. See also A. Setaioli, 'L'epodo XVI di Orazio nella critica del dopoguerra', *A&R* 19 (1974), 9–25.

[2] See notes on *E.* 4. 4, 21–2, 24, 30, 34–5, 38.

[3] *N. Jahrb.* 23 (1909), 28–33 = *Kleine Schriften* (Leipzig, 1914), 370–5.

congress of domestic and predatory animals is an *adynaton.*[4] Horace's 'credula nec rauos timeant armenta leones' is therefore traditional, Virgil's 'nec magnos metuent armenta leones' an innovation. Which is more probable: that Horace, noticing Virgil's novelty, reconstituted it as an *adynaton* in his poem, where it fits perfectly, or that Virgil, who had no use for *adynata* in his poem (there being no 'impossibilities' in the Golden Age), abstracted Horace's *adynaton* from its context and, with an easy modification, adapted it to his description of the Golden Age? And is not such a bold act of appropriation characteristic of Virgil?[5]

B. Snell,[6] who was concerned that Virgil should not be dependent upon Horace, begins by citing *E.* 8. 26–8:

> Mopso Nysa datur: quid non speremus amantes?
> iungentur iam grypes equis, aeuoque sequenti
> cum canibus timidi uenient ad pocula dammae.

—'mirus amor' and innocent congress, apparently, as in *Epode* 16. 30–3; yet Snell finds in these lines not a reminiscence of Horace but an imitation of Theocritus: 'Diese Adynata (die in 52–56 fortgesetzt werden) sind, wie die Kommentatoren hervorheben, von Theokr. I, 132 ff. abhängig' (237). The reader will look in vain for any such comment in Voss, Forbiger, Conington, and T. E. Page; Heyne cautiously remarks 'color ducitur'. Snell further maintains that Horace is imitating *E.* 8. 26–8, and therefore describes *E.* 8 as one of Virgil's earlier *Eclogues.*[7] In any case, imitation of Theocritus would not preclude imitation of Horace, for Virgil constantly practices multiple imitation.[8]

[4] Cf. Hom. *Il.* 22. 262–7, Ar. *Peace* 1075–6 *οὐ γάρ πω τοῦτ' ἐστὶ φίλον μακάρεσσι θεοῖσιν, | φυλόπιδος λῆξαι, πρίν κεν λύκος οἶν ὑμεναιοῖ*, 'For it is not pleasing to the blessed gods that we cease from strife until wolf mates with sheep'. An apparent exception is Theocr. 24. 86–7 *ἔσται δὴ τοῦτ' ἆμαρ ὁπηνίκα νεβρὸν ἐν εὐνᾷ | καρχαρόδων σίνεσθαι ἰδὼν λύκος οὐκ ἐθελήσει*, 'That day shall be when the jag-toothed wolf shall find the couching fawn and shrink from harming it' (Gow); but these lines are interpolated, and come, as Gow suggests in an informative but confused note, 'from some prophecy of a future Golden Age such as is foreseen at *Isaiah* 11. 6, 65. 25, Lactant. *Inst. Diu.* 7. 24'. Cf. also *E.* 8. 26–8, Hor. *Carm.* 1. 33. 7–9 'sed prius Apulis / iungentur capreae lupis / quam turpi Pholoe peccet adultero.'

[5] Of countless instances the most notorious, perhaps, is Catull. 66. 39 'inuita, o regina, tuo de uertice cessi' ∼ *A.* 6. 460 'inuitus, regina, tuo de litore cessi.'

[6] *Hermes*, 73 (1938), 237–42 = *Gesammelte Schriften* (Göttingen, 1966), 192–8.

[7] On the contrary, it must be one of his latest; see *E.* 8, Introduction, pp. 238–9.

[8] See *E.* 8, Introduction, n. 26.

Snell's main argument involves the similarity of *Epode* 16. 49–50:

> illic iniussae ueniunt ad mulctra capellae,
> refertque tenta grex amicus ubera.

and *E*. 4. 21–2:[9]

> ipsae lacte domum referent distenta capellae
> ubera.

Recognizing that no decision about priority can be made if these passages are considered in isolation, Snell adduces Theocr. 11. 12–13 (the Cyclops in love):

> πολλάκι ταὶ ὄιες ποτὶ τωὔλιον αὐταὶ ἀπῆνθον
> χλωρᾶς ἐκ βοτάνας.

Often would his sheep return of their own accord from the green pastures to the fold.

Snell remarks that Virgil is more likely than Horace to have imitated Theocritus: true, but what prevented Virgil from imitating both Horace and Theocritus? Snell is convinced, however, of Horace's dependence upon Virgil: 'Die Abhängigkeit des Horaz von Vergil zeigt sich dann aber darin, daß Horaz die *lacte distenta ubera* übernimmt, die Vergil für seine Schilderung des Goldenen Zeitalters selbständig eingeführt hat' (240)—that is, Virgil has imitated himself, *E*. 7. 3 'distentas lacte capellas', and *E*. 7 accordingly becomes, like *E*. 8, an early *Eclogue*.[10] This elaborate and

[9] These lines are curiously intrusive; were they absent, the repetition 'fundet' (20), 'fundent' (23) would have point: Earth uncultivated will pour forth its plants and flowers for you, your very cradle will pour forth flowers. Cf. 'At tibi' (18), 'ipsa tibi' (23), and 'occidet' (24), 'occidet' (25). And yet ll. 21–2 are necessary where they stand, for otherwise 'occidet et serpens' would have no point: lions will become mild and the snake will perish. Recognizing the difficulty, Snell adopts Klouček's transposition of l. 23, for which see ad loc. A certain oddness, a difficulty, may indicate an imperfect imitation, and Virgil evidently had some trouble arranging his unconventional cradle.

[10] Cf. Fraenkel (above, n. 1), 51 n. 3: '*Ecl*. 7. 3 *distentas lacte capellas*. Virgil's seventh eclogue, probably one of his earliest, is doubtless earlier than the Fourth'. But *E*. 7 is generally considered one of Virgil's latest (see above, p. xxii n. 34), and the phrase 'distentas lacte capellas' appears to be a reworking of 4.21 'lacte . . . distenta capellae.'

improbable reconstruction collapses once a relevant passage in Lucretius, which Snell overlooks, is noticed:

> hinc fessae pecudes pingui per pabula laeta
> corpora deponunt et candens lacteus umor
> uberibus manat distentis. (1. 257–9)

Horace evidently imitated Lucretius,[11] and was in turn imitated by Virgil; or rather, Virgil imitated Horace, Lucretius, and Theocritus simultaneously.

Snell's argument, accepted by Fraenkel and R. Syme,[12] has been influential; it will not, however, bear sustained scrutiny. In all probability, Horace wrote his poem some time after the Battle of Philippi and before the Pact of Brundisium, perhaps during the dark days of the Perusine War, which ended in February 40 BC with the massacre of Perusia, and Virgil, writing in a happier time soon after the Pact of Brundisium, imitated Horace's poem,[13] possibly the poem of a friend, certainly a poem he admired.[14]

[11] Cf. *Epode* 16. 31 'tigris subsidere ceruis', Lucr. 4. 1197–8 'nec ratione alia uolucres armenta feraeque / et pecudes et equae maribus subsidere possent'. In this sexual sense, the verb *subsido* does not occur elsewhere; see W. Wimmel, *Hermes*, 81 (1953), 237, who consulted the editor of *TLL*.

[12] *Sallust* (Berkeley, 1964), 284–6. Syme detects in *Epode* 16 an allusion to the romantic story of Sertorius and supposes that Horace found it in Sall. *Hist.* 1, which was not published before 37 or 36 BC. Cf. Fraenkel (above n. 1), 49: 'possible, but not provable' (quoted by Syme).

[13] See A. D. Nock, *The Cambridge Ancient History*, x (Cambridge, 1934), 472–3. For the priority of Horace's *Epode* see also C. J. Classen, 'Romulus in der römischen Republik', *Philologus*, 106 (1962), 200 n. 6.

[14] Cf. F. Skutsch (above, n. 2), 363: 'Horaz' sechzehnte Epode gehört zu seinen schönsten Gedichten.' The priority of the Fourth *Eclogue* is, to some extent, an emotional as well as a philological problem, for there are those who would rather not imagine Virgil in a relation subordinate to Horace. The controversy over the Hylas story in Ap. Rhod. 1. 1207–72 and Theocr. 13 offers an instructive parallel.

ECLOGUE 5

Introduction

Two shepherds meet, a younger and an older, and the older suggests ('since you are good at piping, I at singing verses', 1–2) that they sit down together among the hazels and elms and make music. Mopsus, for he is the younger, agrees, though not without expressing a preference for a cave near by. Suspecting that Mopsus may resent his casual assumption of superiority and wishing to put the younger man at his ease, Menalcas accepts the cave (as the reader learns in l. 19) and pays him a compliment: 'In our hills only Amyntas competes with you' (8). But Mopsus is not appeased and replies with youthful petulance: 'He might as well compete to excel Phoebus in singing' (9). Menalcas tactfully changes the subject and invites Mopsus to begin, suggesting three themes, all within the range of a Theocritean singer (10–11). But Mopsus has other ideas: he will try out a song he has composed recently (13 'nuper'), words and music; 'then', he adds, 'bid Amyntas compete' (15). Whereupon Menalcas, with a most Theocritean deference, concedes: 'As the willow yields to the olive . . .' (16–18).

The subject of Mopsus' song, of which the reader has had no indication, is startling, *The Death of Daphnis*. Menalcas praises both the song and the singer, 48 'nec calamis solum aequiperas sed uoce magistrum'.[1] Menalcas, too, will sing of Daphnis—a song, as it turns out, which corresponds to that of Mopsus.[2] Mopsus professes himself delighted, nothing could please him more (53), and remarks that Stimichon praised the song to him some time ago

[1] Thus recalling and amending his invidious comparison, 1–2 'boni quoniam conuenimus ambo, / tu calamos inflare leuis, ego dicere uersus'. Who is Mopsus' master? The accepted answer is Daphnis: '"Magistrum" can hardly be anyone but Daphnis, whose minstrelsy is praised by Theocr. l.c.' (Conington on l. 48). But Lee 62–3, pointing out that Virgil says nothing about Daphnis as a musician and much about him that is not in Theocritus, argues that Stimichon (55) is meant. But there seems to be an allusion to Daphnis the unparalleled pastoral musician in l. 49, where see note.

[2] Notably in structure; each song consists of 25 lines divided into sections of 4, 5, 7, 4, 5 lines. See O. Skutsch, 'Symmetry and Sense in the Eclogues', *HSCP* 73 (1969), 157–8, Lee 65–6.

(55 'iam pridem'). In other words, Mopsus already knew Menalcas' song and modelled his own after it.[3] Menalcas is, under the circumstances, exceedingly polite, 51 'Daphninque *tuum* tollemus ad astra'. Mopsus' praise of *The Apotheosis of Daphnis*[4] is elaborate (82–4), as elaborate as was Menalcas' praise of *The Death of Daphnis* (45–7) and couched in similar terms, yet pervaded by a curious dissonance.[5]

Daphnis is dead, destroyed by a cruel fate, and the Nymphs mourned for him, 'Exstinctum Nymphae crudeli funere Daphnin / flebant' (20–1)[6]—so Mopsus begins, taking up, it might seem, where Theocritus left off. In the First *Idyll*, in the song of Thyrsis, Daphnis is represented as mysteriously dying of love: Hermes comes, the herdsmen come, Priapus comes, but Daphnis will not be comforted. Last comes Aphrodite, the mortally offended goddess, to taunt him; Daphnis insults her, then bids the pastoral world farewell, consigns his pipe to Pan, and dies: 'and Daphnis went to the stream. The waters closed over him, the man whom the Muses loved, whom the Nymphs did not hate.'[7]

Virgil had no wish, as may be inferred, to compete directly[8] with a masterpiece of pastoral song, and his Daphnis bears little or no resemblance to the Daphnis of Theocritus.[9] There are, however, in

[3] How else is l. 55 to be interpreted? A line that arrests attention with a unique name (Stimichon, see ad loc.) can hardly be dismissed as an inert detail—supposing there to be such in Virgil. In 9.21–5, one herdsman steals another's song.

[4] Some ancient readers were of the opinion that the deified Daphnis represented Julius Caesar, *Diuus Iulius*; DServ. on l. 56: 'et quibusdam uidetur per allegoriam Caesarem dicere, qui primus diuinos honores meruit et diuus appellatus est'. Cf. B. Otis, *Virgil: A Study in Civilized Poetry* (Oxford, 1963), 135: 'The deified Daphnis of 5 is thus the rather thin mask of the deified Julius'. But Virgil is never so simple, and such an identification, grotesque if insisted upon, would be an inadequate response to the allusiveness and complexity of his poem. According to the *Vita Donati* 14, Daphnis represents Virgil's brother Flaccus, 'cuius exitum sub nomine Daphnidis deflet'.

[5] See notes on ll. 82 and 83–4.

[6] Cf. 34 'postquam te fata tulerunt'. The death of Daphnis is variously reported: that he fell from a cliff, that he was translated to heaven, that he was turned to stone (Gow on Theocr. 1.140). Virgil was only concerned that his death should be mysterious.

[7] 140–1 *χὠ Δάφνις ἔβα ῥόον. ἔκλυσε δίνα | τὸν Μοίσαις φίλον ἄνδρα, τὸν οὐ Νύμφαισιν ἀπεχθῆ*. For the interpretation of this difficult passage see Gow.

[8] Indirectly: 10.9–23, where see notes.

[9] Whose Daphnis in turn bears little or no resemblance to the Daphnis of tradition—insofar as the tradition is extant; see Gow's Preface to the First *Idyll*.

the song of Mopsus allusions to Theocritus, and chiefly to the song of Thyrsis.

24–8: the animals grieve for Daphnis. Cf. Thyrsis 71–5.

32–4: a fourfold comparison in Theocritean style (see ad loc.); 'answered' by Menalcas with a fourfold comparison (76–7) which owes nothing, except perhaps its pastoral compatibility, to Theocritus.

35–9: Pales and Apollo have abandoned the countryside, and natural processes are perverted: instead of barley come up darnel and wild oats; instead of the soft violet and the glowing narcissus, thistles and thorns. Cf. Thyrsis 132–3 *νῦν ἴα μὲν φορέοιτε βάτοι, φορέοιτε δ' ἄκανθαι, | ἁ δὲ καλὰ νάρκισσος ἐπ' ἀρκεύθοισι κομάσαι*, 'Now violets bear, ye brambles, and, ye thorns, bear violets, and let the fair narcissus bloom on the juniper' (Gow).

43–4: Daphnis' self-epitaph, 'Daphnis ego . . .', imitating Thyrsis 120–1 *Δάφνις ἐγὼν* . . . .

Menalcas, too, imitates Theocritus, though sparingly and in a manner Mopsus fails to appreciate or understand. Mopsus wishes to be like Theocritus and therefore borrows from him, while Menalcas transforms the little that he borrows.

67–8: Daphnis will share the altars with Apollo, and Menalcas will establish a cult in his honour, with annual offerings of milk and olive oil. Theocritus 5. 53–4: a *locus amoenus*, with bowls of milk and olive oil for the Nymphs.

72–3: two herdsmen will sing and a third will dance—something may be owing here to Theocritus 7.71–2,[10] but the rest of Menalcas' song owes nothing to Theocritus. (Mopsus' song, by contrast, ends with a precise imitation of Theocritus.)

The main reference of the Fifth *Eclogue*, as the names Amyntas (8, 15, 18) and Antigenes (89) imply, is to the Seventh *Idyll*. Simichidas, the narrator (Theocritus?), goes out of the town with two friends, Eucritus and Amyntas, to take part in a harvest festival on the farm of Phrasidamus and Antigenes. Along the way they encounter the goatherd Lycidas, an older singer who enjoys a high reputation among the herdsmen and the reapers (27–9). Although younger (Lycidas calls him a 'sapling', 44) and as yet no

[10] Rohde's assertion, p. 130: 'Dazwischen steht wie ein Block eine Folge von sieben Versen (67–73), die Theokrit nachahmen', to which Klingner 92 defers, is unfounded.

match for the great master-singers Sicelidas of Samos and Philitas, Simichidas has some reputation and fancies himself the equal of Lycidas, whom he challenges to a country singing-match (36 *βουκολιασδώμεσθα*). Lycidas good-naturedly assents—'see, friend, if this pleases you, a little song I worked out lately on the hill'—and sings a song of thirty-eight lines (52–89). Simichidas, too, has learnt songs on the hill, from the Nymphs—good songs, the very best of which he now sings in honour of Lycidas, who is loved by the Muses, a song of thirty-two lines (96–127). Similar in length, similar in subject-matter (homosexual love), but in composition quite different—that the two songs are equally fine seems to be taken for granted. Laughing pleasantly, Lycidas gives his shepherd's crook to Simichidas as 'a token of friendship in the Muses' (129 *ἐκ Μοισᾶν ξεινήιον*) and proceeds on his way.

Virgil also had in mind another *Idyll* of Theocritus, the Sixth, to which he is indebted for the idea of his ending. Two young herdsmen, Daphnis and Damoetas, meet at a spring, and Daphnis suggests a singing-match. They seat themselves and Daphnis, for the suggestion was his, begins. He addresses his song to Polyphemus and teases him for ignoring the advances of the coquettish Galatea (6–19). Damoetas replies by impersonating Polyphemus:[11] his indifference is feigned, he means to excite her passion; she spies on him from the sea, and perhaps will send a messenger . . . (21–40). The vague fiction of a singing-match is maintained throughout, and the two songs are related, if unobtrusively, in length and in subject-matter, but neither singer gains the victory. Damoetas kisses Daphnis and gives him a pipe, and Daphnis gives Damoetas a pretty flute. Then together they make music, Damoetas on the flute, Daphnis on the pipe, invincible both, while the heifers skip in the soft grass (42–6).

Similarly, at the end of the Fifth *Eclogue*, though in a scene less charmingly idyllic, Menalcas gives Mopsus a pipe, and Mopsus gives Menalcas a shepherd's crook (85–90). The pipe is special; it taught Menalcas 'formosum Corydon ardebat Alexin' and 'cuium pecus? an Meliboei?'[12] His gesture, so qualified, seems to suggest that Menalcas (Virgil?) has reached a crucial stage in his career and is now, as his song indicates, moving away from dependence on

[11] Cf. 7.41–4.

[12] Partial quotations of the opening lines of the Second and Third *Eclogues*, probably Virgil's earliest *Eclogues* and notably Theocritean.

Theocritus, whereas Mopsus, with some way still to go, as his song indicates, can do with the Theocritean pipe.

Bibliography

G. Rohde, *Studien und Interpretationen zur antiken Literatur, Religion und Geschichte* (Berlin, 1963), 121–39.

Klingner 90–9.

G. Lee, 'A Reading of Virgil's Fifth Eclogue', *PCPS* 203, NS 23 (1977), 62–70. (The interested reader will find that I have appropriated Lee's 'reading' almost entirely.)

1. **Mopse:** not a pastoral name before V.; perhaps borrowed from Ap. Rhod. 3.916–18 *Μόψος | Ἀμπυκίδης, ἐσθλὸς μὲν ἐπιπροφανέντας ἐνισπεῖν | οἰωνούς, ἐσθλὸς δὲ σὺν εὖ φράσσασθαι ἰοῦσιν*, 'Mopsus, son of Ampycus, good at interpreting the appearance of birds, good too at advising travellers'. This is the first occurrence of *bonus* with an infinitive (*TLL* s.v. 2098.33); for such adnominal infinitives, a Graecizing construction, see Hofmann–Szantyr 350–1. V. was thinking here, as he was in 7.4–5, of [Theocr.] 8.4 *ἄμφω συρίσδεν δεδαημένω, ἄμφω ἀείδεν*, 'both skilled in piping, both in singing'.

2. **dicere:** 'sane pro *canere*' (DServ.); see 3.55 n.

3. **corylos:** 1.14 n.

consedimus: 3.55 n.

4. **maior:** cf. Plaut. *Stich.* 41 (a younger sister about to disagree with an elder) 'tametsi es maior', Hor. *Epist.* 1.17.16 (addressing a younger man) 'uel iunior audi'. Respect for seniority extends even to the gods, Hom. *Il.* 21.439 (Poseidon to Apollo) *ἄρχε· σὺ γὰρ γενεῆφι νεώτερος*, 'Begin, for you are younger born'.

Menalca: 2.15 n.

5–6. **siue . . . / siue . . . potius:** the uncertain, wind-shifted shade, or rather the cave overspread with clusters of wild grapes. As in Theocr. 1.12–14, 21–3, two places for music-making are suggested and the second chosen as being the more attractive—'let us sit beneath the elm, facing Priapus and the springs, there where the shepherds' seat is and the oaks.' Cf. also Theocr. 5.44–61.

5. **motantibus:** a verb rarely found in classical Latin—here, in 6.28, Ov *Met.* 4.46, *Priap.* 19.3—which acquired a certain vogue in later Latin (*TLL* s.v.).

6–7. No ancient poet, it seems, could describe a cave without recalling Calypso's cave, *Od.* 5.68–9 ἡ δ' αὐτοῦ τετάνυστο περὶ σπείους γλαφυροῖο | ἡμερὶς ἡβώωσα, τεθήλει δὲ σταφυλῇσι, 'And there about the hollow cave was spread a flourishing vine, laden with clusters'.

6. siue: in eighteen of twenty-seven instances in V. *-e*, as here, is elided; see Soubiran 152.

antro: 1.75 n.

succedimus: normally construed with the dative, but in Caesar with *sub* and the accusative (*BG* 1.24.5, *BC* 1.45.2). Here an effect of artful variation is intended.

aspice, ut: 4.50–2 n.

7. siluestris ... labrusca: the wild vine, uncultivated.

raris: the scattered clusters let some light into the cave (Heyne); cf. 7.46.

8. certet: first attested here with the dative, by analogy with *contendo* (Lucr. 3.6–7) or *pugno* (Catull. 62.64); see Pease on *A.* 4.38, Kühner–Stegmann i. 319. Cf. 8.55, *G.* 2.99.

Amyntas: 3.66 n.

9. quid si idem certet: 'What if he should compete ...?', that is, 'He might just as well compete ...'. A colloquial expression, here ironic; cf. Plaut. *Poen.* 728 'quid si recenti re aedis pultem?', 1162 'quid si eamus illis obuiam?', Mart. 1.35.6–7 'quid si me iubeas talassionem / uerbis dicere non talassionis?'

canendo: both singing and piping; see below, 48 n.

10–11. incipe ... si quos ... ignis / ... habes: 9.32 n.

Phyllidis ignis / aut Alconis ... laudes aut iurgia Codri: objective genitives, that is, love of Phyllis or praise of Alcon or abuse of Codrus.

10. Phyllidis ignis: cf. Ov. *Met.* 10.252–3 'haurit / pectore Pygmalion simulati corporis ignes'. For the name see 7.14 n.

11. Alconis: a convenient Greek name, with no especial resonance (*TLL* s.v.); found only here in V.

Codri: a close friend of Corydon in 7.22.

12. pascentis seruabit Tityrus haedos: an echo of Theocr. 3.1–2 Κωμάσδω ποτὶ τὰν Ἀμαρυλλίδα, ταὶ δέ μοι αἶγες | βόσκονται κατ' ὄρος, καὶ ὁ Τίτυρος αὐτὰς ἐλαύνει, 'I am off to serenade Amaryllis, and my goats are grazing on the hill, and Tityrus herds them'. Tityrus is a much less important figure in Theocritus than he is in V.; see 1.1 n.

13. in uiridi . . . cortice fagi: 'ubi enim debuit magis rusticus scribere?' asks Servius. Mopsus has recently inscribed the words and musical notation of his song on a beech tree (as an aid to memory?); in spite of his boastfulness he seems a little unsure of himself (15 'experiar'). But Menalcas, the older and more experienced singer (below, 85–7), has no need of a 'text'; he has got his song by heart. It would be no easy task, Heyne remarks, to carve the twenty-five lines of Mopsus' song in the bark of a beech tree; but such practical considerations do not obtain in the pastoral world. Cf. Calp. Sic. 1.19–88, 3.43–91. In Thomson, *Summer*, 1363–70, Musidora carves four lines on a spreading beech with the 'sylvan pen of rural lovers', that is, a knife (J. Sambrook, ed. (Oxford, 1972) ad loc.: 'The implausibility of Musidora's feat of carving suggests . . .); and in Tasso, *Gerusalemme Liberata* 7. 19, Erminia inscribes Tancredi's name and the sad story of her love on a thousand trees. See 10.53–4.

14. modulans alterna notaui: Mopsus has set his lines to music and marked the interludes of piping; cf. 10.50–1 and see Tarrant on Sen. *Ag.* 672.

15. iubeto certet: so read with P, generally a better MS than R. N.-O. Nilsson, *Eranos*, 53 (1955), 199–200, has shown that elision at this point in the line (*iubeto ut* R) is virtually unparalleled; *ut* was inserted to make the construction clearer. Cf. *A.* 10.53–4 (the only other instance of the future imperative of *iubeo* in V.) 'magna dicione iubeto / Karthago premat Ausoniam'. See also A. Szantyr, *MH* 23 (1966), 212–15.

16. lenta salix: 3.83.

pallenti: 6. 54 n.

oliuae: V. uses *oliua* and *oliuum* at the end of the line, but *olea* and *oleum* within the line, for metrical convenience; see Norden on *A.* 6.224 ff.

17. puniceis . . . rosetis: cf. Hor. *Carm.* 4. 10. 4 'puniceae flore . . . rosae', 3.15.15 'flos purpureus rosae'.

saliunca rosetis: for the conjunction of the name of a plant or tree with that of places where others grow cf. *G.* 2.12–13 'molle siler lentaeque genistae, / populus et glauca canentia fronde salicta' (for metrical convenience; cf. *G*. 2.434 'salices humilesque genistae'), 110–12 'fluminibus salices crassisque paludibus alni / nascuntur, steriles saxosis montibus orni;/litora myrtetis laetissima' (cf. *G*. 4.124 'amantis litora myrtos'). But not merely for metrical convenience; cf. Cic. *De nat. deor.* 2.156 'quid de uitibus oliuetisque dicam?', where Pease cites Sall. *Iug*. 48.3 'collis . . . uestitus oleastro ac myrtetis' and Hor. *Carm*. 2.15.5–6 'uiolaria et / myrtus', where Nisbet–Hubbard cite Mart. 3.58.2–3 'non otiosis ordinata myrtetis / uiduaque platano tonsilique buxeto'.

saliunca: Celtic nard, a species of valerian, with an exquisite perfume but useless for garlands (Pliny, *NH* 21.40). See Sargeaunt 117.

19. sed tu desine plura, puer: cf. 9.66. For the rhythm see 2.53 n.

20. extinctum . . . Daphnin: Daphnis is dead—that is the supreme fact. A slow, solemn line, with a spondee (*flebant*) beginning the next line; cf. *A*. 6.212–13 'nec minus interea Misenum in litore Teucri / flebant' and see Norden 435–6.

crudeli funere: reused in *G*. 3.263 and *A*. 4.308.

21. uos coryli testes et flumina: 'Tritus hic mos Latinis vatibus . . . ut confugiant ad testimonia mutorum' (La Cerda). Cf. e.g. Enn. *Sat*. 10–11 V.[2] 'testes sunt / lati campi quos gerit Africa terra politos', Catull. 64.357 'testis erit magnis uirtutibus unda Scamandri', Tib. 1. 7. 11 'testis Arar Rhodanusque celer magnusque Garunna', Hor. *Carm*. 4.4.38 'testis Metaurum flumen'.

22–3. complexa . . . / atque deos atque astra uocat: critics are divided here: (i) some—Heyne, Conington rather doubtfully, T. E. Page, Klingner, Coleman, Lee—take *atque . . . atque* as meaning 'both . . . and', while (ii) others—Lachmann (see Ribbeck ad loc.), Wagner, *Quaest. Virg*. 35.23, Klotz in *TLL* s.v. 1055.33, Hofmann–Szantyr 516—understand *est* with *complexa* and take the first *atque* as connecting the two verbs. Wagner compares *G*. 3.256–7 'fricat arbore costas / atque hinc atque illinc umeros ad uulnera durat', but, as Mynors there remarks, 'the parallel with *E*. 5.23 is not absolute, because a phrase like *hinc atque illinc* may be almost stereotyped into a unity'. It should also be noticed that elsewhere in V. *complexus* is a participle attached to a verb: *A*. 2.513–14 'fuit . . . complexa', 673–4 'complexa . . . haerebat', 5.766

'complexi ... morantur', 6.785–6 'inuehitur ... complexa', 8.260 'corripit ... complexus'. The fact that *atque ... atque* meaning 'both ... and' is first securely attested in Silius, 1.93–4 'atque Hennaeae numina diuae / atque Acheronta uocat', suggests that Silius so interpreted *atque ... atque* here. Tibullus might be thought to offer a parallel, 2.5.73 'atque tubas atque arma ferunt'; but Conington is unconvinced, K. F. Smith non-committal, and the first *atque* is questioned by M. Platnauer, *Latin Elegiac Verse* (Cambridge, 1951), 79. While some uncertainty may remain, most readers will probably find themselves agreeing with T. E. Page—'the parallelism of *atque ... atque* is so marked that it seems impossible to take one as joining two verbs, the other two nouns'—and prefer to regard this as one of the youthful poet's bolder experiments.

23. astra: a poetic plural; the star under which a man was born influenced his destiny.

crudelia: referring to both *deos* and *astra*.

mater: Daphnis was the son of Hermes and an anonymous Nymph; hence the grief of the Nymphs. Rohde 125–6 suggests, with some probability, that V. was thinking of the death of Orpheus; cf. Antipater of Sidon (available to V. in Meleager's *Garland*) 10.5–6 G.–P. (= *AP* 7.8) *ὤλεο γάρ, σὲ δὲ πολλὰ κατωδύραντο θύγατρες | Μναμοσύνας, μάτηρ δ' ἔξοχα Καλλιόπα*, 'For you are dead, and you were much lamented by the daughters of Mnemosyne, above all by your mother Calliope'. Note, however, the emphasis on *nati* and *mater*; this Pietà-like scene can only be Virgilian.

A similar invention, and not unrelated (for Ovid seems to have been thinking of V.), may be observed in Ov. *Met.* 14.698–758: the story of Iphis and the stony-hearted Anaxarete. His passion unrequited, Iphis hangs himself, and his servants bring his body to his mother, 742–4 '... referunt ad limina *matris*; / accipit illa sinu *complexa*que frigida *nati* / membra sui'. Ovid's story is based on a poem by Hermesianax, of which only a prose version is extant in Antoninus Liberalis, *Metamorphoses* 39 (Hermesianax, fr. 4 Powell). The townspeople, pitying Arceophon (the lover's name in Hermesianax), mourned for his death, and on the third day his relatives (5 *οἱ προσήκοντες*) brought his body out into the open; his mother is not explicitly mentioned.

24–5. non ulli pastos ... egere / ... boues ad flumina: cf. 9.24. The ordinary routine of pastoral life is interrupted by so great a calamity.

24. non ulli: preferred by poets of the Augustan age to *haud ullus*; see W. D. Lebek, *Verba Prisca* (Hypomnemata, 25; Göttingen, 1970), 130–1. In Ovid *haud ullus* is found only in *Met.* 13.460, a suspect line.

25–6. nulla neque amnem / ... nec: an emphatic form of negation found in Cicero, e.g. *De diu.* 2.20 'nulla nec arte nec sapientia prouideri possunt', and Livy, and not therefore colloquial, as has been supposed; see Housman, *CR* 48 (1934), 137 = *Class. Papers*, iii. 1235. Cf. 4.55–6, *A.* 9. 428–9 'nihil iste nec (MP: *neque* R) ausus / nec potuit', 11. 228–9 'nil dona nec (R ω Serv.: *neque* MP) aurum / nec magnas ualuisse preces', 801–2 'nihil ipsa nec (MPR: *neque* ω Serv.) aurae / nec sonitus memor'. V. tends to avoid elision of any sort before a final spondee, and he may have avoided the elision of *neque* altogether before a final spondee beginning with *a*; Soubiran 116 identifies *neque amnem* as unique. This being the case, the variant in P assumes significance: NECQ· with a horizontal line drawn through the *C*, either by the scribe himself (Ribbeck) or by an ancient corrector (Geymonat); 'duplex in exemplo lectio' (Sabbadini).

26. libauit ... attigit: sympathetic animals, grieving for Daphnis, refuse to drink or eat; cf. Theocr. 4. 20 (quoted above, 3. 100 n.), [Moschus], *Epitaph. Bion.* 23–4 καὶ αἱ βόες αἱ ποτὶ ταύροις | πλαζόμεναι γοάοντι καὶ οὐκ ἐθέλοντι νέμεσθαι, 'and the heifers that wander by the bulls lament and refuse their pasture' (Lang). A pastoral motif therefore; but V. may also be alluding to an event of the recent past: the horses which Caesar had consecrated and released at the Rubicon refused to eat shortly before his death and wept copiously (Suet. *Iul.* 81.2).

graminis ... herbam: a curiously rare phrase, found before V. only in Livy's description of the Fetial rite (1.24.5) 'Fetialis ex arce graminis herbam puram attulit'. See *TLL.* s.v. *gramen* 2166.45, Wissowa 551 n. 7, Bömer on Ov. *Met.* 10.87 'graminis herbae'.

27. Daphni ... Poenos ... leones: balanced by *Daphnis ... Armenias ... tigris* (29). African lions and Armenian tigers here make their first appearance in Roman poetry. Cf. Theocr. 1.71–2 (Daphnis) *τῆνον μὰν θῶες, τῆνον λύκοι ὠρύσαντο, | τῆνον χὠκ δρυμοῖο λέων ἔκλαυσε θανόντα,* 'For him the jackals, for him the wolves howled, for him dead even the lion of the forest lamented'. Sicilian lions are imaginary; V. adds a touch of reality by making his African.

ingemuisse: here first with the accusative (*TLL* s.v. 1516.42).

28. montesque feri siluaeque: cf. Lucr. 5.201 'montes siluaeque ferarum'.

feri: in effect modifying both nouns; cf. Hor. *Serm.* 2.6.92 'feris . . . siluis'.

loquuntur: cf. 8.22 'argutumque nemus pinusque loquentis'. The hills and woods, which might be expected to utter their own grief, tell instead how even Punic lions groaned at the passing of Daphnis. Cf. Theocr. 7.74–5 (Tityrus will sing of Daphnis) *χὡς ὄρος ἀμφεπονεῖτο καὶ ὡς δρύες αὐτὸν ἐθρήνευν | Ἱμέρα αἵτε φύοντι παρ' ὄχθαισιν ποταμοῖο,* 'and how the hill was troubled and the oaks that grow upon the banks of the river Himeras mourned for him'. The general grief of nature for a dead singer became a convention of post-Theocritean pastoral; cf. [Moschus], *Epitaph. Bion.* 1–7, [Bion], *Epitaph. Adon.* 31–4.

29–30. Daphnis et Armenias curru subiungere tigris / instituit: 2.32–3 n.

29. Armenias . . . tigris: Kiessling's note on Hor. *Carm.* 3.3.13–14 'hac te merentem, Bacche pater, tuae / uexere tigres' is essentially right: 'für den Panther, der in griechischer Kunst und Poesie das Tier des Dionysos ist, tritt bei den augusteischen Dichtern der armenische, erst seit Alexander im Westen bekannte Tiger ein, als die größere und wildere Bestie (*qui vivus capi adhuc non potuit*, Varr. l.l. V 100)'. But Varro's comment (no doubt V.'s source; see 1.62 n.) should have been quoted in full: 'tigris qui est ut leo uarius, qui uiuus capi adhuc non potuit, uocabulum e lingua Armenia' (one of Varro's few correct explanations; see Ernout–Meillet, *Dict. étym.* s.v.). And of Armenian tigers it would be truer to say that they are Virgilian, introduced here, then recurring in Prop. 1.9.19, Lygdamus 6.15, Ov. *Am.* 2.14.35 'in Armeniis tigres . . . latebris', *Met.* 8.121, 15.86. For a comprehensive discussion of ancient tigers, both literary and real, see Pease on

A. 4.367: 'Tigers were perhaps first seen by Romans in 20 BC (Dio Cass. 54, 9, 8) and first exhibited at Rome in the consulship of Q. Tubero and Fabius Maximus in 11 BC (Plin. *N.H.* 8, 65)'.

curru: 'pro *currui*' (Serv.), a metrical necessity, but the form occurs also in prose; see Austin on *A.* 1.156.

30. thiasos inducere Bacchi: 'accipio pro simplici *ducere* choros Bacchantium, quales saepe in anaglyphis et gemmis' (Heyne), to which Wagner objected: 'sed *Bacchi thiasos inducere* est: thiasorum, iam inventorum, usum introducere aliquo'; and commentators have followed Wagner (so also *TLL* s.v. 1237.76), thus making *inducere* virtually equivalent to *instituit*; see above, 29–30 n. Heyne must be right: *thiasos inducere* is V.'s translation of *θιάσους ἄγειν*. Dionysiac companies (*θίασοι*) always had a female leader (in myth, the god himself; see Dodds on Eur. *Bacch.* 115 *Βρόμιος ὅστις ἄγῃ θιάσους*, 'whoever leads the companies is Bromius', A. Henrichs, *HSCP* 82 (1978), 135–6) and were always led during public appearances; cf. Eur. *Bacch.* 680–2 *ὁρῶ δὲ θιάσους τρεῖς γυναικείων χορῶν, | ὧν ἦρχ' ἑνὸς μὲν Αὐτονόη, τοῦ δευτέρου | μήτηρ Ἀγαύη σή, τρίτου δ' Ἰνὼ χοροῦ*, 'I saw three companies of female revellers, one of which Autonoe led, the second your mother Agave, and Ino the third', Theocr. 26.1–2 *Ἰνὼ καὐτονόα χἀ μαλοπάραυος Ἀγαύα | τρεῖς θιάσως ἐς ὄρος τρεῖς ἄγαγον αὐταὶ ἐοῖσαι*, 'Ino, Autonoe, and white-cheeked Agave, three themselves, led three companies to the mountain', Catull. 64.390–1 'saepe uagus Liber Parnasi uertice summo / Thyiadas effusis euantis crinibus egit', Prop. 2.3.18 'egit ut euantis dux Ariadna choros', *A.* 6.517–18 (Helen) 'illa chorum simulans euhantis orgia circum / ducebat Phrygias'. The right interpretation of *inducere* is to be found in *OLD* s.v. 3.

31. foliis lentas...mollibus hastas: cf. *A.* 7.390 'mollis... thyrsos', 396 'pampineas ... hastas'. The classic thyrsus, or mystic wand of Dionysus, consisted of a stalk of fennel with a bunch of leaves, ivy or vine, attached to the top; see Dodds on Eur. *Bacch.* 113. V. calls the wands *hastae*, presumably from their spear-like appearance; see *TLL* s.v. 2552.43. Although the thyrsus might be employed as a weapon (Eur. *Bacch.* 761–4, where see Dodds, Ov. *Met.* 3.712, 11.27–8), there is no suggestion of violence here; rather, an indication of the thyrsus in its vegetative aspect (see 1. 4 n.).

intexere: V. was evidently thinking of stem-wreathed thyrsi, which 'appear first in Hellenistic art and Roman poetry' (Dodds on Eur. *Bacch.* 1054–5).

32–4. uitis ut arboribus decori est...: 'locus Theocriti est' (Serv.). The figure is Theocritean (see Gow on [Theocr.] 9.31–5), and V. may be indebted here to [Theocr.] 8. 79–80 *τᾷ δρυῒ ταὶ βάλανοι κόσμος, τᾷ μαλίδι μᾶλα, | τᾷ βοῒ δ' ἁ μόσχος, τῷ βουκόλῳ αἱ βόες αὐταί*, 'Acorns are the pride of the oak, apples of the apple tree, her calf of the cow, and of the cowherd his cows alone'. As usual, however, V.'s imitation is more elegant than the original: note the varied repetition of *ut* ('cum summa elegantia', La Cerda), the chiastic arrangement of the comparisons, the succinct final clause ('tu ... tuis'). See 2.63–5 n.

34–5. postquam te fata tulerunt...: cf. Meleager 126 G.–P. (= *AP* 7.535): Pan will quit the countryside now that Daphnis is dead and live in a town.

34. tulerunt: 'pro *abstulerunt*' (Serv.); the simple form instead of the compound, as often in poetry; see 1.3 n.

35. Pales ... Apollo: the repetition 'ipsa ... ipse' seems to insist upon the parity of these two unequal deities, the one Roman, the other Greek. Pales and Apollo are also joined together in *G.* 3.1–2 'Te quoque, magna Pales, et te memorande canemus / pastor ab Amphryso'.

36–9. See above, p. xxx.

36. hordea: barley; cf. *G.* 1.210, 317, and see Abbe 30. The plural first occurs here and appears to be V.'s innovation, for it was ridiculed by one of his ancient critics: 'hordea qui dixit, superest ut tritica dicat'—a verse attributed by DServ. on *G.* 1.210 to the poetasters Bavius and Mevius (*E.* 3.90), but by Cledonius, *GLK* v. 43.2, to a Cornificius Gallus; hence Wissowa's conjecture (*RE* iv. 1629) that, like other such verses of the *obtrectatores Vergilii*, it had been circulating anonymously. *Hordea* was a metrical necessity, as Servius here recognizes, and the poetic plural can be defended by analogy; see Maas 494–5 = 540. The adjective *grandia* also helps; cf. the *rusticum uetus carmen* quoted by Macrobius, *Sat.* 5.20.18 'hiberno puluere, uerno luto, grandia farra, camille, metes', and see Löfstedt i. 49–50. Wheat and barley, *triticum* and *hordeum*, were naturally associated, as by the *obtrectator* quoted above; cf. Cato, *De agr*. 54.2 'paleas triticeas et hordeaceas', Varro, *RR* 2.5.17 'furfures triticeos et farinam hordeaceam', Livy 22.37.6 'trecenta milia modium tritici, ducenta hordei'; but V. did not, nor did any other poet, it seems, attempt *tritica* (explicitly condemned by Caesar, *De analogia*; see Gellius 19.8.3), devising instead the periphrasis 'triticeam in messem' (*G* 1.219), whence Ov. *Met*. 5.486 'triticeas messes'.

37. lolium: darnel, a noxious weed that grows everywhere in wheatfields; cf. *G*. 1.153–4 'interque nitentia culta / infelix lolium et steriles dominantur auenae'.

auenae: the wild oat, another such weed, 'frumenti uitium' (Pliny, *NH* 18.149).

38. molli uiola: 'the softness and delicacy of this sweet flower is opposed to the sharpness of the prickly plants mentioned presently after' (Martyn).

purpureo narcisso: 'duo homoeoteleuta' (DServ.); for V.'s practice see Norden 405–7. The grammarian Diomedes seems to have read *purpurea narcisso*, *GLK* i.453.34: 'primus modus soloecismi fit per immutationem generum nominis, cum dicimus 'atra silex' (*A*. 6.602) aut 'amarae corticis' (*E*. 6.62–3) aut 'purpurea narcisso', cum *ater silex* debeat dici et *amari corticis* et *purpureus narcissus*'. Ribbeck adopted *purpurea narcisso*, and Hofmann–Szantyr 11 list *narcissus* along with *cortex* and *silex* as feminine in V. Although a rarity in Greek, the feminine occurs in Theocr. 1.133 and Meleager 46.2 G.–P. (see 2.46–50 n.). The rhythmical pattern is Greek, and appears to have been introduced into Latin poetry by Catullus, 64.96 'Idalium frondosum', 291 'aerea cupressu'. See 2.24 n.

purpureo: 9.40 n.

narcisso: 2.48 n.

39. carduus: a thistle of any kind; see Abbe 186–7. Cf. *G*. 1.151–2 'segnisque horreret in aruis / carduus'.

spinis surgit paliurus acutis: cf. *A*. 3.46 'iaculis increuit acutis'. Leonidas of Tarentum 18.3 (G.–P. (= *AP* 7.656) describes a grave overgrown 'with sharp paliurus', ὑπ' ὀξείης παλιούρου; see E. A. Schmidt, *Hermes*, 96 (1968–9), 637–8.

paliurus: identified with Christ's Thorn; because of its large sharp thorns recommended by Columella 11.3.4 for a quickset hedge. Sargeaunt 93–5 has a vivid description; see above, p. xxix.

40. Cf. 9.19–20.

foliis: not leaves but flowers, like φύλλα in Theocr. 11.26 ὑακίνθινα φύλλα, 'hyacinth-flowers' (Gow), and [Theocr.] 9.4 ἐν φύλλοισι, 'among the flowers' (Gow); a sense indicated by 9.19 'florentibus herbis' and better suited to the context.

41. mandat fieri sibi talia Daphnis: as if Daphnis had made a will; cf. Ulpian, *Dig.* 17.1.12.17 'si, ut post mortem sibi monumentum fieret, quis mandauit', and see *TLL* s.v. *mando* 263.64, 266.2.

42. tumulum ... tumulo: cf. *A.* 6.380 'et statuent tumulum et tumulo sollemnia mittent', where, for the repeated noun, Norden compares *A.* 1.325, 3.607, 10.149, 202, 11.140. For graves placed near a spring see Gow–Page on Nicias 5 (= *AP* 9.315).

carmen: 'duos uersus carmen uocauit, nec mirum, cum etiam de uno carmen dixerit' (Serv., quoting *A.* 3.288). Very likely V. was thinking of epitaphs composed of a single elegiac couplet; see Clausen 149 n. 76.

43–4. Daphnis ego in siluis ...: an epitaph suggested by a 'couplet' in Theocr. 1.120–1 Δάφνις ἐγὼν ὅδε τῆνος ὁ τὰς βόας ὧδε νομεύων, | Δάφνις ὁ τὼς ταύρως καὶ πόρτιας ὧδε ποτίσδων, 'I am that Daphnis who herded here his cows, that Daphnis who watered here his bulls and calves'.

43. hinc usque ad sidera notus: cf. Hom *Od.* 9.20 καί μευ κλέος οὐρανὸν ἵκει, 'and my fame reaches the heavens', *A.* 1.379 'fama super aethera notus'. Judged incongruous by Heyne ('epicum est') but evidently anticipating the elevation of Menalcas' song (56–7).

44. formosi: an adjective often applied to animals; see 1.5 n., *TLL* s.v. 1111.25.

45–7: cf. Theocr. 1.7–8 ἅδιον, ὦ ποιμήν, τὸ τεὸν μέλος ἢ τὸ καταχές | τῆν' ἀπὸ τᾶς πέτρας καταλείβεται ὑψόθεν ὕδωρ, 'Sweeter, shepherd, is your song than the sound of that water where it pours down from the rock'.

45. diuine poeta: 'inspired poet'; so also of Gallus in 10.17. Cf. 3.37, 6.67.

46–7. in gramine ... / ... aquae ... riuo: cf. Lucr. 2.29–30 (= 5.1392–3) 'in gramine molli / propter aquae riuum' and see *TLL* s.v. *gramen* 2167.55. Cf. also 8.87.

47. aquae saliente ... riuo: cf. Varro, *RR* 1.13.3 'lacum ubi aqua saliat', Columella 1.2.4 (*colles*) 'riuos decurrentes in prata et hortos et salicta uillaeque aquas salientes demittant'.

48. nec calamis solum ...: since Menalcas compliments Mopsus on both his piping and his singing, intervals of piping must be imagined in his song (and that of Menalcas); T. E. Page on 8.21 *mea tibia*: 'the address to his pipe clearly indicates that after each stanza he plays on it'. See above, 9 n., 14 n., and cf. 3.21–2. Obviously the pastoral musician could not be blowing into his pipe and singing at the same time; see Gow on Theocr. 11.39.

aequiperas: a verb found in archaic poetry—in Plautus, Ennius, Pacuvius—and very occasionally in late poetry, but in classical poetry only here and twice in Ov. *Ex Pont.* (*TLL* s.v.). And here it was suggested, in all probability, by ἰσοφαρίζειν in Theocr. 7.27–31 Λυκίδα φίλε, φαντί τυ πάντες | ἦμεν συρικτὰν μέγ' ὑπείροχον ἔν τε νομεῦσιν | ἔν τ' ἀματήρεσσι ... | ... καίτοι κατ' ἐμὸν νόον ἰσοφαρίζειν | ἔλπομαι, 'Friend Lycidas, they all say that among the herdsmen and the reapers you are by far the best piper ... and yet, in my own mind, I think myself your equal'. ἰσοφαρίζω has a history not unlike that of *aequipero*: it is found in archaic poetry—in Homer, Hesiod, Simonides—here in Theocritus, twice in Apollonius of Rhodes, and several times in late poetry.

magistrum: see Introduction, n. 1.

49. fortunate puer: cf. 1.46, 51.

tu nunc eris alter ab illo: apparently a reference to death and pastoral succession; see 2.38 n. and cf. [Moschus], *Epitaph. Bion.* 95–7 ἀλλ' ἄντε διδάξαο σεῖο μαθητάς | κλαρονόμος μοίσας τᾶς Δωρίδος, ᾇ με γεραίρων | ἄλλοις μὲν τεὸν ὄλβον ἐμοὶ δ' ἀπέλειπες ἀοιδάν, 'but (I am) heir of the Doric Muse, which you taught your pupils. With this you honoured me; to others you left your wealth, to me your song'.

50. quocumque modo: a modest disclaimer, contrasting with the arrogance of Mopsus (9).

uicissim: 3.28 n.

51–2. For the epanalepsis see 6.21 n.

52. A carefully elaborated line, divided into two cola of eight syllables each, the first beginning, as the second ends, with Daphnis' name. The verse-technique is Greek (2.6 n.); ictus and accent coincide (1.70 n.).

nos quoque Daphnis: 2.26 n.

55. **Stimichon:** a name not attested before V.; used by Calp. Sic. 6.83, 7.9, 13. Ancient readers were puzzled by Stimichon; DServ. ad loc.: 'quidam per Stimichonem Maecenatem accipiunt; nonnulli Stimichonem patrem Theocriti dicunt'.

56–7. **Candidus ... / ... Daphnis:** a more impressive opening than that of Mopsus, 'Exstinctum ... Daphnin' (20). Cf. *A.* 8.82–3, 12.473–4.

56. **candidus:** radiant, like a star; cf. Plaut. *Rud.* 3–4 'splendens stella candida / signum', *G.* 1.217–18 'candidus auratis aperit cum cornibus annum / Taurus' (*TLL* s.v. 239.60).

insuetum miratur limen: Housman on Luc. 5.163 'insuetus dicitur qui nunc primum aliquid facit patiturue, ut Verg. Aen. VIII 92 et buc. V 56'.

limen Olympi: cf. Hom. *Il.* 1.591–3 (Hephaestus) ἀπὸ βηλοῦ θεσπεσίοιο | ... | κάππεσον ἐν Λήμνῳ, 'from the threshold of the gods ... I fell upon Lemnos', Accius, *Philocteta* 525–6 R.[3] 'Lemnia praesto / litora', 529–31 'Volcania iam templa sub ipsis / collibus, in quos delatus locos / dicitur alto ab limine caeli'.

Olympi: Varro, *LL* 7.20 'caelum dicunt Graeci Olympum'; for Olympus in this sense see Fordyce on *A.* 8.280, Hopkinson on Callim. *Hymn* 6.58.

57–8. **sub pedibusque ... | ... uoluptas:** cf. Lucr. 3.27–8 'sub pedibus quaecumque infra per inane geruntur. / his ibi me rebus quaedam diuina uoluptas / percipit atque horror'. These lines are preceded by Lucretius' rendering (18–22) of Homer's description of Olympus (*Od.* 6.42–6).

58. **alacris:** modifies *uoluptas*; cf. Plaut. *Amph.* 245 'impetu alacri', Cic. *Ad Att.* 1.16.7 'alacris ... improbitas'. Only here in the *E.*, not in the *G.*, in the *A.* applied (as usually) to persons.

siluas et cetera rura: the woods and the rest of the countryside; see above, p. xxvi.

59. **Panaque ...:** cf. *G.* 2.494 'Panaque Siluanumque senem Nymphasque sorores'.

Dryadasque puellas: *G.* 1.11; cf. 10.9–10 'puellae / Naides'.

60–1. **nec lupus insidias pecori ...:** inversions of the natural order, suggesting the Golden Age; cf. 4.22.

nec retia ceruis / ulla dolum meditantur: cf. Plaut. *Pseud.* 941 'meditati sunt mihi doli', Hor. *Epod.* 2.33–4 'tendit retia, / turdis edacibus dolos'. La Cerda compares Pind. *Nem.* 3.51 κτείνοντ' ἐλάφους ἄνευ κυνῶν δολίων θ' ἑρκέων, 'killing stags without hounds or deceptive nets'.

61. ulla: so placed for emphasis; see Norden 399.

bonus: often applied to deities; cf. below, 65, and see *TLL* s.v. 2086.15.

otia: 1.6 n.

62–4. ipsi . . . | . . . ipsae . . . | ipsa: 1. 38–9 n.

62–3. laetitia uoces ad sidera iactant / intonsi montes: cf. Lucr. 2.327–8 'clamoreque montes / icti reiectant uoces ad sidera mundi' and see 6.84 n.

63. intonsi: of men and gods, unshorn or unshaven, from Accius onwards; here only of mountains (*TLL* s.v. *mons* 1433.29)—a bold metaphor which serves to personify the mountains.

64. arbusta: 1.39 n.

deus, deus ille: cf. Lucr. 5. 8 'deus ille fuit, deus' and see *E.* 1.6 n.

65. sis bonus o felixque tuis: cf. *A.* 1. 330 (Aeneas to the disguised Venus) 'sis felix nostrumque leues, quaecumque, laborem', 12.646–7.

en: *en* with the accusative is first attested in Cic. *Verr.* 2.1.93 'en cui tuos libros committas, en memoriam . . ., en metum'; elsewhere Cicero uses the nominative. In his later poetry V. prefers the nominative (*A.* 1.461, 4.597, 5.639–40 'en quattuor arae / Neptuno', 7.452), but once uses the accusative (*A.* 12.359). *Ecce* never takes the accusative in classical Latin; here the accusatives *duas . . . duas* are determined by the case of *aras*. See A. Köhler, *ALL* 5 (1888), 23–4, Kühner–Stegmann i.273–4.

65–6. quattuor aras . . .: cf. Theocr. 26.5–6 δυοκαίδεκα βωμώς, | τὼς τρεῖς τᾷ Σεμέλᾳ, τὼς ἐννέα τῷ Διονύσῳ, 'twelve altars, three for Semele, nine for Dionysus'.

66. altaria: superstructures for burnt offerings ('alii *altaria* eminentia ararum', DServ.), here in apposition to *aras*; see *OLD* s.v. 1. As a rural deity Daphnis will receive no burnt offerings, but libations of fresh milk and olive oil—fresh milk in the spring and olive oil in the fall (Wagner).

67–8. Cf. Theocr. 5.53–4 *στασῶ δὲ κρατῆρα μέγαν λευκοῖο γάλακτος | ταῖς Νύμφαις, στασῶ δὲ καὶ ἁδέος ἄλλον ἐλαίω*, 'I will set out a great bowl of white milk for the Nymphs, and set out another too of sweet olive oil', 58–9 *στασῶ δ' ὀκτὼ μὲν γαυλὼς τῷ Πανὶ γάλακτος, | ὀκτὼ δὲ σκαφίδας μέλιτος πλέα κηρί' ἐχοίσας*, 'I will set out eight pails of milk for Pan, and eight bowls with combs full of honey'.

67. bina: There seems to be no difference between 'pocula bina' and 'crateras . . . duo', as Theocritus (preceding note) indicates. Apparently V. was thinking of a pair of cups; see 3.44 n. See also *TLL* s.v. *bini* 1997.62.

quotannis: 1. 42 n.

68. crateras . . . statuam: V. uses only the Greek form of the noun, as Norden observes on *A.* 6.224 ff., but he is wrong to argue that V. avoids the feminine *cratera* merely for metrical convenience: every poet avoids *cratera*; see Clausen, *CQ*, NS 13 (1963), 85–7. *Pocula* is joined to *statuam* by a slight zeugma.

duo: scanned as an iamb only here; elsewhere in V., and elsewhere in classical poetry, as a pyrrhic (22 times, elided in *A.* 10.124). The anomaly was noticed by Wagner, *Quaest. Virg.* 12.14, who therefore preferred *duos*, a form never used by V. See L. Mueller, *De re metrica*² (Leipzig, 1894), 390, and Housman on Manil. 1.792; that is, *-o* is lengthened here before *st-* as e.g. *-a* is lengthened in Catull. 63.53 'gelida stabula'.

69. in primis (imprimis): rare in poetry after Lucretius, yet not quite so 'unpoetic' as Axelson 94–5 would have it be: it occurs in *G.* 1.338 and *A.* 1.303.

hilarans: a verb neither very old nor very common; see *TLL* s.v.

Baccho: 6.15 n.

71. An intricate line, apparently designed to set off the last two words and call attention to their novelty ('nouum'?): 'Ariusia nectar'.

uina: in the nominative and accusative V. uses only the plural, in the dative and ablative only the singular; that V.'s usage is not idiosyncratic was demonstrated by Löfstedt i. 48.

calathis: here drinking-cups in the shape of an inverted bell; see K. D. White, *Farm Equipment of the Roman World* (Cambridge, 1975), 72; also 2.46 n.

Ariusia: the name of a rough and harbourless region on the north-western coast of Chios where the best Greek wine was produced: Strabo 14.135 *ἡ Ἀριουσία χώρα τραχεῖα καὶ ἀλίμενος ... οἶνον ἄριστον φέρουσα τῶν Ἑλληνικῶν*; also Athenaeus 1.32F *χαριέστατος δ' ἐστὶν ὁ Χῖος καὶ τοῦ Χίου ὁ καλούμενος Ἀριούσιος*, 'the pleasantest is the Chian, and of the Chian that called Ariusian'.

nectar: of wine again in *G*. 4.384. Suggested here by Theocr. 7.153 (where see Gow), but cf. Callim. fr. 399 Pf. (=68 G.–P. = *AP* 13.9) *ἔρχεται πολὺς μὲν Αἰγαῖον διατμήξας ἀπ' οἰνηρῆς Χίου | ἀμφορεύς, πολὺς δὲ Λεσβίης ἄωτος νέκταρ οἰνάνθης ἄγων*, 'Comes many a two-handled vessel cleaving the Aegean from Chios rich in wine, and many an earless one bringing the nectar of the Lesbian vine'.

72–3. Cf. *G*. 1.350 (the rustic worshipper of Ceres) 'det motus incompositos et carmina dicat'.

72. Cf. Theocr. 7.71–2 *αὐλησεῦντι δέ μοι δύο ποιμένες, εἷς μὲν Ἀχαρνεύς, | εἷς δὲ Λυκωπίτας*, 'Two shepherds shall pipe to me, one from Acharnae, and one from Lycope'. There follows, in ll. 74–5, a reference to Daphnis dying of love; see above, 28 n.

Aegon: cf. 3.2. But why should Aegon be called Lyctian, that is, Cretan? Cf. *A*. 3.401 'Lyctius Idomeneus'. The adjective is not attested in the singular before Callimachus, *Epigr*. 37.1 Pf. (=17 G.–P. = *AP* 13.7) *Ὁ Λύκτιος Μενίτας*, and is attached to a proper name only in Callimachus and Nonnus, *Dionys*. 32.187 *Λύκτιος Ἀνθεύς*. Antheus became one of Aeneas' followers; see Clausen 135 n. 1.

73. Lines of four words are rare in V., only one elsewhere in the *E*., 8.34, and only six in the *G*., 1.27, 470, 502, 3.550, 4.111, 336. See S. E. Bassett, 'Versus Tetracolos', *CP* 14 (1919), 216–33. Here perhaps the rhythm is intended to suggest a certain rustic clumsiness (Holford-Strevens).

saltantis Satyros imitabitur: 2.31 n. La Cerda compares Nonnus, *Dionys*. 15.71 *μιμηλὴν Σατύροισι συνεσκίρτησε χορείην*, '(another) leapt with the Satyrs in a mimic dance'.

Alphesiboeus: here first and in 8.1, 5, 62, from *ἀλφεσίβοιος*, an epithet as old as Homer, *Il*. 18.593 of virgins who gain cattle for their parents as wedding-gifts; in later Greek simply 'cattle-fattening', 'fruitful', an appropriate name for a herdsman. In the feminine, however, the name occurs in several Greek legends (*RE*

i. 1636) and Alphesiboea, daughter of Bias and Pero, appears in Theocr. 3.45. Cf. 1.6 n. (*Meliboee*).

75. reddemus: as commonly of discharging a vow, e.g. Cic. *De leg.* 2.22 'caute uota reddunto', Catull. 36.16 'acceptum face redditumque uotum', Hor. *Carm.* 2.7.17 'ergo obligatam redde Ioui dapem'.

lustrabimus agros: 3.76–7 n.

76–8. Cf. *A.* 1.607–9.

77. pascentur apes: 1.54 n. (*florem depasta salicti*).

rore cicadae: the belief that cicadas lived on dew is first attested in [Hes.] *Scut.* 393; see Gow on Theocr. 4.16 for later references and the truth of the matter.

79. Baccho Cererique: the two deities who preside over the countryside, *G.* 1.7 'Liber et alma Ceres'. *Bacchus*, here necessitated by metre, is novel in this conjunction, *Liber* traditional; cf. Lucr. 5.14 'Ceres . . . Liberque', Varro, *RR* 1.1.5 'Cererem et Liberum', and see *TLL* Onomast. s.v. *Ceres* 341.22. Cf. also *G.* 1.343 'Cererem' (the goddess), 344 'Baccho' (the metonym). See 6.15 n.

80. damnabis tu quoque uotis: anyone who had made a vow to a god was condemned to discharge it if the request was granted, and remained 'guilty of his vow' (*A.* 5.237 'uoti reus') until he had done so. The usual construction is with the genitive singular (hence *uoti* in R) and the perfect passive participle, *uoti damnatus* (*TLL* s.v. *damno* 20.31).

81. quae tibi, quae . . .: an impassioned question; for the position of *tibi* see Wackernagel, *IF* 1 (1892), 409–11 = *Kleine Schriften* (1953), i. 77–9. Cf. *A.* 5.670 'quo nunc, quo . . .', 9.252–3 'quae uobis, quae . . .', Soph. *Phil.* 1348 τί με, τί . . . .

82. nam neque me tantum . . .: negative comparisons, as DServius observes, quoting *A.* 2.496, where see Austin, and 5.146. Cf. also Callim. *Hec.* fr. 69.11–14 Hollis (with his note on l. 11), [Moschus], *Epitaph. Bion.* 37–44, Catull. 68.119–28.

uenientis: cf. *A.* 10.99 'uenturos . . . uentos'.

sibilus Austri: the whistling of the South Wind, which is frequently stormy and violent; cf. Enn. *Ann.* 432–4 Skutsch 'concurrunt ueluti uenti, quom spiritus Austri / imbricitor Aquiloque suo cum flamine contra / indu mari magno fluctus extollere certant', Cic. *Arat.* 101 'uiribus erumpit qua summis spiritus Austri', *G.* 3.278–9 'unde nigerrimus Auster / nascitur et pluuio contristat frigore caelum', *A.* 2.110–11 'saepe illos aspera ponti / interclusit hiems et terruit Auster euntis'. See also 2.58 n. and Introduction, p. 152.

83–4. Cf. 9.43, Quintilian 9.4.7 'ceterum quanto uehementior fluminum cursus est prono alueo ac nullas moras obiciente quam inter obstantia saxa fractis aquis ac reluctantibus, tanto quae conexa est et totis uiribus fluit fragosa atque interrupta melior oratio'.

84. saxosus: 1.5 n.

85. ante: 'bene anticipat et offert munus' (Serv.). V. was thinking of Theocr. 6.42–4; see Introduction, p. 154. Here Mopsus' mention of gifts (81 'dona') prepares for the exchange.

cicuta: 2.36 n.

86–7. The Second and Third *Eclogues* are identified by partial quotations of the first lines. Literary works were sometimes identified in antiquity by the opening words of the text: thus 'Vtinam ne in nemore', Cic. *De fin.* 1.5; 'O Tite, si quid', Cic. *Ad Att.* 16.3.1; 'Aeneadum genetrix', Ov. *Trist.* 2.261; 'Cynthia', Prop. 2.24.2, Mart. 14.189.1; 'Arma uirumque', Pers. 1.96, Mart. 8.55.19, 14.185.2. See E. J. Kenney, 'That Incomparable Poem the "Ille ego"', *CR*, NS 20 (1970), 290.

88. pedum: a word found only here in literature; cf. Festus, p. 292 L. '*pedum* est quidem baculum incuruum, quo pastores utuntur ad comprehendendas oues, aut capras, a pedibus'.

89. Antigenes: 3.66 n.

et: introducing a parenthesis; cf. e.g. Cic. *Pro Flacc.* 71 '(et magis erat tuum)', Lucr. 4.1174 '(et scimus facere)', Sall. *Iug.* 52.3 '(et iam die uesper erat)', A. 11.364–5 '(et esse / nil moror)'. See *TLL* s.v. 891.80.

90. formosum paribus nodis atque aere: 'et ab arte et a natura laudauit: *paribus nodis* id est natura formosum, *atque aere* hoc est pulchrum aere artificium' (DServ.). Cf. [Theocr.] 9.23–4 *Δάφνιδι μὲν κορύναν, τάν μοι πατρὸς ἔτραφεν ἀγρός, | αὐτοφυῆ, τὰν οὐδ' ἂν ἴσως μωμάσατο τέκτων*, 'to Daphnis a staff grown on my father's farm, self-shaped, yet even a craftsman, maybe, would have found no fault in it' (Gow), Wordsworth, *Michael* 190–3 'Then Michael from a winter coppice cut / With his own hand a sapling, which he hooped / With iron, making it throughout in all / Due requisites a perfect Shepherd's Staff'.

ECLOGUE 6

Introduction

Virgil's Muse, his Thalea, first condescended to play with Syracusan verse and dwelt unembarrassed in the woods (1–2). Then, abruptly, Virgil tells how he came to write pastoral poetry.

cum canerem reges et proelia, Cynthius aurem
uellit et admonuit: 'pastorem, Tityre, pinguis
pascere oportet ouis, deductum dicere carmen.'
nunc ego (namque super tibi erunt qui dicere laudes,
Vare, tuas cupiant et tristia condere bella)
agrestem tenui meditabor harundine Musam. (3–8)

With the decline of letters in late antiquity, when Greek poetry, and especially Hellenistic poetry, ceased to be read in the West, Apollo's epiphany as literary critic could only be interpreted in an autobiographical sense[1]—that is, it could no longer be recognized for what it was, a literary allusion, Virgil's pastoral rendering of Callimachus' rejection of epic.

καὶ γὰρ ὅτε πρώτιστον ἐμοῖς ἐπὶ δέλτον ἔθηκα
γούνασιν, Ἀπόλλων εἶπεν ὅ μοι Λύκιος·
'.......]... ἀοιδέ, τὸ μὲν θύος ὅττι πάχιστον
θρέψαι, τὴ]ν Μοῦσαν δ' ὠγαθὲ λεπταλέην.'

(*Aetia* fr. 1.21–4 Pf.)

For, when I first placed a writing-tablet on my knees, Lycian Apollo said to me: 'Poet, feed your burnt offering to be as fat as possible, but your Muse, my friend, keep her slender'.

[1] Thus *Vita Donati* 19 'mox cum res Romanas incohasset, offensus materia ad Bucolica transiit'; DServ. on l. 5: 'quidam uolunt hoc significasse Vergilium, se quidem altiorem de bellis et regibus ante bucolicum carmen elegisse materiam, sed considerata aetatis et ingenii qualitate mutasse consilium et arripuisse opus mollius, quatenus uires suas leuiora praeludendo ad altiora narranda praepararet'; Serv. on l. 3: 'et significat aut Aeneidem aut gesta regum Albanorum, quae coepta omisit nominum asperitate deterritus'. Since 1927, when Callim. *Aet.* fr. 1 Pf. was published by A. S. Hunt (P. Oxy. XVII 2079), there has been no excuse for this misinterpretation; but venerable error dies hard, e.g. B. Otis, *Virgil: A Study in Civilized Poetry* (Oxford, 1963), 33: 'Virgil had originally, it seems, thought of writing a Roman epic (*res Romanae*) but gave up the attempt and turned instead to the *Bucolics*'; similarly R. D. Williams, *Virgil: The Eclogues and Georgics* (New York, 1979), on l. 3.

'Agrestem tenui meditabor harundine Musam': no attentive reader will fail to hear the echo of the First *Eclogue*, 2 'siluestrem tenui Musam meditaris auena'. There *tenui* may have seemed ornamental, contributing rather to the shape and balance of the line than to its sense.[2] Here, however, *tenui* is defined by its context: it is the equivalent of λεπταλέην[3] and signifies a concept of poetry, poetry as conceived by Callimachus.[4] His pastoral poetry, Virgil implies, though ostensibly Theocritean, is essentially Callimachean.

After paying an artful compliment to Varus, to whom the poem is addressed, and, in so doing, declining to celebrate his military exploits in an epic (6–12), Virgil introduces the figure of Silenus. Early in the morning two young shepherds, Chromis and Mnasyllos, discover Silenus asleep in a cave, drunk as usual. The garlands have fallen from his head and lie on the ground near by, yet even in his drunken stupour he clings to his beloved tankard. The shepherds bind him with his garlands, hoping thus to extort a song, of which he had often cheated them in the past. Aegle, loveliest of the Naiads, joins them and boldly (for Silenus is now awake and watching) daubs his face with mulberry juice. Laughing, Silenus promises a song to the shepherds, but to the Naiad—something else.

Unknown to the poetry of Theocritus, Silenus is an intractable figure whom Virgil hardly confines within a pastoral frame of reference (13–30, 85–6). Traditional features remain:[5] he is the ancient drunkard still, the lover of Nymphs and music, and possessed of secret wisdom, which he must be constrained, here playfully constrained, to reveal. Yet Virgil's Silenus—and the Silenus of the Sixth *Eclogue* can only be Virgil's[6]—is wondrously changed: the forest seer has undergone a Callimachean transformation, he has become a literary critic.

[2] No doubt Virgil was thinking of Callimachus when he wrote 'tenui . . . auena', but could he have expected his reader, on a first reading of the Book of *Eclogues* and as yet ignorant of the Sixth *Eclogue*, to be aware of the literary implications of *tenui*?

[3] A metrical convenience; the usual form is λεπτός. See Pfeiffer ad loc.

[4] For a detailed description of Callimachus' style see A. W. Bulloch, *CHCL* i. 549–70.

[5] See *OCD* s.v. 'Satyrs and Sileni', T. H. Carpenter, *Dionysian Imagery in Archaic Greek Art* (Oxford, 1986), 76–7.

[6] F. Skutsch (1901), 48–9 thinks Virgil found him in the poetry of Gallus.

The song of Silenus originates in Apollonius' song of Orpheus,[7] as Virgil wishes his reader to notice, for he describes Silenus as singing even more enchantingly than Orpheus.[8]

῎Ηειδεν δ' ὡς γαῖα καὶ οὐρανὸς ἠδὲ θάλασσα,
τὸ πρὶν ἐπ' ἀλλήλοισι μιῇ συναρηρότα μορφῇ,
νείκεος ἐξ ὀλοοῖο διέκριθεν ἀμφὶς ἕκαστα·
ἠδ' ὡς ἔμπεδον αἰὲν ἐν αἰθέρι τέκμαρ ἔχουσιν
ἄστρα σεληναίης τε καὶ ἠελίοιο κέλευθοι·
οὔρεά θ' ὡς ἀνέτειλε, καὶ ὡς ποταμοὶ κελάδοντες
αὐτῇσιν Νύμφῃσι καὶ ἑρπετὰ πάντ' ἐγένοντο.
ἤειδεν δ' ὡς πρῶτον 'Οφίων Εὐρυνόμη τε
'Ωκεανὶς νιφόεντος ἔχον κράτος Οὐλύμποιο.

(Ap. Rhod. 1. 496–504)

He sang how earth and sky and sea, joined together of old in one form, were separated from each other by a deadly quarrel; and how for ever in the heavens the stars and the paths of the moon and the sun keep their steadfast place; how the mountains rose up, and how sounding rivers with their Nymphs and all the animals came to be. He sang how first Ophion and Eurynome, the daughter of Ocean, held sway over snowy Olympus.

Silenus, like Orpheus, begins with the creation of the world and the emergence of living things, primeval figures, Pyrrha, Saturn, Prometheus; and his song is similarly if less regularly articulated: 31 'Namque canebat uti',[9] 41–2 'hinc . . . refert', 43 'his adiungit', 61 'tum canit', 62 'tum . . . circumdat', 64 'tum canit', 74–8 'Quid loquar aut . . . aut ut', 82–4 'omnia . . . ille canit'. But while Orpheus 'stayed his lyre and his immortal voice' (Ap. Rhod. 1. 512) with Zeus still a child, still thinking childish thoughts, in the Dictaean cave, Silenus continues on, singing as if, in the plenitude of song, his song could have no ending, singing 'until the Evening Star advanced in the unwilling sky' (86). He sings distractedly, it seems, touching on various themes; in fact, his song is a dense and harmonious composition, a neoteric *ars poetica* artfully concealed,

[7] The song of Silenus, like the song of Orpheus, like the song of Demodocus (*Od.* 8. 500–20; see below, n. 9), is reported. Cf. *A.* 1. 742–6 (the song of Iopas), 8. 287–302 (the song of the men of Pallanteum, of which, however, only the first part is reported).

[8] Notice how Orpheus' name immediately precedes Silenus' song, 30–1 'Orphea. / Namque canebat uti'.

[9] So begins the song of Orpheus, *῎Ηειδεν δ' ὡς*. Cf. *Od.* 8. 514 *ἤειδεν δ' ὡς*.

with but a single subject: poetry, poetry as conceived by Callimachus (and poets after Callimachus) and now embodied in Gallus.[10] That Pasiphae's perverse passion, her ἐρωτικὸν πάθημα (45–60), and the literary initiation of Gallus (64–73) occupy so much of the song is indicative of contemporary taste, Virgil's and that of his friends.

Although the Sixth *Eclogue* is addressed to Varus, the central figure is evidently Gallus, and Virgil's readers have therefore sensed a certain inconsistency or failure of design in it.[11] The fault lies not with Virgil but with his readers, unschooled in Callimachean poetics. In refusing to write an epic a poet undertook to incorporate his refusal in a poem that should exemplify a contrasting idea of poetry.[12] Apollo's magisterial rebuke to the poet and the poet's initiation on Helicon—these scenes are related and complementary, the one being implicitly, the other explicitly, programmatic, and are found in the second and first preface to the *Aetia*.

One of the Muses conducts Gallus to the top of Helicon, where Linus, the divinely inspired shepherd, gives him the pipes of Hesiod, with these words:

> hos tibi dant calamos (en accipe) Musae,
> Ascraeo quos ante seni, quibus ille solebat
> cantando rigidas deducere montibus ornos.
> his tibi Grynei nemoris dicatur origo,
> ne quis sit lucus quo se plus iactet Apollo. (69–73)

Apollo will be pleased by Gallus' epyllion about the Grynean Grove; and the reader may recall how displeased Apollo was by an epic about kings and battles. Now, with a dismissive 'Quid loquar?' (74), Virgil hurries the song of Silenus (and his own) to a pastoral close. The deliberate abruptness of the question underlines the reference to Gallus' epyllion, and Virgil speaks again, as

[10] The hypothesis of F. Skutsch (1901), 38–48, that the song of Silenus is an allusive catalogue of the epyllia of Gallus, is perceptive but extreme. In *G.* 4. 345–7, Clymene is described as singing of the love-affairs of the gods and, like Silenus and Orpheus, beginning from Chaos, 347 'aque Chao densos diuum numerabat amores'. As Hollis observes on Ov. *Ars* 1. 283–342, the song of Silenus owes something to the catalogue-poetry of the Hellenistic period.

[11] Thus Servius, concerned that Varus should not simply disappear from the poem, identifies Chromis and Mnasyllos as Virgil and Varus, with Silenus representing their teacher Siro, and Aegle Epicurean *uoluptas*.

[12] See Clausen, 'Callimachus and Latin Poetry', *GRBS* 5 (1964), 189 = K. Quinn (ed.), *Approaches to Catullus* (Cambridge, 1972), 277.

he did at the beginning, in his own person, 'cum canerem reges et proelia'.

Bibliography

F. Skutsch, *Aus Vergils Frühzeit* (Leipzig, 1901), 28–49.
—— *Gallus und Vergil* (Leipzig, 1906), 128–55.
G. Jachmann, 'Vergils sechste Ekloge', *Hermes*, 58 (1923), 288–304.
O. Skutsch, 'Zu Vergils Eklogen', *RhM* 99 (1956), 193–5.
Z. Stewart, 'The Song of Silenus', *HSCP* 64 (1959), 179–205.
J. P. Elder, '*Non iniussa cano*: Virgil's Sixth Eclogue', *HSCP* 65 (1961), 109–25.
G. Williams, *Tradition and Originality in Roman Poetry* (Oxford, 1968), 243–9.
W. Spoerri, 'Zur Kosmogonie in Vergils 6. Ekloge', *MH* 27 (1970), 144–63, 265–72.
Ross (1975), 18–38.
Clausen, 'Theocritus and Virgil', *CHCL* ii. 317–19.
P. E. Knox, 'In Pursuit of Daphne', *TAPA* 120 (1990), 183–202.

1–2. Prima . . . / nostra . . . Thalea: V. has recreated Theocritean pastoral in Latin; an assertion of primacy (within the limits of pastoral decorum) tempered by an awareness that the appropriated genre is relatively minor. Greek poets from Homer onward boast of the novelty of their songs, but this possessive sense of the tradition is peculiarly Roman; for references see Nisbet–Hubbard on Hor. *Carm.* 1. 26. 10. V. was not the first Roman poet to make a claim of this sort; Ennius had done so, to judge from Lucr. 1. 117–18 'Ennius ut noster cecinit qui primus amoeno / detulit ex Helicone perenni fronde coronam'. (Perhaps V. had 'Ennius . . . noster . . . primus' in mind when he wrote 'Prima . . . nostra . . . Thalea'.) And Lucretius makes a similar claim—intense and personal, yet remotely Callimachean (see E. J. Kenney, 'Doctus Lucretius', *Mnemosyne*, 4th ser., 23 (1970), 369–70)—for his own poetry, 1. 926–30 (=4. 1–5) 'auia Pieridum peragro loca nullius ante / trita solo. iuuat integros accedere fontis / atque haurire, iuuatque nouos decerpere flores / insignemque meo capiti petere inde coronam / unde prius nulli uelarint tempora Musae'. Cf. *G.* 3.

10–11 'primus ego in patriam mecum . . . / Aonio rediens deducam uertice Musas', Hor. *Carm.* 3. 30. 13–14 'princeps Aeolium carmen ad Italos / deduxisse modos', *Epist.* 1. 19. 23–4 'Parios ego primus iambos / ostendi Latio', Prop. 3. 1. 3–4 'primus ego ingredior puro de fonte sacerdos / Itala per Graios orgia ferre choros'.

1. Syracosio . . . uersu: '*Syracosio* autem Graece ait, nam Latine *Syracusanus* dicimus' (Serv.). The Greek form, first attested here, is rare and almost exclusively poetic; prose writers and Plautus (*Men.* 1069, 1109) use *Syracusanus*. Cf. 10. 50 'Chalcidico . . . uersu'.

ludere: cf. 1. 10; here opposed to *canerem* (3), which connotes epic.

2. erubuit: here first with an infinitive (*TLL* s.v. 822. 66).

siluas habitare: 2. 29 n.

Thalea: her first appearance in Latin poetry. The names of the Muses, a bevy of goddesses only vaguely realized in Homer (but see Heubeck on *Od.* 24. 60), appear to have been invented by Hesiod; see West on *Theog.* 75–9. Propertius to the contrary, 3. 3. 33 'diuersaeque nouem sortitae iura Puellae', it was not until late antiquity that a definite province of poetry was allotted to each Muse, to Thalea comedy and light verse; see Roscher's *Lexicon* s.v. *Thaleia, Thalia* 450. According to Rhianus, it makes no difference which Muse you invoke, because 'when you speak the name of one, all listen', fr. 19 Powell *πᾶσαι δ' εἰσαΐουσι, μιῆς ὅτε τ' οὔνομα λέξῃς*. See also Nisbet–Hubbard on Hor. *Carm.* 1. 24. 3.

3–5. See Introduction, p. 174.

3. Cynthius: 'Apollo a Cyntho monte Deli, in quo natus est' (Serv.). Greek poets apply the adjective to the hill sacred to the god (*The Homeric Hymn to Apollo* 17 *Κύνθιον ὄχθον*, Eur. *IT* 1098, Ar. *Clouds* 596–7) but not to the god himself, with a singular exception: Callimachus, *Hymn* 4. 9–10 *Ἀπόλλων* | *Κύνθιος*, where see Mineur, *Aet.* fr. 67. 5–6 Pf. *ἄναξ* . . . | *Κύνθιε*, and 114. 8 *Κύνθιε* in the same metrical position as *Cynthius* here; cf. C. Weber, *HSCP* 91 (1987), 268: 'the aesthetic implicit in *cum canerem reges et proelia* is undermined by the *mollities* of the following diaeresis'. (On the infrequency of the bucolic diaeresis in the *E.* as compared with

Theocritus—there are at least five times as many bucolic diaereses in *Idylls* 1 and 5 (302 lines) as in all the *E.* (828 lines)—see ibid. n. 39.) *Cynthius* as an epithet must be Callimachus' invention, that is, a literary epithet unhallowed by cult or tradition, and as such perceived by V., who introduces it into Latin poetry and uses it only once again, in an intensely Callimachean context, *G.* 3. 36. See Clausen, *AJP* 97 (1976), 245–7 and 98 (1977), 362.

3–4. **aurem / uellit:** 'tweaked my ear', as a reminder, a proverbial expression, e.g. *Copa* 38 'Mors aurem uellens "uiuite," ait, "uenio"', Sen. *Epist.* 94. 55 'sit ergo aliquis custos et aurem subinde peruellat'. See Otto, no. 212. 4. Hence a plaintiff would call a bystander to witness a summons (*antestari*) by tweaking his ear; cf. Pliny, *NH* 11. 251 'est in aure ima memoriae locus, quem tangentes antestamur', Plaut. *Persa* 745–8, Hor. *Serm.* 1. 9. 76–7.

4. **pinguis:** proleptic, 'ut pinguescant' (DServ.); cf. Hor. *Serm.* 2. 6. 14–15 'pingue pecus domino facias et cetera praeter / ingenium'.

5. **oportet:** a very prosaic word, here only in V., and generally avoided in the higher forms of poetry; see Axelson 13–14, Ross (1969), 69–70. But why should V. have used such a word here, in his elegant imitation of Callimachus? Was he thinking of Catullus, who had used it in his imitation of Callim. *Epigr.* 25 Pf. (= 11 G.–P. = *AP* 5. 6)? See Fordyce on Catull. 70, Ross (1969), 152–3. 'Pascere oportet ouis' is metrically equivalent, even in the detail of an elision, to Catull. 70. 4 'scribere oportet aqua', that is, to the second half of a pentameter; cf. Prop. 3. 7. 72 'condar oportet iners', 4. 1. 70 'sudet oportet equus', Lygdamus 1. 14 'mittere oportet opus', Ov. *Ex Pont.* 3. 1. 6 'condar oportet humo', 144 'tu quoque oportet eas'. Placed as it is here, in what may be described as its elegiac position, *oportet* is unique; elsewhere, *Ciris* 262 'oportet amari' and Juv. 14. 207 'oportet habere' excepted, *oportet* always stands at the end of the hexameter: Lucr. 1. 778, Catull. 90. 3, Hor. *Serm.* 1. 6. 17, 2. 6. 52, *Epist.* 1. 2. 49, 10. 12, *Dirae* 36, Prop. 2. 4. 1, 8. 25, Ov. *Her.* 1. 83, *Ars* 1. 699, *Rem.* 23, *Trist.* 1. 5. 51, *Ex Pont.* 3. 1. 35, Pers. 5. 155, *Priapea* 38. 1.

deductum . . . carmen: 'tenue; translatio a lana, quae deducitur in tenuitatem' (Serv.). The metaphor is explicit in Hor. *Epist.* 2. 1. 225 'tenui deducta poemata filo', implicit in Hor. *Serm.* 2. 1. 4 'mille die uersus deduci posse' and here.

dicere: 3. 55 n.

6. super tibi erunt: the future tense as in Hor. *Carm.* 1. 7. 1 'Laudabunt alii'; others, but not Horace, will praise the famous cities of Asia and Greece. For the tmesis cf. Plaut. *Curc.* 85 'si quid super illi fuerit', Lucr. 3. 878 'esse sui quiddam super', *A.* 7. 559 'si qua super fortuna laborum est', and see Wackernagel ii. 175.

6–7. laudes, / Vare, tuas: conventional and convenient flattery; cf. Cic. *Pro Marcell.* 9 'itaque, C. Caesar, bellicae tuae laudes celebrabuntur'. But the exploits of P. Alfenus Varus (*cos. suff.* 39 BC), such as they may have been, can hardly have been a fit subject for epic celebration; see R. Syme, *The Roman Revolution* (Oxford, 1939), 235, 245 n. 4, Nisbet–Hubbard on Hor. *Carm.* 1. 18. Not surprisingly, readers in late antiquity were puzzled and identified Varus as P. Quinctilius Varus, who perished with his legions in Germany in AD 9: 'alii Varum eum dicunt qui in Germania cum tribus legionibus interiit' (DServ.). Servius invents (or inherited) a somewhat more complicated fiction: 'hic autem Varus Germanos uicerat et exinde maximam fuerat et gloriam et pecuniam consecutus, per quem Vergilius meruerat plurima'. According to the *Vita Donati* 19, V. wrote his *E.* 'maxime ut Asinium Pollionem, Alfenum Varum, et Cornelium Gallum celebraret, quia in distributione agrorum . . . indemnem se praestitissent'; cf. DServ. here and on 9. 27.

7. tristia: 'epitheton bellorum perpetuum' (Serv.); cf. *A.* 7. 325 'tristia bella', Hor. *Ars* 73–4 'res gestae regumque ducumque et tristia bella / quo scribi possent numero monstrauit Homerus'.

condere bella: a variation of the phrase *carmen* (*carmina*) *condere*; cf. 10. 50–1 'condita . . . / carmina' and see *TLL* s.v. *condo* 153. 74.

8. agrestem tenui meditabor harundine Musam: cf. 1. 2 'siluestrem tenui Musam meditaris auena' and see Introduction, p. 175. Had V. written *siluestrem* here, readers would doubtless refer it to 'neque erubuit siluas habitare Thalea'. But he wished to vary the phrase, after the fashion of Hellenistic poets, while simultaneously imitating Lucretius: 4. 589 'fistula siluestrem ne cesset fundere Musam', 5. 1398 'agrestis enim tum Musa uigebat'. Neither phrase occurs elsewhere in either poet; for such precise imitation see 2. 69 n.

9. non iniussa cano: modelled on the old ablative *iniussu* (Ter. *Hec.* 562 'iniussu meo', Cato, *De agr.* 5.3 'iniussu domini'), *iniussus* is first attested in Horace, *Epod.* 16.49 'illic iniussae ueniunt ad mulctra capellae', *Serm.* 1. 3. 3 'iniussi' (opposed to 'rogati'). V. uses the word three times: *G.* 1.55–6 'iniussa uirescunt / gramina', where it means 'unbidden', *A.* 6. 375 'ripamue iniussus adibis', where it means 'forbidden', and here. Here Heyne and others take *non* with *iniussa* as a kind of litotes; but then *tamen* has no force (see below). V. will not sing of what Apollo has forbidden, of kings and battles, high heroic themes—yet he ventures to hope (note the graceful diffidence of 'si quis . . ., si quis') that some reader may be captivated by his pastoral song. *Non* must therefore be taken with *cano*; so Voss, Conington, O. Skutsch 193, Elder 110. Similarly, Apollo scolds Propertius for having attempted an epic in the style of Ennius, 3.3.15–16 'quis te / carminis heroi tangere iussit opus?'

si quis . . ., si quis: repetition of a word, for *si quis* counts rhythmically as a word, with, as usually in V., a shift of the ictus.

tamen: goes with what follows: see Housman on Juv. 6. 640 'facinus tamen ipsa peregi', and his note on p. l, where he cites this passage ('where editors try to refer *tamen* to what precedes').

10. captus amore: cf. *G.* 3. 285 'singula dum capti circumuectamur amore' ('capti harum rerum studio', Heyne). 'The address to Varus is eminently beautiful' (Samuel Johnson, *The Adventurer*, No. 92, 22 Sept. 1753).

myricae: 4.2 n.

13. Pierides: 3. 85 n.

Chromis: from Theocr. 1. 24 ὡς ὅκα τὸν Λιβύαθε ποτὶ Χρόμιν ᾆσας ἐρίσδων, 'as once you sang in a match with Chromis from Libya'. Chromis appears in Homer, together with Ennomos, as a chief of the Mysians (*Il.* 2. 858); and Theocritus qualified him for pastoral employment with the addition τὸν Λιβύαθε. The pastoral renown of Libya, where lambs were horned from birth, is as old as Homer (*Od.* 4. 85–9); see Mynors on *G.* 3. 339 'pastores Libyae'.

Mnasyllos: 'Rarae in hominum nominibus Graecae formae' (Wagner, *Quaest. Virg.* 4). Mnasyllos is a real name and in the feminine occurs in a poem V. had probably read: Perses 7. 1 G.–P. (= *AP* 7. 730) Μνάσυλλα. The epigram is one of nine by Perses included in Meleager's *Garland*; see Meleager 1. 26 G.–P. (= *AP*

4.1). The first of these concerns three marvellous pairs of antlers dedicated to Apollo (cf. 7. 29–30)—antlers of bucks taken on Mt. Maenalus in Arcadia, the landscape of V.'s pastoral imagination in its latter development; see 8. 22 n. The Greek form of the name may have been determined by the character of the line: four of the words in it are 'Greek'.

antro: 1. 75 n.

14. pueri: shepherds, *pastores*, like Tityrus (4) and Linus (67), as V. indicates at the end (85). Nymphs will consort with shepherds—'Nymphs and Shepherds, dance no more . . .'—in spite of scholiasts: 'isti pueri satyri sunt' (Serv.); for modern scholiasts see C. Segal, *AJP* 92 (1971), 56–61.

somno . . . iacentem: again in *G.* 4. 404, of Proteus; a phrase derived ultimately from Enn. *Ann.* 288 'nunc hostes uino domiti somnoque sepulti', where see Skutsch. Cf. also *A.* 4. 527 'somno positae', Ov. *Am.* 1.4.53 'si bene compositus somno uinoque iacebit'.

iacentem: 16 *iacebant*; Housman, *Lucan*, p. xxxiii: 'Horace was as sensitive to iteration as any modern . . . Virgil was less sensitive, Ovid much less; Lucan was almost insensible'. Cf. above, 5 *dicere*, 6 *dicere*, below, 84 *referunt*, 85 *referre*, and see Norden on *A.* 6. 423, Austin on *A.* 2. 505, Fordyce on *A.* 7. 491.

15. inflatum . . . uenas: cf. Lucr. 3. 476–7 'hominem cum uini uis penetrauit / acris et in uenas discessit diditus ardor', Hor. *Serm.* 2. 4. 25–6 'quoniam uacuis committere uenis / nil nisi lene decet', Mart. 5. 4. 4 'rubentem prominentibus uenis'. For the construction see 1. 54 n.

Iaccho: 'uino, a Libero patre, qui etiam Iacchus uocatur' (Serv.). This metonym, modelled on *Bacchus*, does not occur in Greek poetry, nor in Latin poetry before V. Iacchus seems to have been a minor Eleusinian deity or *daimon* who became confused with Bacchus by the middle of the fifth century BC, if not earlier; see *RE* ix. 613–22, *OCD* s.v., F. Graf, *Eleusis und die orphische Dichtung Athens in vorhellenistischer Zeit* (Berlin, 1974), 51–8. For the Latin poets he is hardly more than a metrical convenience, as in Catull. 64. 251, here, and in 7. 61 (but not in *G.* 1. 166). Metonymy of this kind, that is, the name of the deity put for the element or activity with which the deity was intimately associated or identified (see Dodds on Eur. *Bacch.* 274–85), is as old as Homer, and no doubt much older, e.g. *Il.* 2. 426, 440, *Od.* 22. 444—Hephaestus, Ares, Aphrodite, but not Bacchus, because the wine-god was unknown to Homer. Bacchus as a metonym first occurs in Euripides (*IT* 164, *IA* 1061); thereafter in Hellenistic poetry, e.g. Antipater of Sidon 15. 7 G.–P. (= *AP* 7. 27), 27. 5–6 G.–P. (= *AP* 7. 353), Philodemus, *AP* 11. 34. 7; and frequently in Latin poetry, beginning with Lucr. 3.221 'Bacchi cum flos euanuit', where Lucretius might have written *uini* instead of *Bacchi* (cf. Plaut. *Curc.* 96 'flos ueteris uini') since he had criticized such metonymies, 2. 655–7 'hic si quis mare Neptunum Cereremque uocare / constituet fruges et Bacchi nomine abuti / mauult'. For Lucretius the genitive of *Liber*—the name 'abused' by earlier poets—was metrically impossible; cf. Liv. Andr. *trag.* 30 R[3] 'florem anculabant Liberi ex carchesiis', Plaut. *Cas.* 640 'flore Liberi', *Cist.* 127 'flore Liberi' (in Plautus *Liber* is always a metonym). See O. Gross, *De metonymiis sermonis Latini a deorum nominibus petitis* (Diss. Philol. Halenses, 19; Halle, 1911), 342–9, Fordyce on *A.* 7. 113, Pease on Cic. *De nat. deor.* 2. 60, Wackernagel ii. 62–3. See also 10. 5 n.

16. procul: with *iacebant*, 'his garlands lay close by just fallen from his head' (T. E. Page). *Procul* signifies separation without regard to distance, which may be small, as here, or great, as in Ov. *Trist.* 4. 2. 17 'nos procul expulsos'.

tantum capiti delapsa: 'intulit . . ., ut ostenderet non longius prouolutam coronam' (Serv.). Wearing a garland was associated with drunkenness, e.g. Plaut. *Amph.* 999 'capiam coronam mi in caput, adsimulabo me esse ebrium', *Pseud.* 1287 'cum corona ebrium', Hor. *Serm.* 2. 3. 255–6 'potus ut ille / dicitur ex collo furtim carpsisse coronas'.

capiti: an archaic ablative in long *i*, of which there are a number (but not *capiti*) in Lucretius; see Munro on Lucr. 1. 978. *Capiti* is found occasionally in later poetry where, as here, it is convenient: Prop. 2. 30. 39, Ov. *Ars* 1. 582, 2. 528, *Met.* 15. 610, Silius 16. 434–5; in all these places the reference is to a garland. Cf. also Catull. 68. 124 'suscitat a cano uolturium capiti'.

17. pendebat: 'manibus non emissum significat' (DServ.); 'he is too drunk to sustain it, and too fond of it, even in this almost senseless condition, to let it go out of his hand' (Martyn).

cantharus: a deep two-handled drinking-cup, here with well-worn handles, 'frequenti scilicet potu' (Serv.). In Greek and

Roman comedy the *cantharus* (*κάνθαρος*) is simply an article of everyday use; see Athenaeus 11. 473 D–474 D, *TLL* s.v. 280. 71. But from the sixth century BC onward it was associated with Dionysus, and no doubt that is the suggestion here (Macrob. *Sat.* 5. 21. 14 'aptissime proprium Liberi patris poculum adsignat Sileno'); see T. H. Carpenter, *Dionysian Imagery in Archaic Greek Art* (Oxford, 1986), 117–23 and pls. 2, 9A, 14A and B, 17, 20B, 21, 24A and B, 25, in all of which Dionysus is shown holding the *cantharus* by one of its handles. It is reported of Gaius Marius that, in the insolence of victory, he would drink from a *cantharus*: Val. Max. 3. 6. 6 'iam C. Marii paene insolens factum. nam post Iugurthinum Cimbricumque et Teutonicum triumphum cantharo semper potauit, quod Liber pater Indicum ex Asia deducens triumphum hoc usus poculi genere ferebatur, ut inter ipsum haustum uini uictoriae eius suas uictorias compararet'. See also Nisbet–Hubbard on Hor. *Carm.* 1. 20. 2.

18. adgressi: V. imagines Silenus as a Proteus-like figure, an ancient wizard, reluctant and evasive; cf. *G.* 4. 402–4 'cum sitiunt herbae et pecori iam *gratior* umbra est [a reminiscence of l. 11 'nec Phoebo gratior ulla est', the comparative of *gratus* not being found elsewhere in the *E.* and *G.*], / in secreta *senis* ducam, quo fessus ab undis / se recipit, facile ut *somno adgrediare iacentem*'.

18–19. spe ... / luserat: cf. *A.* 1. 352 'uana spe lusit'.

19. ipsis ex uincula sertis: see below, 33 n.

20–2. The shepherd boys are abashed at their own boldness, but the forward Nymph daubs the face of Silenus, now awake and watching, with mulberry juice; cf. *A.* 9. 345 'Rhoetum uigilantem et cuncta uidentem'. A common trick, as it must have been, described in graphic detail by Petronius 22. 1 'cum Ascyltos grauatus tot malis in somnum laberetur, illa quae iniuria depulsa fuerat ancilla totam faciem eius fuligine larga perfricuit et non sentientis labra umerosque sopionibus pinxit'; see Lindsay on Plaut. *Capt.* 656, B. Brotherton, *The Vocabulary of Intrigue in Roman Comedy* (Menasha, Wis., 1926), 85–9; W. M. Thackeray, *The Memoirs of Barry Lyndon, Esq.*, ch. 19: 'We painted the parson's face black, when his reverence had arrived at his seventh bottle, and his usual insensible stage'. A scene of this sort might have been expected in Plautus, but there is only the faded metaphor *os alicui sublinere* 'to trick, cheat someone'; cf. Nonius p. 65 L. (on Plaut. *Aul.* 668) '*subleuit* significant inlusit et pro ridiculo habuit, tractum a genere ludi, quo dormientibus ora pinguntur'. Plautus has the phrase thirteen times; but for the refined taste of Terence even the metaphor was too crude. In fact, only one writer after Plautus appears to have used the phrase, Symmachus, *Epist.* 4. 18. 1 'ne mihi os subleueris!' (La Cerda). Aegle daubs Silenus' face not with soot, of which, dramatically, there could be none to hand (and which, besides, would be inappropriate in a pastoral setting), but with the blood-red juice of the mulberry (Nicander, *Alex.* 69 μορέης ... φοινηέσσης). Line 22 is related to 10. 27 (Pan) 'sanguineis ebuli bacis minioque rubentem'. Gods of the countryside customarily had their faces raddled (see S. Weinstock, *Divus Julius* (Oxford, 1971), 68 n. 5), and rustic worshippers of Bacchus painted their own faces with vermilion for his holidays, Tib. 2. 1. 55 'agricola et minio suffusus, Bacche, rubenti'.

20. addit se sociam: cf. *A.* 2. 339 'addit se socios', 9. 149–50 'addant se . . . / . . . socios'.

addit se: Maas 515 n. 24 = 557 n. 24, observes that the second syllable of trochaic words (except those ending in *-m*) is seldom lengthened in the first foot of the hexameter. Here *se* is enclitic; in 2. 39, 7. 58, and 8. 83 proper names are involved.

superuenit: 3. 38 n.

Aegle: not a pastoral name; it may have been suggested by an episode in Ap. Rhod. 4. 1393–1431. The Argonauts, desperately thirsty after portaging the Argo across the Libyan desert, search for a spring; startled by their sudden arrival, the Hesperides turn into dust on the spot; Orpheus prays and is heard: Hespere becomes a poplar, Erytheis an elm, and Aegle a willow, and Aegle speaks to the Argonauts.

21. Aegle Naiadum pulcherrima: such repetition, or epanalepsis, which is as old as Homer, became a feature of Hellenistic poetry, and of pastoral especially; see Norden on *A.* 6. 164, McLennan on Callim. *Hymn* 1. 91, R. Gimm, *De Vergilii stilo bucolico quaestiones selectae* (diss. inaug. Leipzig, 1910), 87–8. Cf. 6. 33–4, 55–6, 9. 27–8, 47–8, 10. 72–3; somewhat different though not unrelated are 1. 27–9, 4. 55–7, 5. 51–2, 7. 2–3, 8. 55–6, 9. 64–5, 10. 31–3.

Naiadum: like the *Pierides* (3. 85 n.), the *Naiades* are confined to the *E*.; plural here and in 10. 10 (*Naides*), singular in 2. 46.

pulcherrima: here and in 7. 65 before a bucolic diaeresis, elsewhere, except *A*. 1. 72 'pulcherrima Deiopea', occupying the fifth foot of the line (once in the *G*., ten times in the *A*.).

24. satis est potuisse uideri: 'ut uideamini me uincire potuisse' (Heyne). Forbiger compares *A*. 5. 231 'possunt, quia posse uidentur', and this seems to be the simplest interpretation of this rather cryptic remark.

25. carmina . . .; carmina uobis: the repetition and rhythm suggest a pastoral refrain; see 8. 68 n.

cognoscite: cf. Lucr. 1. 921 'nunc age quod superest cognosce et clarius audi'.

26. huic aliud mercedis erit: 'Nymphae minatur stuprum latenter, quod uerecunde dixit Vergilius' (Serv.); see 3. 8 n.

simul incipit ipse: cf. *G*. 4. 386 (Cyrene) 'sic incipit ipsa', *A*. 10. 5 (Jupiter) 'incipit ipse'.

27–8. Like Orpheus, Silenus charms the world of nature with his song. Description of its effect precedes the song itself, a late example of an 'archaic sequence' (West on Hes. *Theog*. 43).

27. tum uero: only here in the *E*. and only once in the *G*., 3.505, where again *-o* is elided; see Axelson 86–7.

in numerum: keeping time; cf. *A*. 8. 452–3, Lucr. 2. 630–1 (Curetes) 'inter se . . . / ludunt in numerumque exsultant'.

Faunos: Skutsch on Enn. *Ann*. 207: 'Faunus, divinity of the woods, to whom the mysterious voices of the forest are ascribed . . . exists properly only in the singular . . . a pluralization is not unnatural and has certainly occurred, in the same way as probably that of Pan, and later that of Silvanus (Ov. *Met*. 1. 193), under the influence of the Satyrs'. Cf. Lucr. 4. 580–3 'haec loca capripedes Satyros Nymphasque tenere / finitimi fingunt et Faunos esse loquuntur / quorum noctiuago strepitu ludoque iocanti / adfirmant uulgo taciturna silentia rumpi', *A*. 8. 314 'haec nemora indigenae Fauni Nymphaeque tenebant', and see Mynors on *G*. 1. 10.

-que . . . -que: only here attached to successive nouns in the *E*.; cf. 8. 22, 10.23. For a brief history of the idiom see Skutsch on Enn. *Ann*. 170. See below, 32 n.

28. rigidas . . . quercus: the tree obedient to Orpheus' lyre is usually the tough and unyielding oak; the manna-ash, below, 71 'rigidas . . . ornos', is simply a metrical convenience. The adjective is not found elsewhere in the *E.*, nor is it elsewhere applied to the manna-ash (*TLL* s.v. *ornus*).

motare: 5. 5 n.

29–30. nec tantum . . . / nec tantum: Ursinus compares [Moschus], *Epitaph. Bion.* 89–90 *οὐ τόσον Ἀλκαίῳ περιμύρατο Λέσβος ἐραννά, | οὐδὲ τόσον τὸν ἀοιδὸν ὀδύρατο Τήιον ἄστυ*, 'Not so much did lovely Lesbos mourn for Alcaeus, nor so much did Teos town grieve for her poet'.

29. Parnasia rupes: a neoteric phrase, modelled on Catull. 68. 53 'Trinacria rupes' and Theocr. 7. 148 *Παρνάσιον αἶπος*, 'the steep of Parnassus'. Like *αἶπος*, for which see Fraenkel on Aesch. *Ag.* 285, *rupes* is an old poetic word, first attested in Accius 505 R.[3], then occurring in Lucretius, in the longer poems of Catullus, 61. 27–8 'Thespiae / rupis', 64. 154, 68. 53, and in Augustan poets, especially V. (30 times), though not in Tibullus; found in prose in Varro, Caesar, and Sallust. See Tränkle 13.

30. Rhodope: a mountain-range in Thrace—'where Woods and Rocks had Ears / To rapture'—associated with Orpheus; cf. Ov. *Met.* 10. 11–12 'Rhodopeius . . . / . . . uates', with Bömer's note.

Ismarus: in Homer, a town of the Cicones, a Thracian tribe, sacked by Odysseus (*Od.* 9. 39–40); in V., as in other Latin poets, a mountain in Thrace, *G.* 2. 37–8 'iuuat Ismara Baccho / conserere', Prop. 3. 12. 25 'Ciconum mons Ismara'. (The town still existed in V.'s day: Strabo 7, fr. 43, says that it is now called Ismara.) Lucretius seems to have been—but can he have been?—the first poet after Homer to mention Ismarus, 5. 31 'Ismara propter'. The neuter plural is the usual form in Latin poetry: *Ismarus* here is unique. This variation in the form of place-names, Greek in origin, was elaborated by the Latin poets as a metrical convenience, e.g. 8. 22, 10. 15 *Maenalus* but 10. 55, *G.* 1. 17 *Maenala* (cf. Theocr. 1. 124 *μέγα Μαίναλον*, 'mighty Maenalus', Theocritus' only reference to this mountain haunt of Pan); *A.* 6. 577 *Tartarus* but elsewhere in V. *Tartara*; *G.* 3. 44 *Taygeti* but *G.* 2. 488 *Taygeta*.

Orphea: a synizesis, here first in Latin poetry and extremely rare; cf. *G.* 1. 279 *Typhoea*, whence Ov. *Met.* 3. 303 *Typhoea* and

Silius 8. 540 *Typhoea*, and see Leumann, *Lat. Laut- und Formenlehre*², 120.

31. namque canebat uti: here, as in *A.* 1. 466 'namque uidebat uti', *namque* gives the reason, there for Aeneas' tears, here for the effect produced on the natural world. The archaic form *uti*, frequent in Lucretius, is not found elsewhere in the *E.*

magnum per inane: a Lucretian phrase (1. 1018, 1103, 2. 65, 105, 109).

31–2. coacta / semina: cf. Lucr. 2. 1059–60 'semina rerum / . . . coacta'. *Semina* is the commonest term for atoms in Lucretius.

32. terrarum: 'earth', often plural in Lucretius, especially in the genitive, accusative, and ablative cases; see Munro on 1. 3 'terras frugiferentis' and cf. 5. 446 'a terris altum secernere caelum'.

-que . . . -que . . . -que: cf. 4. 51; a sequence not found elsewhere in the *E.*

animae: 'air'; cf. Lucr. 1. 714–15 (referring to Empedocles) 'et qui quattuor ex rebus posse omnia rentur / ex igni terra atque anima procrescere et imbri'.

33. liquidi . . . ignis: Macrob. *Sat.* 6. 5. 4 'illud audaciae maximae uideri possit quod ait in Bucolicis "et liquidi simul ignis" pro puro uel lucido seu pro effuso et abundanti, nisi prior hoc epitheto Lucretius usus fuisset in sexto [205] "deuolet in terram liquidi color aureus ignis"'. The last item in the enumeration is carried over into the next line and weighted with an adjective; cf. Lucr. 6. 529–30 (next note), 5. 68–9 'terram caelum mare sidera solem / lunaique globum', *A.* 6. 724–5 'caelum ac terras camposque liquentis / lucentemque globum lunae', Milton, *Paradise Lost* 4. 721–3 (Adam and Eve) 'and under op'n Sky ador'd / The God that made both Sky, Air, Earth and Heav'n / Which they beheld, the Moon's resplendent globe'. Stewart 184 notes that 'the slight distinction given to fire' is Empedoclean.

his ex omnia primis: of the two ancient MSS which contain this line, P has *ex omnia*, R *exordia*; the latter being the vulgate, for *exordia* is found in DServ., Macrob. *Sat.* 6. 2. 22, and in every medieval MS that has been examined. *EXOMNIA*, the *lectio difficilior*, was altered, whether by accident or design, to *EXORDIA*, a fairly frequent Lucretian term for atoms (2. 333, 1062, 3. 31, 380, 4. 45, 114, 5. 430, 471, 677). The reading of P was discovered by Ribbeck, *Prolegomena critica ad P. Vergili Maronis opera maiora* (Leipzig, 1856), 225, and published in the apparatus criticus of his first edition (1859); shortly thereafter Peerlkamp, who apparently had not seen Ribbeck's edition, published *ex omnia* as a conjecture (*Mnemosyne*, 10 = NS 1 (1861), 23–4). Since Ribbeck retained *exordia* in the text of both his first and second (1894) editions, it was left for Sabbadini (1937) to restore *ex omnia* to V. The word-order is distinctively Lucretian; cf. Lucr. 2. 731–2 'albis ex alba rearis / principiis', 3. 10 'tuisque ex inclute chartis', 4. 829 'ualidis ex apta lacertis', 6. 788 'terris ex omnia surgunt', and see Munro on Lucr. 1. 841. By 'his ... primis' V. must mean the four elements, each composed of its proper atoms (*semina*), thus using a Lucretian term in an un-Lucretian sense; cf. Lucr. 1. 61 (the atoms) 'corpora prima, quod ex illis sunt omnia primis'. Finally, the epanalepsis 'omnia primis, / omnia', frequent in Hellenistic poetry (see above, 21 n.), is also Lucretian, 2. 955–6 'uincere saepe, / uincere', 6. 528–30 'et quae concrescunt in nubibus, omnia prorsum, / omnia, nix uenti grando gelidaeque pruinae / et uis magna geli' (for the punctuation see G. B. Townend, *CQ*, NS 19 (1969), 337).

34. ipse tener mundi ... orbis: the upper sky, the bright cloudless ether (so Voss, Conington, Munro on Lucr. 5. 468); cf. Lucr. 5. 498–9 'inde mare, inde aer, inde aether ignifer ipse / corporibus liquidis sunt omnia pura relicta', 510 'magnus caeli ... orbis'.

tener: 'recens factus' (DServ.), and here contrasted with the earth, 35 'tum durare solum'.

concreuerit: a Lucretian verb (18 times), here agreeing with its nearer subject. A. Ernout, *Philologica* (Paris, 1946), 93, notes that this is the first instance of *concreui* (again in *A.* 12. 905); Cicero uses only *concretus sum*.

35. tum: with *coeperit*, not, like *tum* in ll. 61 and 64, introducing another part of Silenus' song.

durare: intransitive; cf. *Aetna* 497 'flumina ... frigore durant' and see *TLL* s.v. 2296. 17; also, for verbs ordinarily transitive used intransitively, Fordyce on *A.* 7. 27. The spongy earth, natal wet, begins to exhale its moisture and grow hard.

discludere: Macrob. *Sat.* 6.4.11 'ferit aures nostras hoc uerbum *discludere* ut nouum, sed prior Lucretius in quinto [437–8] "diffugere inde loci partes coepere paresque / cum paribus iungi res et discludere mundum"'.

Nerea ponto: to shut off the sea-god in the sea. Neoteric wit? Nereus was the eldest child of Pontos, Hes. *Theog.* 233 *Νηρέα δ' ἀψευδέα καὶ ἀληθέα γείνατο Πόντος*, 'Pontos begat undeceitful and truthful Nereus'. 'Sed video in Virgilio quid peculiare. Nereus ab Hesiodo in Theog. signate dicitur filius Ponti, non maris, aut aequoris. Itaque coniungitur filius cum patre' (La Cerda).

ponto: local ablative unmodified, as in *G.* 1. 372, *A.* 1. 40, 70. *Πόντος* was Latinized by Ennius, *Ann.* 217 Skutsch 'urserat huc nauim compulsam fluctibus pontus', and remained exclusively poetic. See Skutsch ad loc., Tränkle 40.

36. paulatim: a Lucretian adverb (23 times).

37. stupeant: earth stands in awe of the sun, now rising for the first time ever. The natural world is imagined as animate and capable of emotion; cf. 29 'gaudet', 30 'miratur', 40 'ignaros', 82 'beatus'. *Stupeo* first occurs here with an accusative and infinitive (Hofmann–Szantyr 358).

lucescere solem: an innovation, since the verb is impersonal before V., e.g. Plaut. *Amph.* 533 'prius quam lucescat', or virtually so, Plaut. *Amph.* 543 'lucescit hoc iam' (*TLL* s.v. 1703. 5, 21). But the compound verbs *inlucesco* and *dilucesco* furnished an analogy, e.g. Plaut. *Persa* 712 'hic tibi dies inluxit lucrificabilis', *G.* 2. 337 'inluxisse dies', Lucr. 5. 176 'donec diluxit rerum genitalis origo'.

38. altius atque: several editors put a comma after *altius*, for the reason that '*atque* is never second word in a clause in Virgil' (T. E. Page); but, taken with *lucescere, altius* makes little or no sense. The new sun—new, strange, never before seen, *nouum* is emphasized by its separation from *solem*—now begins to shine, and rain falls from the clouds because the earth's moisture has moved higher up and no longer clings to it like a mist. *Altius* belongs to *summotis*, 'nubibus in altum leuatis' (Serv.; so also F. Skutsch (1901), 46); cf. Luc. 3. 401 'alte summotis solibus'—or rather, to both *summotis* and *cadant*; cf. 1. 83 'maioresque cadunt altis de montibus umbrae'. That a very confident young poet should have employed a verbal sophistication which he afterwards avoided is hardly surprising; thus *atque* is postponed four times in the first book of Horace's *Sermones*, 5. 4, 6. 111, 131, 7. 12, a book published at about the same time as V.'s book of *E.*, but nowhere in his later hexameters.

39. cum . . . cumque: reproducing the sound, if not the sense, of Lucr. 2. 114 'cum solis lumina cumque'.

40. rara per ignaros errent animalia montis: cf. Lucr. 2. 532 'nam quod rara uides magis esse animalia quaedam'. If, as Conington suggests, V. was thinking of Lucr. 5. 823–4 (*terra*) 'animal prope certo tempore fudit / omne quod in magnis bacchatur montibus passim', he imagines a very different scene: his lonely animals roam over the mountains unknowing and unknown. Cf. *A.* 10. 706 'ignarum Laurens habet ora Mimanta', quoted by Gellius 9. 12. 20, with the comment: '*ignarus* aeque utroqueuersum dicitur non tantum qui ignorat sed et qui ignoratur'. It is unusual for a true reading to be preserved in R alone; P (M is defective here) and the Carolingian MSS have *ignotos*, inharmonious and banal (cf. Livy 29. 32. 5, 31. 42. 8, 36. 19. 10, 38. 2. 14, 39. 45. 6), a *lectio facilior*.

41–4: brief, allusive references in the Alexandrian manner to well-known stories. The lack of chronology was noted in antiquity; 'quod autem dicit "Saturnia regna", fabularum ordinem uertit' (Serv.). But Silenus is offering a history of the world in brief: first the creation of man, symbolized by Pyrrha, because the Ancients, having no Adam and Eve, referred their legendary origin to Deucalion and Pyrrha; then the Golden Age, the end of which is marked by the introduction of technology, symbolized by Prometheus' theft of fire, while the new art of navigation, auspicated by the Argonauts, is symbolized by Hylas (Mynors).

41. lapides Pyrrhae iactos: here, as La Cerda remarks, Pyrrha is mentioned without her husband, but in *G.* 1.62 'Deucalion uacuum lapides iactauit in orbem' Deucalion without his wife. *Pyrrhae* is dative of agent. The story is told in Ov. *Met.* 1. 313–415.

Saturnia regna: 4. 6 n.

42. Caucasias . . . uolucris: cf. Ap. Rhod. 3. 851–3 *καταστάξαντος ἔραζε | αἰετοῦ ὠμηστέω κνημοῖς ἐνὶ Καυκασίοισιν | αἱματόεντ' ἰχῶρα Προμηθῆος μογεροῖο*, 'when the ravenous eagle let fall to earth on the slopes of the Caucasus the bloody ichor of wretched Prometheus'; also 2. 1247–50. *Caucasius* is first attested here in Latin. The tradition before V. is unanimous: it was a single bird, an eagle, that came daily to feed on Prometheus' liver. There may be some contamination with the story of Tityos, *Od.* 11. 578 *γῦπε δέ μιν ἑκάτερθε παρημένω ἧπαρ ἔκειρον*, 'and two vultures sat, one on either side, tearing his liver', Lucr. 3. 984 'nec Tityon uolucres ineunt Acherunte iacentem', *A.* 6. 595–8 'nec non et Tityon . . . / . . . per tota nouem cui iugera corpus / porrigitur, rostroque immanis uultur obunco / immortale iecur tondens'. See K. F. Smith on Tib. 1. 3. 75–6 'porrectusque nouem Tityos per iugera terrae / adsiduas atro uiscere pascit aues'.

Promethei: the genitive *-ei* of Greek proper names in *-eus* is monosyllabic in V.: cf. below, 78 *Terei*, *A.* 1. 41 *Oilei*, 120 *Ilionei* (also 9. 501), 8. 383 *Nerei*, 11. 262 *Protei*, 265 *Idomenei*, and see Leumann, *Lat. Laut- und Formenlehre*², 120.

43–4: an allusion to the story of Hylas, a minor episode of the voyage to Colchis. The Argonauts bivouac on the shore of the Propontis and Hylas, the beautiful boy, the beloved of Heracles, goes looking for water. He finds a spring, reaches down, and (the enamoured Nymphs clasp his hand) disappears into the dark water. Wild with grief, Heracles searches for him, shouting his name three times, to which the lost boy, as though from far away, faintly answers (Theocr. 13. 58–60). Unlike the stories alluded to in 41–2, the Hylas story—a local legend given literary form by Apollonius 1. 1207–72 and Theocritus 13—is Hellenistic and must therefore have seemed, to V. and his readers, relatively modern. Hylas became popular as a subject of poetry (and art: *LIMC* v/1. 574–9, v/2. 396–9)—too popular in V.'s opinion, *G.* 3. 6 'cui non dictus Hylas puer?' The story had an unmistakable literary connotation, however, and here serves to introduce (45 '*et* fortunatam') Pasiphae, whose story V. conceives as an epyllion; see below, 45–60 n.

43. quo fonte: Ap. Rhod. 1. 1221–2 *αἶψα δ' ὅ γε κρήνην μετεκίαθεν ἣν καλέουσι | Πηγὰς ἀγχίγυοι περιναιέται*, 'Hylas soon came to the spring which the inhabitants near by call the Streams', whence Prop. 1.20.33 'hic erat Arganthi Pege sub uertice montis'; profusely described but not named in Theocr. 13. 40–2. No doubt Apollonius' naming of the spring was controversial; see F. Skutsch (1901), 96 n. 2. What song did the Sirens sing? In which hand was Aphrodite wounded by Diomedes? What was the name of Proteus' father? of Hecuba's mother? of Anchises' nurse? Such fantastic literary questions, ζητήματα, were attributed to the followers of Callimachus, 'super-Callimachuses', by Philip (61 Gow–Page, *The Garland of Philip* (Cambridge, 1968) = *AP* 11. 347). See Courtney on Juv. 7. 234.

44. clamassent: there is no reference here to a treble calling or last conclamation; see Gow on Theocr. 13. 58.

Hyla, Hyla: for the prosody see 3. 79 n.

45–60. V.'s miniature of an epyllion: 45–6 elliptical narrative, assuming the familiar story; 47 the poet's sympathetic apostrophe; 48–51 Proetus' daughters—a story within the story, as in Callimachus' *Hecale* (Erichthonius), Moschus' *Europa* (Io), Catullus 64 (Ariadne); 52 repetition of the apostrophe to frame the inner story; 52–5 the poet's feeling comment; 55–60 the heroine's lovely, sad, and disproportioned speech.

45. fortunatam, si: cf. *G.* 2. 458 'O fortunatos nimium, . . . si', *A.* 4. 657 'felix, heu nimium felix, si'.

46. Pasiphaen . . . solatur amore: when Minos failed to sacrifice the beautiful bull that Poseidon sent him from the sea, keeping it for himself and sacrificing another bull instead, the angry sea-god visited Minos' wife, Pasiphae, with a monstrous passion for the animal. The poet—here Silenus—is said to do what he describes being done, a well-attested form of expression; cf. below, 62 and 63, 9. 19–20, *G.* 3. 386–7, [Moschus], *Epitaph. Bion.* 82 (the poet Bion) σύριγγας ἔτευχε καὶ ἁδέα πόρτιν ἄμελγε, 'fashioned pipes and milked the sweet heifer', Hor. *Serm.* 1. 10. 36 'turgidus Alpinus iugulat dum Memnona', *Carm.* 2. 1. 17–18 'iam nunc minaci murmure cornuum / perstringis auris', Ov. *Am.* 3. 12. 25 (we poets) 'per spatium Tityon porreximus ingens', *Trist.* 2. 439 'is quoque, Phasiacas Argon qui duxit in undas', and see R. Kassel, *RhM* 109 (1966), 8–10. Cf. also Gibbon, *Decline and Fall of the Roman Empire*, ch. 68 n. 59 (differing accounts of the death of Constantine Palaeologus): 'Ducas kills him with two blows of Turkish soldiers; Chalcocondyles wounds him in the shoulder, and then tramples him in the gate'.

For the rhythm of this line—strong caesura in the third foot, weak in the fourth—see Norden 427–9 and cf. 10. 10.

niuei: precedes, as *niueum* (53) follows, 'a, uirgo infelix' (47, 52), to heighten the symmetrical effect.

47. a, uirgo infelix: 'Caluus in *Io* "a, uirgo infelix, herbis pasceris amaris"' (DServ. = *FPL* fr. 9 Büchner). For the exclamation see 10. 47 n.; for the sympathetic apostrophe, Norden on *A*. 6. 14 ff. Twice below, as if thinking of 'herbis pasceris amaris', V. uses the adjective *amarus* (62, 68). Cf. Ovid's exuberant yet precise imitation, *Met*. 1. 632–4 'frondibus arboreis et *amara pascitur herba*, / proque toro terrae non semper gramen habenti / incubat *infelix*'.

uirgo: apart from this repeated quotation of Calvus and a reference to the constellation (4. 6), *uirgo* does not appear in the *E*. Even the chaste Atalanta is called, like the provocative Galatea in 3. 64, *puella* (below, 61). Three times in the *G*., forty-three times in the *A*., *uirgo* belongs to the higher style of poetry; see Axelson 58, P. Watson, 'Puella and Virgo', *Glotta*, 61 (1983), 119–43.

quae te dementia cepit: Corydon's self-reproach; see 2. 69 n.

48. Proetides: daughters of Proetus, king of Argos, who were afflicted with madness by Hera, according to one tradition, because they had mocked her wooden image; fancying themselves to be transformed into cows, they wandered about the countryside until cured of their hallucination by the seer Melampus; see G. Radke, *RE* xxiii. 117–25, Maehler on Bacchyl. 11 (pp. 196–202). Since 1962 they have been happily reunited in a euphonious line of the Hesiodic *Catalogue* fr. 129. 24 M.–W.[3] [Λυσίππην τε καὶ Ἰφι]νόην καὶ Ἰφιάνασσαν; cf. 'Lysippe, Iphinoe, Iphianassa' (DServ.). Did Calvus somehow involve Proetus' daughters in his *Io*? V. implies as much, and Calvus had before him the example of Moschus' *Europa* with the story of Io (like Io, Proetus' daughters had offended Hera) worked in gold on Europa's golden flower-basket (44–9).

falsis mugitibus: 'their imitated lowings' (Dryden).

50. concubitus: used only of animals by V. (*TLL*. s.v. 100. 63). The hysterical girls imagined themselves cows; they did not, however, behave like cows.

quamuis: with the pluperfect subjunctive first in Cicero, e.g. *Pro Mil*. 21 'multa etiam alia uidit, sed illud maxime, quamuis atrociter ipse tulisset'; again in *A*. 8. 379–80.

collo: dative; cf. *A*. 2. 130, 729.

51. leui . . . fronte: 'humana scilicet' (DServ.), and perhaps that is all the humane V. wished his reader to understand here; cf. Ov.

Ars. 1. 308 (Pasiphae) 'quam cuperes fronti cornua nata tuae'. But V. certainly knew that Proetus' daughters had been described as losing their hair—a symptom of their morbidly excited condition—in the Hesiodic *Catalogue* fr. 1. 33. 4–5 M.–W.[3] *αἱ δέ νυ χαῖται | ἔρρεον ἐκ κεφαλέων, ψίλωτο δὲ καλὰ κάρηνα*, 'now the hair was falling from their heads, and their lovely heads became bald'. Similarly, the lovesick Simaetha loses her hair, Theocr. 2. 89 *ἔρρευν δ' ἐκ κεφαλᾶς πᾶσαι τρίχες*, 'and all the hair was falling from my head'.

52. in montibus erras: like a grazing animal, e.g. 2. 21 'errant in montibus agnae', *G.* 4. 11 'errans bucula campo', or possibly a distracted lover. Wandering aimlessly is a symptom of passion in Hellenistic poetry; see 10. 55–6 n.

53. latus . . . fultus: 1. 54 n.

niueum: in contrast with the dark green foliage of the ilex beneath which he lies, hardly with 'the deep purple of the hyacinths' (T. E. Page); 'molli . . . hyacintho', a delicious soft bed merely. *Niueus* first occurs in Cic. *Progn.* fr. 3. 3 Soubiran, then eight times in Catullus, with the exception of 58b. 4—an apparent exception only, for the first four lines of this poem are a parody of the Alexandrian style—in his long poems: 61. 9–10 'niueo gerens / luteum pede soccum' ('*niueo luteum*, der seit hellenistischer Zeit beliebte Farbenkontrast', Kroll), 63. 8, 64. 240, 303, 309 'at roseae niueo residebant uertice uittae', 364, 68. 125. See Clausen 103–4.

molli . . . hyacintho: 2. 50 n.

fultus hyacintho: DServius compares Lucil. 138 M. 'et puluino fultus' and *A.* 7. 94–5 'atque harum effultus tergo stratisque iacebat / uelleribus'. For the prosody cf. Catull. 66. 11 'auctus hymenaeo' and see Norden 451, Fordyce on *A.* 7. 398.

54. pallentis ruminat herbas: *pallentis* presumably represents *χλωρός*, pale greenish-yellow, as grass becomes during the Mediterranean summer; Longus 1. 17. 4 *χλωρότερον τὸ πρόσωπον ἦν πόας θερινῆς*, 'his face was paler than summer grass' (cf. Sappho fr. 31. 14–15 L.-P.). Used of violets (2. 47, where see note), ivy (3. 39, *G.* 4. 124), and the olive tree (5. 16), *pallens* is, as Conington observes, 'an unusual epithet of grass'; it is, in fact, unique (*TLL* s.v. *herba* 2620. 78). This being so, the interpretation of Servius should not, perhaps, be excluded: 'reuomit ac denuo consumit . . . *pallentis* autem . . . quae uentris calore propria uiriditate caruerunt'. Cf.

[Ov.] *Am.* 3. 5. 17–18 'dum iacet et lente reuocatas ruminat herbas / atque iterum pasto pascitur ante cibo' (the pentameter is quoted by DServ. here), [Ov.] *Hal.* 119 'ut Scarus, epastas solus qui ruminat escas', Calp. Sic. 3. 17 (*taurus*) 'matutinas reuocat palearibus herbas'. Is *pallentis* a touch of Hellenistic realism?

ruminat: here first in its literal sense. Four earlier instances are cited by Nonius, all deponent and all metaphorical, 'to turn over in the mind, recall', the earliest from Livius Andronicus (p. 245 L.), the other three from Varro (pp. 245, 770 L.). The word had not ceased to be used in its literal sense (cf. Columella 6. 6. 1 'bos neque ruminat', Pliny, *NH* 9. 62, 11. 161); it was simply too uncouth for literary use. But why so exquisite a coarseness here? Is V. again imitating Calvus?

55. Pasiphae's speech begins after a bucolic diaeresis, pauses, then continues (58 'forsitan' . . .) after a bucolic diaeresis.

55–6. Nymphae, / Dictaeae Nymphae: *Νύμφαι, | Δικταῖαι Νύμφαι.* Cf. Theocr. 13. 43–4 *ὕδατι δ' ἐν μέσσῳ Νύμφαι χορὸν ἀρτίζοντο, | Νύμφαι ἀκοίμητοι*, 'And in the water Nymphs were arraying the dance, the sleepless Nymphs' (Gow), Nonnus, *Dionys.* 17. 310–11 *Νύμφαι, | Νύμφαι Ἁμαδρυάδες*, 'Nymphs, Hamadryad Nymphs'.

56. nemorum iam claudite saltus: Enn. *Ann.* 580 Skutsch 'siluarum saltus', with no apparent distinction of meaning; cf. 10. 9, *G.* 1. 16, 3. 40, 4. 53, *A.* 4. 72. Legally defined, *saltus* is pasturage among trees; Festus, p. 392 L. 'saltus est, ubi siluae et pastiones sunt'. Sometimes (a favourite word in Livy) it is a defile or narrow valley with a clear bottom and wooded sides; cf. *A.* 11. 904–5 'Aeneas saltus ingressus apertos / exsuperatque iugum siluaque euadit opaca' (Mynors). The bull will be found in the upland pastures, but only by ringing them with Nymphs playing the part of huntsmen; cf. 10. 56–7, *G.* 1. 140, *A.* 4. 121 'dum trepidant alae saltusque indagine cingunt'.

57. si qua forte: 'if by some chance'; cf. Livy 6. 3. 7 'si qua forte se in agros eicere possent', 1. 4. 4 'forte quadam diuinitus super ripas Tiberis effusus', *A.* 1. 377 (*nos*) 'forte sua Libycis tempestas appulit oris', 2. 94 'fors si qua tulisset', and see *TLL* s.v. *fors* 1128. 36, 1130.45.

58: errabunda . . . uestigia: cf. Catull. 64. 113 (Theseus in the Labyrinth) 'errabunda regens tenui uestigia filo'; *errabundus* is found elsewhere in poetry only in Lucr. 4. 692 (*sonitus*) 'errabundus enim tarde uenit'. In general, poets avoid such adjectives when formed on verbs of the first conjugation; cf. M. Zicàri, 'Moribunda ab sede Pisauri', *Studia Oliveriana*, 3 (1955), 63–4 = *Scritti Catulliani* (Urbino, 1978), 193–4: 'L'aggettivo appartiene all'esiguo manipolo di deverbali in *i/ebundus* accolti dalla poesia, mentre il solo Lucrezio usò *versabundus* (6, 438; 6, 582), ed *errabundus* non si trova in poeti dopo Verg. *ecl.* 6, 58. Questa limitazione, oltre che negli epici e in Ovidio, si osserva anche in Cicerone, che peraltro non influisce sui prosatori posteriori, nei quali o il tipo manca del tutto, o spesseggiano proprio le forme in *abundus*'. For a list of such adjectives see P. Langlois, *REL* 39 (1961), 128–34, and E. Pianezzola, *Gli aggettivi verbali in* -bundus (Florence, 1965), 239–40. V. was careful to avoid *errabundus* in A. 6. 30 'caeca regens filo uestigia', while retaining the bucolic diaeresis; see Clausen 113–14.

60. Gortynia: the topographical reference is (as so often) to a place in a poem: Catull. 64. 75 'Gortynia templa'. Minos' capital was at Cnossus. See 9. 13 n.

61. puellam: Atalanta, in this sophisticated poetry denied 'the simplicity of her own name'—so Ross (1975), 62, on Prop. 1. 1. 15 'ergo uelocem potuit domuisse puellam' (a different version of the story, but with this fleeting reference to V. and the commoner version; see Gow on Theocr. 3. 40). There is also a metrical reason for Atalanta's anonymity; see Norden on *A*. 6. 28. The story of Atalanta and the deadly footrace is as old as the Hesiodic *Catalogue* (frs. 74–6 M.–W.[3]). The golden apples Hippomenes cast in her way were given to him by Aphrodite—apples from the garden of the Hesperides; see Bömer on Ov. *Met.* 10. 644. Cf. Lucr. 5. 32 'aureaque Hesperidum . . . fulgentia mala'.

62. Phaethontiadas: Phaethon's sisters; a curious use of the patronymic form, though Meleager's sisters were called *Meleagrides* after being turned into guinea-fowl; see *RE* xv. 445–6, Bömer on Ov. *Met.* 8. 533–46, Coleman here, and J. Huyck, *HSCP* 91 (1987), 217–28.

circumdat: 'mira autem est canentis laus, ut quasi non factam rem cantare, sed ipse eam cantando facere uideatur' (Serv.); see above, 46 n.

62–3. amarae / corticis: 'epitheton naturale' (DServ.); perhaps, but see above, 47 n. *Cortex* is feminine in Lucr. 4. 51, but elsewhere in V., where its gender can be determined, masculine: *G*. 2. 74, *A*. 7. 742, 9. 743. See 5. 38 n.

63. proceras . . . alnos: the phrase has a faint literary resonance: Catull. 64. 289–91 'proceras stipite laurus / non sine nutanti platano lentaque sorore / flammati Phaethontis'. In all other accounts, Phaethon's sisters, weeping over their brother's charred remains on the banks of the Eridanus, are turned into poplars distilling tears of amber. Ancient readers, understandably, were puzzled: 'quidam alnos poetica consuetudine pro populis accipiunt' (DServ.). But V. remembered the alder-fringed Po, the mythical Eridanus, of his youth, *G*. 2. 451–2 'nec non et torrentem undam leuis innatat alnus / missa Pado'. It is rare for a Latin poet to prefer personal experience to literary tradition, and V. may have regretted his youthful originality, for he later took pains to 'correct' it, *A*. 10. 189–91 'namque ferunt luctu Cycnum Phaethontis amati / populeas inter frondes umbramque sororum / *dum canit* et maestum Musa *solatur amorem*'. See Clausen 148 n. 61.

64. tum canit, errantem Permessi ad flumina: thus, unobtrusively, before he can quite be identified, Gallus is introduced into the world of pastoral fantasy. He wanders by a river, as shepherds do with their flocks, e.g. Ap. Rhod. 2. 502–3 (Cyrene) *Ἀπόλλων | τήν γ' ἀνερειψάμενος ποταμῷ ἔπι ποιμαίνουσαν*, 'Apollo snatched her up while she was tending her flock by the river', *E*. 10. 18 'ouis ad flumina pauit Adonis', and as might a disconsolate lover, like Antimachus by the Pactolus (Hermesianax fr. 7. 41–2 Powell) or Orpheus 'deserti ad Strymonis undam' (*G*. 4. 508). But the Permessus is no vulgar stream; it has a Hesiodic, a Callimachean resonance, and only a poet-shepherd would be found wandering along its banks.

Owing to Pfeiffer's patient research, the topography of Callimachean Helicon is much less obscure than it once was. Callimachus distinguishes two springs as sources of poetry:

(i) Hippocrene, the Horse's Spring, near the summit, where the Muses met Hesiod, *Aet*. fr. 2. 1–2 Pf. *Ποιμένι μῆλα νέμοντι παρ' ἴχνιον ὀξέος ἵππου | Ἡσιόδῳ Μουσέων ἑσμὸς ὅτ' ἠντίασεν*, 'when a bevy of Muses met the shepherd Hesiod tending his flock by the hoofprint of the fiery horse'. Hesiod himself is less precise; he places his encounter with the Muses 'under holy Helicon', *Theog.* 23 Ἑλικῶ-νος ὕπο ζαθέοιο, and assigns no symbolic value to the Permessus or to Hippocrene.

> *Μουσάων Ἑλικωνιάδων ἀρχώμεθ᾽ ἀείδειν,*
> *αἵ θ᾽ Ἑλικῶνος ἔχουσιν ὄρος μέγα τε ζάθεόν τε,*
> *καί τε περὶ κρήνην ἰοειδέα πόσσ᾽ ἁπαλοῖσιν*
> *ὀρχεῦνται καὶ βωμὸν ἐρισθενέος Κρονίωνος·*
> *καί τε λοεσσάμεναι τέρενα χρόα Περμησσοῖο*
> *ἢ Ἵππου κρήνης ἢ Ὀλμειοῦ ζαθέοιο*
> *ἀκροτάτῳ Ἑλικῶνι χοροὺς ἐνεποιήσαντο.* (*Theog.* 1–7)
>
> Let us begin our song from the Heliconian Muses,
> who inhabit Helicon, the great and holy mountain,
> and dance soft-footed around the dark spring
> and the altar of the mighty son of Cronus;
> and, with tender bodies bathed in the Permessus
> or Hippocrene or the Olmeius, holy stream,
> on the very top of Helicon make their dances.

Evidently Callimachus has remodelled Helicon to suit himself, imposing upon the antique poet; see *CHCL* ii. 183.

(ii) Aganippe, about which nothing is known before Callimachus (Pfeiffer on fr. 696), although Gow–Page on the epigrammatist Alcaeus 12. 5–6 (= *AP* 7. 55) remark that Hesiod's κρήνην ἰοειδέα is 'probably Aganippe'. Aganippe derives its origin from the Permessus and is therefore the daughter of the Permessus, 'filia Permessi . . . per Musarum vallem fluentis' (see Pfeiffer, ii. 102–3, on fr. 2a. 16–19, 20–5); Campbell on Soph. *Trach.* 14: 'The well-springs in the neighbourhood of a river were regarded in Greek mythology as the offspring of the river. Thus Callirhoë is the daughter of Scamander, and Achelöus too has a daughter Callirhoë'. Cf. Callim. *Hymn.* 4. 76–7 Δίρκη τε Στροφίη τε μελαμψήφιδος ἔχουσαι | Ἰσμήνου χέρα πατρός, 'Dirce and Strophie, holding the hand of their father, black-pebbled Ismenus'.

The only interpretation which V.'s lines (64–73) will bear is the obvious one: Gallus quits the valley (Permessus–Aganippe) for the mountain; he rises, so to speak, from a lower to a higher level of poetry (Hippocrene), from love-elegy and Lycoris to aetiology and the Grynean Grove. Both R. Reitzenstein, *Hermes*, 31 (1896), 194–5, and F. Skutsch (1901), 34 find in these lines a direct reference to Gallus' epyllion—to the proem, that is, in which, presumably, Gallus described his poetic initiation on Helicon.

V. seems to have been the only poet well acquainted with the Callimachean topography, or rather, hydrography, of Helicon; the only poet, at any rate, concerned to represent it accurately. Certainly Nicander was not; see below, 70 n. Propertius understood the significance of Aganippe, 2. 3. 20 (Cynthia) 'par Aganippaeae ludere docta lyrae', 10. 25–6 'nondum etiam Ascraeos norunt mea carmina fontis, / sed modo Permessi flumine lauit Amor'; but he erred strangely, or wilfully, with regard to Hippocrene, which he imagines in 2. 10. 25 and, with considerable detail, in 3. 3. 1–16 as the source of epic poetry. The self-styled 'Roman Callimachus' may have been misled by the proem to *G.* 3 (much of his erudition is borrowed from V.), in which Ennian and Callimachean references are so curiously mingled; see Clausen 13–14. Within a generation of Propertius' death the significance of Callimachus' two springs had been thoroughly confused: Ov. *Fast.* 5. 7 'Aganippidos Hippocrenes'—a confusion, however, for which Ovid is the sole evidence, and which A. Barchiesi, 'Discordant Muses', *PCPS* 37 (1991), 18 n. 5, suggests is intentional.

For the topography of Helicon and the valley that lies below it see P. W. Wallace, 'Hesiod and the Valley of the Muses', *GRBS* 15 (1974), 5–24, bearing in mind that Callimachus had never seen Helicon or the Valley of the Muses; his experience was entirely literary.

Gallum: for his life and poetry see *E.* 10, Introduction.

65. Aonas in montis: 'th'Aonian Mount'; cf. Lucr. 6. 786 'magnis Heliconis montibus'.

Aonas: a barbarian people anciently inhabiting Boeotia (Strabo 9. 401), whose name, however, is not certainly attested before Callimachus (Pfeiffer on fr. 572). The adjectival use of *Aones* must represent a Hellenistic development of such quasi-appositional phrases in older poetry as Hom. *Il.* 1. 594 *Σίντιες ἄνδρες*, 'Sintian men', Eur. *Andr.* 592 *ἀνδρὸς Φρυγός*; 'a Phrygian man'; cf. Nonnus, *Dionys.* 5. 37 *Ἄονι . . . λαῷ*, 'Aonian people', 286 *Ἄονες αὐλοί*, 'Aonian pipes', and see Schwyzer–Debrunner, *Griech. Gramm.* ii. 614. Pertinent also is an epigram by Antipater of Thessalonica, a poet nearly contemporary with V., addressed to a cupbearer named Helicon, who pours 'Ausonian Bacchus', *Αὔσονα Βάκχον* (3.3 Gow–Page, *The Garland of Philip* (Cambridge, 1968) = *AP* 11. 24). As an adjective, *Αὔσων* is rare; the regular adjective *Αὐσόνιος* is, like Ἀόνιος, Hellenistic (Gow–Page ibid., Hollis on Callim. *Hec.* fr. 18. 14) and was destined to have a long success 'in Ausonian land' (Milton); see *TLL* s.v. For *Aonius* see 10. 12 n.

una sororum: for the phrase cf. Hom. *Il.* 14. 267 Χαρίτων μίαν ὁπλοτεράων, 'one of the younger Graces', Eur. *Bacch.* 917 Κάδμου θυγατέρων . . . μιᾷ, 'one of the daughters of Cadmus', Ap. Rhod. 4. 896 Τερψιχόρη, Μουσέων μία, 'Terpsichore, one of the Muses', Enn. *Scen.* 71 V.^{2} = 49 J. 'Furiarum una'. The Muse so indicated must be Calliope, mother of Orpheus (4. 56–7) and eldest of the nine (Hes. *Theog.* 79)—as Propertius inferred, 3. 3. 37–8 'e quarum numero me contigit una dearum / (ut reor a facie, Calliopea fuit)'. In Longus 3. 27. 2, the eldest Nymph speaks for the group.

sororum: cf. [Eur.] *Rhesus* 891 Μοῦσα, συγγόνων μία, 'a Muse, one of the Sisters', 976 ἀδελφαί, 'Sisters'; Callim. *Aet.* fr. 43. 56–7 Pf. Κλειὼ . . . | χεῖρ' ἐπ' ἀδελφειῆς ὦμον ἐρεισαμένη, 'Clio . . . resting a hand upon her sister's shoulder', Ov. *Ars* 1. 27–8 'nec mihi sunt uisae Clio Cliusque sorores / seruanti pecudes uallibus, Ascra, tuis', Naev. *Bell. Poen.* 1 Strz. 'Nouem Iouis concordes filiae sorores', Prop. 2. 30. 27, Ov. *Met.* 5. 255, *Trist.* 5. 12. 45, Stat. *Theb.* 9. 317, Mart. 4. 14. 1, 31. 5, 9. 42. 3. In classical Greek poetry, apart from the *Rhesus*, the Muses are called not sisters but daughters, daughters of Zeus (Hom. *Il.* 2. 491–2) or Zeus and Mnemosyne (see West on Hes. *Theog.* 54) or Uranos (see West on Hes. *Theog.* 78), or simply goddesses.

66. uiro Phoebi: a deliberate juxtaposition—man and god—emphasizing the respect shown to Gallus (for the juxtaposition see Richardson on *Homeric Hymn to Demeter* 111). The rhetorical effect may in part be regarded as the solution to a problem of technique, that is, the avoidance of an oblique case of the pronoun *is*; see Axelson 70–1, Norden on *A.* 6. 174 'inter saxa uirum spumosa immerserat unda', Housman on Luc. 1. 293 'accenditque ducem'.

chorus: here first of the Muses in Latin (*TLL* s.v. 1024. 54); V. was thinking of Hesiod, *Theog.* 7 ἀκροτάτῳ Ἑλικῶνι χοροὺς ἐνεποιήσαντο, 'make their dances on topmost Helicon'.

adsurrexerit: 'in honour rose', an unprecedented gesture of respect by the Muses. Cf. *G.* 2. 98 and see *TLL* s.v. *assurgo* 938. 3.

67. Linus . . . pastor: 'quaeritur cur *pastor* dixerit' (DServ.). The answer must be that Linus, like Gallus, has become a poet-shepherd, Alexandrian and vatic; see Ross (1975), 21–3.

diuino carmine: may be taken with *pastor*, 'shepherd of inspired song', (so T. E. Page and others) or with *dixerit* (so Heyne and others); cf. Catull. 64. 321 'talia diuino fuderunt carmine fata'.

68. floribus atque apio: cf. Lucr. 5. 1399–1400 'caput atque umeros plexis redimire coronis / floribus et foliis' (note 'agrestis . . . Musa' in l. 1398). Celery was used for garlands; see *TLL* s.v. *apium* 240. 22, Gow on Theocr. 3. 23, Nisbet–Hubbard on Hor. *Carm.* 1. 36. 16. *Amarus* ('Vnde Epithetum?', La Cerda) is not elsewhere applied to *apium*; see above, 47 n.

crinis ornatus: 1. 54 n.

69–70. 'The Muses give you the pipes (come, take them), which they once gave to the old man of Ascra'. In fact, they gave him a laurel branch (*Theog.* 30), but giving the pipes of a dead singer to a worthy successor was a pastoral tradition (2. 38 n.).

69. en: here first with the imperative (*TLL* s.v. 547. 28).

70. Ascraeo: not so much a local as a literary reference, to Callimachus and his conception of Hesiod; cf. *G.* 2. 176, Prop. 2. 10. 25–6 (quoted above, 64 n.). The epithet first occurs in Nicander, *Ther.* 10–12 *εἰ ἐτεόν περ | Ἀσκραῖος μυχάτοιο Μελισσήεντος ἐπ' ὄχθαις | Ἡσίοδος κατέλεξε παρ' ὕδασι Περμησσοῖο*, 'if indeed he spoke the truth, Ascraean Hesiod on the steeps of secluded Melisseeis by the waters of Permessus' (Scholfield). According to the scholiast, Melisseeis was the place where Hesiod received instruction from the Muses.

Ascraeo . . . seni: cf. Callim. *Hymn* 4. 304 *Λυκίοιο γέροντος*, 'the old man of Lycia', with Mineur's note, *G.* 4. 127 'Corycium . . . senem'.

71. cantando . . . deducere montibus ornos: 'Novum vero hoc, quod nunc Hesiodo tribuitur, id quod de Orpheo solenne est, silvas eius cantum esse sequutas' (Heyne). K. Ziegler, *RE* xviii. 1249 n. 1, suggests that V.'s innovation, as it appears to be, may be Hellenistic. It is perhaps of some importance that Hesiod was reputedly a descendant of Orpheus (*RE* viii. 1169–70); but V. also attributes an Orphean potency to the song of Silenus (above, 27–30) and to the songs of Damon and Alphesiboeus (8. 1–4).

72. Grynei nemoris . . . origo: a grove consecrated to Apollo 'a Gryno filio; uel a Grynio, Moesiae ciuitate, ubi est locus arboribus multis iucundus, gramine floribusque uariis omni tempore uestitus, abundans etiam fontibus' (DServ.; see Ross (1975), 79–80). 'In quo aliquando Calchas et Mopsus dicuntur de peritia diuinandi inter se habuisse certamen, et cum de pomorum arboris cuiusdam contenderent numero, stetit gloria Mopso, cuius rei dolore Calchas interiit [cf. Hes. fr. 278 M.–W.[3]]. hoc autem Euphorionis continent carmina, quae Gallus transtulit in sermonem Latinum, unde est illud in fine, ubi Gallus loquitur "ibo et Chalcidico quae sunt mihi condita uersu / carmina" [10. 50–1, where see note], nam Chalcis ciuitas est Euboeae, de qua fuerat Euphorion' (Serv.).

The Grynean Grove may have been suggested to Gallus as a subject for an aetiological poem by his teacher Parthenius. Parthenius himself wrote a poem about a more celebrated cult-place of Apollo, Delos, which contained the phrase *Γρύνειος Ἀπόλλων* (*Suppl. Hell.* 620); cf. *A.* 4. 345 'Gryneus Apollo' and see Clausen, *GRBS* 5 (1964), 192 = K. Quinn (ed.), *Approaches to Catullus* (Cambridge, 1972), 280. And Parthenius did make such suggestions; in his *Ἐρωτικὰ παθήματα* he collected and summarized thirty-six stories for Gallus to use, as he writes in the preface, *εἰς ἔπη καὶ ἐλεγείας*—not 'for either epic or elegiac verse' (Gaselee in the Loeb Classical Library) but 'for either hexameter or elegiac verse', that is, epyllia or elegies. See F. Jacoby, 'Zur Entstehung der römischen Elegie', *RhM* 60 (1905), 69 n. 2.

73. ne quis sit lucus quo se plus iactet Apollo: an imitation of Callimachus' hymn to Delos, 4. 269–70 *οὐδέ τις ἄλλη | γαιάων τοσσόνδε θεῷ πεφιλήσεται ἄλλῳ*, 'nor will any other land be so loved by another god' (Heyne).

quis: despite the sigmatism, *quis* must be right here (*quis* P: *qui* MR), for this is the only form of the indefinite used by V.; see Löfstedt ii. 87 n. 3.

iactet: cf. *G.* 1. 102–3 'nullo tantum se Mysia cultu / iactat', *A.* 6. 876–7 'nec Romula quondam / ullo se tantum tellus iactabit alumno' (*TLL* s.v. 61. 21).

74. Scyllam Nisi: cf. *A.* 6. 36 'Deiphobe Glauci' and see Kühner–Stegmann i. 414. V. has deliberately conflated two stories, for he might have written *Scyllam Phorci*; cf. A. 5. 240, 824, 10. 328 and see Livrea on Ap. Rhod. 4. 828. It was Scylla, the daughter of Phorcys, who was transformed into a sea-monster. Scylla, the daughter of Nisus, betrayed the city of Megara to

Minos by cutting off her father's magical lock of hair and was transformed into a sea-bird; see Hollis on Callim. *Hec.* fr. 90. Older commentators supposed that V. had simply made a mistake, e.g. Wagner, *Quaest. Virg.* 40. 3 ('Virgilius dormitans aliquando'). But V. was not 'nodding'; he was concerned, rather, to tell one of the Scylla-stories while reminding his reader of the other, as Propertius was later to do with the story of Atalanta (above, 61 n.). The manner is Alexandrian, consummately so.

The story of Scylla, the daughter of Nisus, is found in its 'pure' form in *G.* 1. 404–9, Prop. 3. 19. 21–6, Ov. *Met.* 8. 1–151 (note, however, Scylla's curse, 120–1 'non genetrix Europa tibi est, sed inhospita Syrtis, / Armeniae tigres Austroque agitata Charybdis'; I owe this observation to J. Huyck), and the *Ciris*; the story of Scylla, the daughter of Phorcys, in Ov. *Met.* 13. 898–14. 74. The conflated story is briefly told in Prop. 4. 4. 39–40, Ov. *Am.* 3. 12. 21–2 'per nos (*poetas*) Scylla patri caros furata capillos / pube premit rabidos inguinibusque canes', and *Ars* 1. 331–2, and alluded to in *Rem.* 737 and *Fast.* 4. 500 'et uos, Nisei, naufraga monstra, canes'.

quam fama secuta est . . .: the poet appeals to tradition for the truth of what he says, a literary convention as old as Homer. For a discussion of such expressions in Greek and Latin poetry see Norden on *A.* 6. 14 'ut fama est'; also Austin on A. 6. 14, Nisbet–Hubbard on Hor. *Carm.* 1. 7. 23, N. Horsfall, *Virgilio: l'epopea in alambicco* (Naples, 1991), 117–33. *Secuta* is unparalleled and curious; had V. vaguely in mind the pursuit of Scylla by Nisus (*G.* 1. 408 'insequitur Nisus')?

75. candida succinctam latrantibus inguina monstris: cf. Lucr. 5. 892–3 'aut rabidis canibus succinctas semimarinis / corporibus Scyllas'.

candida . . . inguina: cf. Ov. *Am.* 2. 16. 23 'quae uirgineo portenta sub inguine latrant'.

succinctam . . . inguina: 1. 54 n.

76. Dulichias: the adjectival use of proper nouns in poetry is as old as Lucilius 676 M. 'Metello . . . munere'; cf. Lucr. 6. 738 'Auerna . . . loca', *A.* 4. 552 'cineri . . . Sychaeo', 6. 876–7 'Romula . . . / . . . tellus', 10. 273 'Sirius ardor', Prop. 1. 11. 30 'Baiae . . . aquae', 2. 1. 76 'esseda . . . Britanna', and see Hofmann–Szantyr 427. Odysseus was lord of Dulichium, Same, Zacynthus, and

Ithaca (*Od.* 1. 246–7, where see S. West), but in Latin poetry he is especially associated with Dulichium; see *TLL* Onomast. s.v. 62.

uexasse: like that of its English derivative 'vex' (Milton, *Paradise Lost*, 1. 305–6 'when with fierce Winds Orion arm'd / Hath vext the Red-Sea Coast'), the force of *uexo* was weakened with the passage of time, and later readers found *uexasse* incongruous here, Gellius 2. 6. 2 '*uexasse* enim putant uerbum esse leue et tenuis ac parui incommodi nec tantae atrocitati congruere'. Wishing to defend *uexasse*, Gellius § 7 cites Cato, *De Achaeis*, fr. 187 M. 'cumque Hannibal terram Italiam laceraret atque uexaret', and Cic. 2 *Verr.* 4. 122 (*Aedis Mineruae*) 'quae ab isto sic spoliata atque direpta est, non ut ab hoste aliquo, qui tamen in bello religionem et consuetudinis iura retineret, sed ut a barbaris praedonibus uexata esse uideatur'. DServius, or rather Donatus, likewise cites Cato and Cicero here, adding 2 *Verr.* 4. 104 'di ablati, fana uexata, direptae (DServ., *nudatae* Cic.) urbes reperiuntur'. Cf. also Hor. *Carm.* 3. 2. 3–4 'Parthos ferocis / uexet eques'. See L. A. Holford-Strevens, *Aulus Gellius* (London, 1988), 150–1.

rates: poetic plural; Odysseus had only one ship left (Wagner).

gurgite: a poetic word, usually occupying the fifth foot of the hexameter (*TLL* s.v. 2360. 3). See Clausen, *HSCP* 90 (1986), 166.

77. a: 10. 47 n.

78–81: an allusive summary of the old and horrible, yet ever popular, tale of Tereus king of Thrace, his wife Procne, her raped and mutilated sister, Philomela, and their transformation into birds; see D'A. W. Thompson, *A Glossary of Greek Birds* (Oxford, 1936), 20–2, Ov. *Met.* 6. 424–674, with Bömer's note on 668–9. In Greek literature, Procne becomes a nightingale, Philomela a swallow; in Latin, however, their roles are usually reversed, and Philomela with her 'melodious pain' becomes a nightingale. Philomela is oddly ascendant in these lines, for rhetoric demands that she be the subject of *pararit*, *petiuerit*, and *uolitauerit* (so T. E. Page and Lee; Klingner 105 makes Tereus the subject of *petiuerit* and *uolitauerit*, Coleman of *uolitauerit*; Conington is uncertain). Line 81 'sua tecta' (although attachment to a dwelling, if not merely pathetic here, indicates a swallow) implies that Philomela, not the unnamed Procne, was Tereus' wife; or rather, as in the case of the two Scyllas, an ambiguous tradition permitted V. to conflate the roles of the two sisters.

79–81. quas . . . dapes, quae dona . . . / quo cursu . . . quibus . . . / . . . alis: eccentric emphasis as above, 43 'quo fonte', a characteristic of Alexandrian narrative; see F. Skutsch (1901), 32–3, 42.

79. dona: 'quod satiato Tereo caput et pedes filii uxor intulerit' (DServ.).

80. quo cursu: 'by what route' (Lee); cf. *A.* 6. 194–5 'cursumque per auras / derigite in lucos'.

deserta: 10. 52 n.

81. infelix: cf. Hor. *Carm.* 4. 12. 5–6 'nidum ponit Ityn flebiliter gemens / infelix auis'.

super uolitauerit: or *superuolitauerit*? The problem is an old one; see Thilo's app. crit. to Servius here. Recently R. O. A. M. Lyne, *Ciris* (Cambridge, 1978), 19 n. 1, has argued for the compound; and see now S. J. Harrison on *A.* 10. 384. But a weak caesura in the third foot unaccompanied by a strong caesura in the fourth ('infelix sua tecta superuolitauerit') would be abnormal: it occurs in only six lines of the *E.*, 2. 24, 4. 16, 34, 5. 52, 9. 60, 10. 12, all of which contain Greek words; see Norden 431–2. Cf., for the rhythm, *A.* 5. 330 'fusus humum uiridisque super madefecerat herbas'. See 3. 38 n.

82–4. These lines have been understood in two different ways: (i) the song of Silenus is not his own composition but a song composed and sung by Apollo; or (ii) the song of Apollo is the last of the themes in the song composed and sung by Silenus. The second interpretation must be right, for, as F. Skutsch (1906), 133–8 observes, Apollo's song could not conceivably include Gallus' initiation on Helicon; 'ille canit' (84) therefore introduces a new theme, like 61 'tum canit', 64 'tum canit'. For a brief history of the question see Knox 185 n. 8.

What, then, was the subject of Apollo's song? Why was Apollo singing by the Eurotas? And why did the Eurotas bid his laurels learn Apollo's song? Cf. DServius on 83: 'nam hunc fluuium Hyacinthi causa Apollo dicitur amasse', the name of the river being thus an allusion to the story of Apollo's love for Hyacinthus; cf. Nicander, *Ther.* 903–5 *ὃν Φοῖβος θρήνησεν ἐπεί ῥ' ἀεκούσιος ἔκτα | παῖδα βαλὼν προπάροιθεν Ἀμυκλαίου ποταμοῖο, | πρωθήβην Ὑάκινθον,* 'over whom Phoebus wept, since without willing it, hard by the river of Amyclae he slew with a blow the boy Hyacinthus in the bloom of youth' (Scholfield). DServius' explanation no doubt represents an ancient consensus, and every modern commentator has accepted it. But Knox argues that V. is alluding not to Hyacinthus but to Daphne. No extant witness to the story of Apollo and Hyacinthus mentions a song, but Apollo's song of courtship to Daphne is attested, if barely, in literature; the main evidence is in art. Daphne's metamorphosis—and metamorphosis is a feature of Silenus' song—into a laurel-tree (*δάφνη*) accounts for the presence of laurel-trees by the Eurotas (see below, 83 n.), where their whispering leaves repeat Apollo's song, learned at the river's bidding. Further summary would only diminish the effect of Knox's elaborate and elegant argument, which should be studied in its entirety.

82. Phoebo . . . meditante: the reminiscence of 1. 8 'agrestem tenui meditabor harundine Musam' is intentional. Apollo, too, is a votary of the slender Muse.

83. lauros: in Theocr. 18. 44, plane-trees grow by the Eurotas, as in fact they do (A. Lindsell, *G&R* 6 (1937), 81), but in Catull. 64. 89, myrtles; Wilamowitz, *Hellenistische Dichtung* (Berlin, 1924), ii. 284: 'Der Myrtenbusch . . . wächst 64, 89 am Eurotas: man sieht, welchen Wert der geographische Name hat'.

84. ille canit, pulsae . . .: the valleys reverberate Silenus' voice to the stars; cf. *G.* 4. 49–50 'ubi concaua *pul*su / saxa sonant uocisque offensa re*sult*at imago', *A.* 5. 149–50 'consonat omne nemus, uocemque in*clu*sa uo*lut*ant / litora, *pulsa*ti colles clamore re*sulta*nt', 7. 701–2 'sonat amnis et Asia longe / *pul*sa *pal*us, 12. 334–5 'gemit *ult*ima *pul*sa / Thraca ped*um* circ*um*que . . .'. See also 1. 5 n. (*Amaryllida siluas*), 5. 62–3 n.

referunt: 'quo uerbo aliter in sequenti uersu utitur' (DServ.); see above 14 n.

85. stabulis: dative, as so often in V. with verbs of motion.

numerumque referre: cf. 3. 34, *G.* 4. 436 'numerumque recenset'.

86. iussit: the subject of the infinitive is to be supplied from the context; cf. 4. 32–3 and the passages collected by Forbiger here.

inuito: La Cerda compares Ov. *Met.* 1. 682–3 'euntem multa loquendo / detinuit sermone diem', Voss Hom. *Il.* 18. 239–40 Ἠέλιον δ' ἀκάμαντα βοῶπις πότνια Ἥρη | πέμψεν ἐπ' Ὠκεανοῖο ῥοὰς *ἀέκοντα νέεσθαι*, 'Queen Hera the ox-eyed sent the unwearied, unwilling sun to sink in the Ocean stream'.

processit: of a star coming into view; Cic. *Arat.* 391 'iam toto processit corpore Virgo'. Cf. 9. 47.

Vesper Olympo: Catull. 62. 1 'Vesper Olympo'; the same cadence in *A.* 1. 374, 8. 280. For Olympus denoting the sky see 5. 56 n.

ECLOGUE 7

Introduction

Discussion of the Seventh *Eclogue* has long been preoccupied with a single question, 'Why does Thyrsis lose the singing-match?' Generations of readers have been surprised, and puzzled, by the ending—as was Samuel Johnson, *The Adventurer*, No. 92, 22 September 1753: 'One of the shepherds now gains an acknowledged victory, but without any apparent superiority, and the reader, when he sees the prize adjudged, is not able to discover how it was deserved'.

La Cerda, scrupulous as always, examines each exchange separately and renders his decision: in the first three (21–4 ~ 25–8, 29–32 ~ 33–6, 37–40 ~ 41–4) Corydon is the winner; the fourth and fifth (45–8 ~ 49–52, 53–6 ~ 57–60) are a draw; but in the sixth and last (61–4 ~ 65–8) Corydon's victory is complete.[1] Cartault[2] finds Corydon superior to Thyrsis in the second exchange, Thyrsis superior to Corydon in the fifth, and both equal in the sixth; as for the first, third, and fourth, 'C'est affaire de goût personnel de préférer l'un ou l'autre' (195). Others[3] have tried to justify Corydon's victory by discovering technical defects in the lines of Thyrsis; but that Virgil—or any Latin poet, for that matter—should have defaced his poem for the sake of an effect is out of the question, and in any case the defects attributed to Thyrsis tend, on closer inspection, to disappear.[4] Those committed to a point of view will hardly be impartial, as Heyne observes in an ironic note on l. 69:

[1] 'Victoria Corydonis est, et ea manifestissima, et multiplex, adeo ut putet Nannius 6. Miscell. ex extrema hac lite adiudicatam a Meliboeo victoriam Corydoni', citing Pieter Nanninck, *Συμμίκτων, siue Miscellaneorum decas una* (Leuven, 1548), 180–1, who offers no opinion on the preceding exchanges.

[2] A. Cartault, *Étude sur les Bucoliques de Virgile* (Paris, 1897), 191–8.

[3] Notably V. Pöschl, *Die Hirtendichtung Virgils* (Heidelberg, 1964), 101–42; refuted in some detail by O. Skutsch, *Gnomon*, 37 (1965), 166–7, and H. Dahlmann, *Hermes*, 94 (1966), 218–29 = *Kleine Schriften* (Hildesheim, 1970), 166–77.

[4] Thus F. H. Sandbach, *CR* 47 (1933), 216–19, finds two metrical 'defects' in the lines of Thyrsis: but the first, a matter of punctuation in l. 35, is more apparent than real (see G. B. Townend, *CQ*, NS 19 (1969), 339); and the second, the elision of a pyrrhic word in l. 41, unexceptionable, as O. Skutsch points out, *BICS* 18 (1971), 28: 'elision of pyrrhic words is rare . . . but it is not therefore objectionable, and *tibi*

Operose docent viri docti per totam Eclogam, quantopere in singulis versibus superior Thyrsi sit Corydon, iis tamen fere argumentis, ut, si poeta Thyrsin victorem pronuntiasset, iidem huius iudicii caussas et probationes ab iisdem locis ducturi fuisse videantur.[5] Hoc facile perspicio, Thyrsin esse perpetuum Corydonis obtrectatorem; id quod ipsa lex carminis amoebaei postulat; huius tamen versus qua tandem in re Thyrsidos numeris praestent, qui tardus meus sensus est, equidem non perspicio.[6]

Critics have been unkind to Thyrsis. True, he is an *obtrectator*, as Heyne remarks, contentious, sardonic, vain; but to some extent this character is imposed by his role as second singer.[7] If Thyrsis is to be judged fairly, he must be judged as a singer subordinate to Corydon but not necessarily inferior; indeed, he is superior to Corydon in at least one exchange,[8] failing only, so that there may

amarior in any case is a deliberate and very effective echo of *mihi dulcior* in line 37'. Curiously, Sandbach overlooks an even worse 'defect', the elision *si ultra* in l. 27 (Norden on *A*. 6. 770 'mit sehr seltner Synaloephe', found once elsewhere in the *E*., 8. 41 'ut uidi, ut perii', but not in the *G*. and only five times in the *A*.). Again, the same explanation obtains: *aut, si ultra* is a deliberate echo of *aut, si non* in l. 23. As to Sandbach's other criticisms, 'subjective, it must be confessed' (218).

[5] Pöschl (above, n. 3) finds expressed in l. 53 'stant et iuniperi et castaneae hirsutae' a teeming abundance and ripeness: 'Der merkwürdige Vers mit seiner Häufung von Spondeen, Hiaten und Diphthongen malt die strotzende, lastende Fülle der in voller Reife stehenden fruchtschweren Bäume' (134). Cf. Skutsch (above, n. 3) 168: 'zwei Hiate und ein spondeischer Schluß, unerhört; schade, daß es Corydon ist und nicht Thyrsis!' And in fact H. Fuchs, *MH* 23 (1966), 218–22, following J. Perret, *Virgile, Les Bucoliques* (Paris, 1961), 82–3, wishes to rearrange the quatrains—an old suggestion, mentioned by Heyne in his app. crit. to l. 53—so as to give 57–60 to Corydon and 53–6 to Thyrsis. Perret: 'Les v. 57–60 ont l'élégance, les images gracieuses, la noblesse qui caractérisent Corydon; le v. 53 est écrit dans le stil âpre, volontiers heurté, de Thyrsis, il évoque l'image d'objets raides et piquants; le v. 56 est prosaïque et sans fantaisie, il termine le quatrain sur la vision désagréable d'une nature rétractée', quoted approvingly by Fuchs.

[6] Cf. T. E. Page on l. 69: 'Of course, however, the defeat of Thyrsis is wholly fictitious, and no one would have been more astonished than Virgil himself if any one had taken him literally and pointed out to him the defects in the lines which he assigns to Thyrsis'.

[7] The second singer need not be disagreeable, however. There is no acerbity in the responses of Daphnis to Menalcas in [Theocr.] 8, which served as V.'s model, and Daphnis wins the prize. Thyrsis' bad temper disposes the reader to acquiesce in his defeat.

[8] The fifth (53–6 ~ 57–60); Klingner 124: 'Endete der Zwiegesang mit diesen Gruppen, so könnte Corydons Recht auf den Sieg zweifelhaft scheinen'.

be a slight pretext for his defeat, at the very end.[9] No doubt Corydon is the more musical of the two, yet the voice of Thyrsis is essential to the harmony of Virgil's composition.

The Seventh is unique among the *Eclogues* in being a reminiscence,[10] the story of an important singing-match between Corydon and Thyrsis as told by Meliboeus, a small farmer. Daphnis was there, under a rustling ilex, where Corydon and Thyrsis had driven their flocks in the heat of the day, and Meliboeus was mulching his myrtles against the danger of a late frost when, unaccountably, his he-goat wandered off to where Daphnis was sitting. Daphnis' invitation to join them in the shade, '... your steers will find their own way through the meadows to drink here' (11), his own hesitation, since he had no Phyllis, no Alcippe, at home to help with the chores—all this Meliboeus remembers vividly and the singing-match word for word, but at the end only that Thyrsis lost, 69 'haec memini, et uictum frustra contendere Thyrsin'. Why so terse and vague at the end?

On being introduced into the poem, Corydon and Thyrsis are described as singers, Arcadians both, both in the flower of youth, well paired and prepared to sing (2–5). The ilex, under which they are resting with their flocks, is a shepherds' tree, a favoured resort for them; but Daphnis happens to be there (1). His fortuitous presence indicates that Daphnis has been chosen umpire;[11] for umpires are chosen—Morson as he cuts the heather near by,[12] the goatherd whose barking dog causes him to be noticed[13]—without forethought, on the spur of the moment. There is no mention of the usual preliminaries to a singing-match, because Meliboeus remembers none; he had been too busy with his myrtles, until distracted by the aberration of his he-goat, to notice what was

[9] Cf. Skutsch (above, n. 4), 28: 'the inferiority of the loser, only faintly indicated elsewhere because otherwise the beauty of the whole composition would be damaged, must be most strongly in evidence immediately before the decision is given'; Nannius (above, n. 1) 180: 'Adfert et ille quidem quatuor arbores, uerum populus quae quarta est, iam a Corydone praeoccupata fuit, quod est signum ingenii in aliqua parte deficientis'.

[10] Only one of Theocritus' *Idylls*, the Seventh, is a reminiscence.

[11] Cf. Dryden's confident translation: 'Daphnis, as Umpire, took the middle Seat'.

[12] Theocr. 5. 61–6.

[13] [Theocr.] 8. 25–7. Cf. *E.* 3. 50 'audiat haec tantum—uel qui uenit ecce Palaemon'.

going on under the ilex. By making his poem dramatic—the imperfect recollection, that is, of a witness to an event and not, like the Third *Eclogue*,[14] an omniscient poet's description of it— Virgil frees himself from the obligation, or possibly the embarrassment, of justifying his umpire's decision.[15] Meliboeus remembers that Thyrsis lost and forgets everything else, for, in comparison with Corydon's victory, nothing else mattered: from that time on Corydon was—Corydon.[16]

Bibliography

Klingner 118–25.

C. Fantazzi and C. W. Querbach, 'Sound and Substance: A Reading of Virgil's Seventh Eclogue', *Phoenix*, 39 (1985), 355–67.

1–5. V. begins with an imitation of the opening lines of two *Idylls*, Theocr. 6. 1–4 *Δαμοίτας καὶ Δάφνις ὁ βουκόλος εἰς ἕνα χῶρον | τὰν ἀγέλαν ποκ', Ἄρατε, συνάγαγον· ἦς δ' ὁ μὲν αὐτῶν | πυρρός, ὁ δ' ἡμιγένειος· ἐπὶ κράναν δέ τιν' ἄμφω | ἑσδόμενοι θέρεος μέσῳ ἄματι τοιάδ' ἄειδον*, 'Damoetas and Daphnis the neatherd once, Aratus, gathered the herd together to the same place. Golden was the chin of one; the other's beard half-grown. And at a spring the pair sat down, in summer at noonday, and thus they sang' (Gow), [Theocr.] 8. 1–4 *Δάφνιδι τῷ χαρίεντι συνάντετο βουκολέοντι | μῆλα νέμων, ὡς φαντί, κατ' ὤρεα μακρὰ Μενάλκας. | ἄμφω τώγ' ἤστην πυρροτρίχω, ἄμφω ἀνάβω, | ἄμφω συρίσδεν δεδαημένω, ἄμφω ἀείδεν*, 'While herding his sheep on the high hills Menalcas met (it is said) the beautiful Daphnis pasturing his cattle. Both were red-haired, both young; both skilled in piping, both in singing'.

1. sub arguta . . . ilice: the holm-oak, *Quercus ilex* L.; 'In a wind there is a harsh rustling in the leaves' (Sargeaunt 62).

consederat: may imply that he was making music; cf. Longus 1. 13. 4 *ὁ μὲν Δάφνις ὑπὸ τῇ δρυῒ τῇ συνήθει καθεζόμενος ἐσύριττε*,

[14] Virgil seems to invite comparison with the Third *Eclogue*: in both the Third and the Seventh the singing-match consists of 48 lines.

[15] See *E.* 3, Introduction, p. 91.

[16] 70 'ex illo Corydon Corydon est tempore nobis', where see note. Cf. [Theocr.] 8. 92 *κὴκ τούτω πρᾶτος παρὰ ποιμέσι Δάφνις ἔγεντο*, 'And from that day was Daphnis first among the herdsmen' (Gow), that is, from the day he defeated Menalcas in their singing-match.

'Daphnis sat down beneath the usual oak and began to play his pipe', and see *E.* 3. 55 n.

2–3. Corydon et Thyrsis: the names of the two herdsmen are repeated in chiastic order with some elaboration; similarly in Theocr. 6. 42–4. The pattern is as old as Homer, cf. *Il.* 2. 870–1 *τῶν μὲν ἄρ' Ἀμφίμαχος καὶ Νάστης ἡγησάσθην, | Νάστης Ἀμφίμαχός τε*, 'of these the leaders were Amphimachus and Nastes, Nastes and Amphimachus', and, for more complex examples, *Od.* 15. 249–53, Ap. Rhod. 1. 71–4, *A.* 5. 294–6, 11. 690–5. See Hopkinson on Callim. *Hymn* 6. 71–70, S. E. Bassett, *The Poetry of Homer* (Berkeley, 1983), 120–8.

2. compulerantque greges: 2. 30 n.

Corydon: 2. 1 n.

Thyrsis: a name borrowed from Theocr. 1. 19; there perhaps with a 'vaguely Dionysiac suggestion' (Gow), here simply pastoral.

in unum: a Lucretian phrase (seven times), which V. was not to use again until late in the *A.* (8. 576, 9. 801, 10. 410, 12. 714); here representing *εἰς ἕνα χῶρον*. Corydon and Thyrsis had come together at the usual tree, a shepherds' tree: Theocr. 1. 21–3 *δεῦρ' ὑπὸ τὰν πτελέαν ἑσδώμεθα τῶ τε Πριήπω | καὶ τᾶν κρανίδων κατεναντίον, ᾇπερ ὁ θῶκος | τῆνος ὁ ποιμενικὸς καὶ ταὶ δρύες*, 'come, let us sit beneath the elm, facing Priapus and the springs, where that shepherds' seat is and the oaks', Leonidas of Tarentum 86. 4 G.–P. (=*A.Pl.* A 230) *πὰρ κείνᾳ ποιμενίᾳ πίτυϊ*, 'by that shepherds' pine', Longus 1. 13. 4 (above, 1 n.), 2. 30. 2 *ἐπὶ τὴν συνήθη φηγόν*, 'to the usual oak', Calp. Sic. 4. 2–3 'quidue sub hac platano . . . / insueta statione sedes?', Milton, *Epitaph. Dam.* 15 'assueta seditque sub ulmo'. In *G.* 4. 433–6, Proteus, who is likened to a shepherd, seeks refuge from the noonday heat for himself and his 'flock' in the usual cave (429 'consueta . . . antra'), so that Cyrene knows where to find him.

3. distentas lacte capellas: cf. Lucr. 1. 258–9 'candens lacteus umor / uberibus manat distentis', Hor. *Epod.* 16. 49–50 'iniussae ueniunt ad mulctra capellae, / refertque tenta grex amicus ubera', *E.* 4. 21–2 'ipsae lacte domum referent distenta capellae / ubera'. The time of day seems to be late afternoon, for though it is still hot (10) the goats' udders are swollen with milk—late afternoon, then, of a day in early spring, when there is some danger of frost and the lambs have just been weaned (below, 15 n.).

4. aetatibus: for the plural (unique in V.) see Kühner–Stegmann i. 77–8. The singular would perhaps have been more normal (cf. Livy 26. 49. 13 'et aetate et forma florentes . . . filiae', 29. 1. 2 'trecentos iuuenes, florentes aetate') but V. was concerned to reproduce the thythm of [Theocr.] 8. 3–4 (above, 1–5 n.).

Arcades ambo: cf. Erucius (1 Gow–Page, *The Garland of Philip* = *AP* 6. 96) *Γλαύκων καὶ Κορύδων οἱ ἐν οὔρεσι βουκολέοντες,* | *Ἀρκάδες ἀμφότεροι, τὸν κεραὸν δαμάλαν* | *Πανὶ φιλωρείτᾳ Κυλληνίῳ αὐερύσαντες* | *ἔρρεξαν καί οἱ δωδεκάδωρα κέρα* | *ἅλῳ μακροτένοντι ποτὶ πλατάνιστον ἔπαξαν* | *εὐρεῖαν, νομίῳ καλὸν ἄγαλμα θεῷ,* 'Glaucon and Corydon, those ox-herds on the hills, Arcadians both, in honour of Cyllenian Pan the mountain-lover drew back the head of a horned calf and sacrificed it, and with a long-tapering nail they fastened its horns of twelve palms' length to a broad plane-tree, a fine ornament for the pastoral god' (Gow–Page). 'Arcades ambo', *Ἀρκάδες ἀμφότεροι*—what is the connection, if any, between V. and Erucius? Three explanations are possible: (i) the phrase is derived independently from a common source: so R. Reitzenstein, *RE* vi. 565; (ii) Erucius is imitating V.: so Wilamowitz, *Die Textgeschichte der griechischen Bukoliker* (Philol. Untersuch. 18; Berlin, 1906), 111 n. 1, and Norden in C. Cichorius, *Römische Studien* (Leipzig, 1922), 306; or (iii) V. is imitating Erucius, a possibility which Norden dismisses without argument. Of these explanations, the first is not improbable; Amaryllis and her resonant name occur in both V. and Longus, and yet neither can be shown to have imitated the other; see 1. 5 n. The second explanation is the least probable, for Greek poets seem never, or almost never, to have imitated Latin poets; see R. Reitzenstein, *Epigramm und Skolion* (Giessen, 1893), 132 n. 2, Clausen 138 n. 27. Norden's point, that Erucius bears a Roman name (see Cichorius 304, R. Syme, *History in Ovid* (Oxford, 1978), 180), is of no consequence: so does Longus. There remains the probability that V. is imitating Erucius; in which case V. combines an imitation of Theocritus with an imitation of Erucius, as in 10. 38–9 he combines an imitation of Theocritus with an imitation of Asclepiades.

Erucius, perhaps a native of Cyzicus, is the author of fourteen extant epigrams, 'fluent, elegant, and lucid' (Gow–Page); his *floruit* may be placed in the second half, perhaps the third quarter, of the first century BC. Erucius was therefore a contemporary of Parthenius, whom he outlived and attacks, even though dead, with an animosity that suggests personal acquaintance (13 Gow–Page = *AP* 7. 377). That Erucius composed an epigram using the pastoral name Corydon should occasion no surprise; the Greek pastoral tradition was continued, if incidentally and on a diminished scale, by the epigrammatists (Reitzenstein, *RE* vi. 565). Cf. Glaucus 3. 1–2 G.–P. (= *AP* 9. 341) *Νύμφαι, πευθομένῳ φράσατ' ἀτρεκὲς εἰ παροδεύων | Δάφνις τὰς λευκὰς ὧδ' ἀνέπαυσ' ἐρίφους*, 'Nymphs, answer me truly, if Daphnis passing by rested here his white goats'; Meleager 29. 1 G.–P. (= *AP* 5. 139) *Ἁδὺ μέλος, ναὶ Πᾶνα τὸν Ἀρκάδα*, 'Sweet is the song, by Pan the Arcadian', 126. 1–3 G.–P. (= *AP* 7. 535) *Οὐκέθ' ὁμοῦ χιμάροισιν ἔχειν βίον, οὐκέτι ναίειν | ὁ τραγόπους ὀρέων Πὰν ἐθέλω κορυφάς· | τί γλυκύ μοι, τί ποθεινὸν ἐν οὔρεσιν; ὤλετο Δάφνις*, 'No more to live with goats, no more do I, goat-footed Pan, wish to dwell upon the mountain tops. What pleasure, what delight have I in the mountains? Daphnis is dead'; Zonas (8. 1–2 Gow–Page, *The Garland of Philip* = *AP* 9. 556) *Νύμφαι ἐποχθίδιαι Νηρηίδες, εἴδετε Δάφνιν | χθιζόν;*, 'Nymphs of the shore, Nereids, did you see Daphnis yesterday?'

Finally, it will be noticed that V.'s 'Arcadians' are metaphorical, while the Arcadians of Erucius are real, not singers but herdsmen; and it is easier to imagine V. translating the Arcadians of Erucius to the Po valley than it is to imagine Erucius returning V.'s 'Arcadians' to the Peloponnesus.

5. et cantare pares et respondere parati: 'qui possent et continuum carmen dicere . . . et amoebaeum referre' (Serv.); a perfectly balanced line consisting of two assonating parallel clauses, the second longer than the first. For 'pares . . . parati' see Norden on *A*. 6. 204 ff.; for the adnominal infinitive, Hofmann–Szantyr 350–1. Cf. 10. 32.

6–7. dum . . . defendo . . . / . . . deerrauerat: for the pluperfect cf. Enn. *Ann*. 356 Skutsch 'missaque per pectus dum transit striderat hasta', Cic. *Ad Att*. 10. 16. 5 'dum redeo, Hortensius uenerat' (Hofmann–Szantyr 613); Heyne compares *A*. 6. 171–4.

6. teneras defendo a frigore myrtos: the normal construction, e.g. Columella, *De arb*. 12. 1 'ea res et a sole et a nebula maxime uuam defendit' (*TLL* s.v. *defendo* 299. 32). Meliboeus had planted some myrtles and wished to protect them from the cold.

teneras: probably young, or frost-tender, as myrtles of any age are (Ov. *Am*. 1. 15. 37 'metuentem frigora myrtum'), and these are growing far inland. The myrtle, 'fearless of the mild sea air', likes to be near the sea (*G.* 4. 124 'amantis litora myrtos'), where Aphrodite rising from the foam first saw it and whose tree it became. The Po valley is cold in winter, too cold even for the olive; see J. Bradford, *Ancient Landscapes* (London, 1957), 163.

7. uir gregis ipse caper: cf. [Theocr.] 8. 49 *ὦ τράγε, τᾶν λευκᾶν αἰγῶν ἄνερ*, 'He-goat, husband of the white nannies' (Gow). See Nisbet–Hubbard on Hor. *Carm.* 1. 17. 7 'olentis uxores mariti', A. S. Hollis, *LCM* 17 (1992), 36.

ipse caper: the he-goat himself (Meliboeus is surprised), who ought to have remained with his 'wives'. See 4. 38 n.

deerrauerat: Velius Longus, *GLK* vii. 65. 2 'in hac tamen *de* praepositione . . . animaduertendum illud, quod imminuitur si quando sequens uox a littera *e* incipit, ut est *derrare*, *desse*'.

atque: Plautine, 'introducing some new person or fact' (Sonnenschein on *Rud.* 492); cf. *Bacch.* 278–9 'forte ut adsedi in stega, / dum circumspecto, atque ego lembum conspicor', *Most.* 487–8 'lucernam forte oblitus fueram exstinguere; / atque ille exclamat derepente maximum'. Here only in the *E.*, possibly in *G.* 1. 203 (Gellius 10. 29. 4), occasionally in the *A.*; see Wagner, *Quaest. Virg.* 35. 22, Norden on *A.* 6. 161 ff.

8. ille ubi me contra uidet: cf. Plaut. *Mil.* 122–3 'uideo illam amicam erilem . . . / ubi contra aspexit me' (*TLL* s.v. *contra* 738. 45).

9. huc ades, o Meliboee: without the pathos of 'o Meliboee' in 1. 6; cf. 9. 39 'huc ades, o Galatea', 2. 45 'huc ades, o formose puer'.

haedi: the kids had wandered off after the he-goat.

11. ipsi: 'sponte sua' (Serv.); cf. 4. 21. His steers will come of their own accord to drink; no need for Meliboeus to drive them.

12–13. hic uiridis tenera praetexit harundine ripas / Mincius: the slow-winding, reed-bordered river; cf. *G.* 3. 14–15 'tardis ingens ubi flexibus errat / Mincius et tenera praetexit harundine ripas'.

13. eque sacra resonant examina quercu: the bees have made their hive in the hollow oak; cf. Hes. *Works and Days* 232–3 *οὔρεσι δὲ δρῦς | ἄκρη μέν τε φέρει βαλάνους, μέσση δὲ μελίσσας*, 'in the mountains the oak bears acorns at the top, bees in the middle', *G.* 4. 44 'exesaeque arboris antro', and see West on Hes. *Theog.* 594.

Some beekeepers, Varro says, build hives of a hollow tree, *RR* 3. 16. 15 'alii ex arbore caua'.

eque sacra: 'et e sacra, ut *aque chao*' (Serv.); curious that Servius should have felt the need to explain *eque*. Cf. Lucr. 1. 37, *A.* 12. 671, Ov. *Met.* 1. 468, 2. 96, 15. 268.

sacra . . . quercu: 'an auncient tree / Sacred with many a mysterie' (Spenser, *The Shepheardes Calender*, *Februarie* 207–8). Cf. *G.* 3. 332 'magna Iouis antiquo robore quercus.'

14. quid facerem?: 1. 40 n.

neque ego Alcippen nec Phyllida habebam: *conseruae*, the only 'wives' permitted to a slave (see 1. 30 n.). Although Theocritean (5. 132, where see Gow), Alcippe is unique in Theocritus, as she is in V.; for Phyllis, no Theocritean, see 3. 76 n. The Graecizing elision *Phyllida habebam* is unparalleled in the *E.* (Norden 455 n. 3).

15. depulsos a lacte: lambs were conceived 'ab Arcturi occasu ad Aquilae occasum' (Varro, *RR* 2. 2. 13; so also Pliny, *NH* 8. 187, who supplies the exact dates: from 13 May to 23 July); the ewes dropped towards the end of autumn after a pregnancy of 150 days (14 'ouis praegnas est diebus CL. itaque fit partus exitu autumnali, cum aer est modice temperatus et primitus oritur herba imbribus primoribus euocata'); the lambs were given a supplementary feeding of ground vetch or tender grass twice daily for four months (16–17) and then weaned, in March or April therefore; when weaned (17 'cum depulsi sunt agni a matribus') they were kept apart from the flock until they had forgotten the taste of milk and no longer missed their mothers (18).

16. certamen . . . Corydon cum Thyrside, magnum: a match, Corydon v. Thyrsis, a big one. The apposition is extraordinary; 'uidetur nominatiuum pro genitiuo posuisse' (DServ.). But for an even more extraordinary apposition cf. Cic. *Phil.* 2. 58 'sequebatur raeda cum lenonibus, comites nequissimi', and see Löfstedt i. 82–3.

17. posthabui: not a common verb in poetry or in prose; found in *A.* 1. 16 and several times in Statius. Cf. Ter. *Phorm.* 908 'omnis posthabui mihi res'.

illorum mea: the genitive plural of *ille* is infrequent in poetry after Lucretius: *A.* 7. 282, Hor. *Serm.* 1. 3. 141, 4. 39–40 'primum ego me illorum . . . / excerpam numero', Prop. 1. 13. 9 (*illarum*), 2. 30B. 15, 3. 6. 34, Ovid *Am.* 3. 3. 47, *Ex Pont.* 4. 6. 43, Juv. 15. 105; see Axelson 72–4. The tone of 'illorum mea' is established by comparison with Plaut. *Cas.* 445 'nam illorum me alter cruciat', *Mil.* 1350 'mea, non illorum dedi', *Most.* 883 'minoris pendo tergum illorum quam meum'.

seria ludo: cf. Hor. *Serm.* 1. 1. 27 'sed tamen amoto quaeramus seria ludo'.

18–20. Cf. [Theocr.] 8. 30–2 *πρᾶτος δ' ὧν ἄειδε λαχὼν ἰυκτὰ Μενάλκας, | εἶτα δ' ἀμοιβαίαν ὑπελάμβανε Δάφνις ἀοιδάν | βουκολικάν*, 'First, then, clear-voiced Menalcas sang, for to him the lot had fallen, and then Daphnis took up the answering strain of pastoral song' (Gow).

18. igitur: though common in Lucretius, a word generally avoided in the higher style of poetry. Only here in the *E.*, not in the *G.*, twice in the *A.* (in speeches: 4. 537, 9. 199). See Axelson 92–3.

19. coepere, alternos: one of the five instances of elided *-ēre* in V. where the penult does not coincide with the ictus: *G.* 4. 272, *A.* 4. 321, 6. 201, 12. 447; see D. W. Pye, *TPhS* (1963), 21.

alternos Musae meminisse uolebant: 'the Muses wished to recall alternate verses'; cf. 3. 59, *A.* 7. 645 'et meministis enim, diuae, et memorare potestis'. It was an ancient belief that the Muses, daughters of Memory, put the poet in mind of his song and enabled him to sing it. Poets thus became 'the holy interpreters of the Muses', Theocr. 16. 29 *Μοισάων . . . ἱεροὺς ὑποφήτας*, where see Gow.

21. Nymphae, noster amor, Libethrides: Corydon begins with a show of Hellenistic learning: 'Nymphae . . . Libethrides', *Νύμφαι Λιβηθρίδες*; see 4. 1 n. and cf. Varro, *LL* 7. 20 (*Musae Olympiades*) 'ita enim ab terrestribus locis cognominatae Libethrides, Pipleides, Thespiades, Heliconides', Strabo 9. 2. 25 (Mt. Helicon) *ἐνταῦθα δ' ἐστὶ τὸ τῶν Μουσῶν ἱερὸν καὶ ἡ Ἵππου κρήνη καὶ τὸ τῶν Λιβηθρίδων Νυμφῶν ἄντρον*, 'Here is the temple of the Muses and Hippocrene and the cave of the Libethrian Nymphs'; the relationship of Nymphs to Muses is, as Gow remarks on Theocr. 7. 92, 'close and not precisely definable'. The *Λιβηθρίδες* are found in Euphorion and in two anonymous poems of the Hellenistic period

(*Suppl. Hell.* 988. 1, 993. 7), but not elsewhere in Latin poetry. For the word-order (parenthetic apposition) see 1. 57 n.

21–2. The similarity of these lines to 10. 1–3 suggests that V. is here alluding to a contemporary poet and a friend; cf. DServ.: 'Codrus poeta eiusdem temporis fuit, ut Valgius in elegis suis refert'. Valgius' poem is preserved, though badly damaged, by the Verona scholiast here: 'Codrum plerique Vergilium accipiunt, alii Cornificium, nonnulli Heluium Cinnam putant, de quo bene sentit. similiter autem hunc Codrum in elegiis Valgius honorifice appellat et quadam in ecloga de eo ait (*FPL* fr. 2 Büchner) "Codrusque ille canit, quali tu uoce canebas, / atque solet numeros dicere, Cinna, tuos ..."'. For C. Valgius Rufus, already an important poet or man of letters when the Book of *Eclogues* was published, as may be inferred from Hor. *Serm.* 1. 10. 81–2 'Plotius et Varius, Maecenas Vergiliusque, / Valgius', see H. Gundel, *RE*, 2nd ser., viii. 272–6, R. Syme, *History in Ovid* (Oxford, 1978), 111–12.

22. meo Codro: cf. 10. 2 'meo Gallo', Catull. 95. 1 'mei Cinnae'. This simple style of reference connotes intimacy and affection ('Positus in hac particula magnus affectus amoris', La Cerda): so Horace refers to Lamia, *Carm.* 1. 26. 8–9 'necte meo Lamiae coronam, / Piplei dulcis', and so, once only, near the end of their long friendship, to Maecenas, *Carm.* 4. 11. 19 'Maecenas meus'; see Fraenkel, *Horace*, 416–17, Clausen 155 n. 37, *TLL* s.v. *meus* 917. 67.

22–3. proxima Phoebi / uersibus: 'subaudis carmina' (Serv.); Burman compares *A.* 8. 427–8 'fulmen erat, toto genitor quae plurima caelo / deicit'. But V. means (*uersus*) *proximos Phoebi uersibus* and employs the neuter plural for metrical convenience (so Heyne); cf. *A.* 6. 169–70 'Dardanio Aeneae sese fortissimus heros / addiderat socium, non inferiora secutus' ('*inferiorem* war für Verg. metrisch unbrauchbar', Norden), 9. 27–8 'Messapus primas acies, postrema coercent / Tyrrhidae iuuenes'. The conceit may owe something to Theocr. 1. 3 *μετὰ Πᾶνα τὸ δεύτερον ἆθλον ἀποισῇ*, 'After Pan you will take the second prize'; cf. Longus 2. 32. 3 *ἐσεμνύνετό τις ὡς λύκον ἀποκτείνας, ἄλλος ὡς μόνου τοῦ Πανὸς δεύτερα συρίσας*, 'One man boasted of having killed a wolf, another of having been second only to Pan in piping'.

23. facit: *ποιεῖ*; cf. 3. 86 'Pollio et ipse facit noua carmina'. For the prosody cf. *A.* 12. 883 'te sine, frater, erit? o quae ...' and see 1. 38 n.

si non possumus omnes: 8. 63 n.

24. arguta ... fistula: cf. Callim. *Hymn* 3. 242–3 *λίγειαι* | ... *σύριγγες*, 'shrill pipes', Ap. Rhod. 1. 577–8 *σύριγγι λιγείῃ* | *καλὰ μελιζόμενος νόμιον μέλος*, 'gaily playing a shepherd's tune on his shrill pipe'.

pendebit: 'It was customary on leaving off any occupation to dedicate some of the instruments connected with it to an appropriate divinity, e.g. a warrior dedicates his arms to Mars, a fading beauty her mirror to Venus' (T. E. Page). So Corydon, if defeated, will hang his pipe on the tree sacred to Pan (Prop. 1. 18. 20 'Arcadio pinus amica deo') as a sign that he has given up music. A common theme; cf. Theocr. *Epigr.* 2 Gow (= *AP* 6. 177), Longus 4. 26. 2, Tib. 2. 5. 29–30 'pendebatque uagi pastoris in arbore uotum, / garrula siluestri fistula sacra deo'.

25. hedera: symbolic of poetic achievement; cf. 8. 13 and see Nisbet–Hubbard on Hor. *Carm.* 1. 1. 29. See also 4. 19 n.

crescentem: *nascentem* has equally good if not better MS authority and is interpreted by Servius here: 'uerecunde locutus est, non iam poetam dicens sed nascentem'. Obviously, Servius is mistaken: Thyrsis boldly claims the recognition due to a poet, a potential bard. And it may be doubted whether 'nascentem ... poetam' could mean anything 'but the poet at his birth', as in 4. 8 'nascenti puero', Hor. *Carm.* 4. 3. 1–2 'Quem tu, Melpomene, semel / nascentem placido lumine uideris'.

25–6. pastores ... / Arcades: for the delayed apposition cf. 10. 70–2 'diuae ... / ... / Pierides'.

25. poetam, 28 uati: 9. 32–6 n.

26. rumpantur: a proverbial expression; cf. Mart. 9. 97. 1–2 'Rumpitur inuidia quidam, carissime Iuli, / quod me Roma legit, rumpitur inuidia' (and the rest of the poem) and see Fedeli on Prop. 1. 8. 27 'rumpantur iniqui'. V. seems to have been thinking of Catull. 11. 20 'ilia rumpens', though there the sense is different; see below, 28 n.

27. ultra placitum: 'quicquid autem ultra meritum laudatur, dicitur fascinari' (DServ.). Praise, even self-satisfaction, the

Ancients felt, could be dangerous; cf. Titinius 109–10 R.[3] 'Paula mea, amabo, pol tuam ad laudem addito "praefiscini"', Plaut. *Rud.* 461–2 (Sceparnio is exceedingly pleased with himself) 'ut sine labore hanc extraxi! praefiscine. / satin nequam sum, utpote qui hodie amare inceperim?', and see Fraenkel on Aesch. *Ag.* 904. Pliny, *NH* 7. 16, reports the existence in Africa of certain families of sorcerers 'effascinantium . . . quorum laudatione intereant probata, arescant arbores, emoriantur infantes'. For the neuter singular of the perfect passive participle used as a substantive see Kühner–Stegmann i. 768–9, adding *G.* 1. 412 'praeter solitum' (also Hor. *Carm.* 1. 6. 20), *A.* 10. 9 'contra uetitum', Hor. *Carm.* 2. 9. 23 'intraque praescriptum'.

baccare: βάκκαρις or βάκχαρις, an unidentifiable plant, on which see P. Wagler, *RE* ii. 2803. Dioscorides describes it as sweet-smelling and suitable for garlands, *De mat. med.* 3. 44. 1 εὐώδης, στεφανωματική. In Latin poetry it is found only here and in 4. 19, in both places ablative, occupying the fifth foot of the line, and linked with ivy. V. is alone in attributing magical properties to it; 'herba est ad depellendum fascinum' (DServ. here) and 'herba est quae fascinum pellit' (Serv. on 4. 19)—both inferences from this text. For the apotropaic qualities of garlands and plants see L. Duebner, *Archiv für Religionswissenschaft*, 30 (1933), 86–90.

28. mala lingua: cf. Catull. 7. 12 'mala fascinare lingua'.

29–32. 'Corydon speaks in the character of Micon' (Conington); cf. below, 37–40, 41–4, and 3. 78–9. These lines resemble a Greek dedicatory epigram, e.g. Rhianus 6 G.–P. (= *AP* 6. 34), available to V. in Meleager's *Garland*.

Τὸ ῥόπαλον τῷ Πανὶ καὶ ἰοβόλον Πολύαινος
τόξον καὶ κάπρου τούσδε καθᾶψε πόδας,
καὶ ταύταν γωρυτὸν ἐπαυχένιόν τε κυνάγχαν
θῆκεν ὀρειάρχᾳ δῶρα συαγρεσίης.
ἀλλ', ὦ Πὰν σκοπιῆτα, καὶ εἰσοπίσω Πολύαινον
εὔαγρον πέμποις υἱέα Σημύλεω.

Polyaenus hung here as a gift to Pan the club, the arrow-shooting bow, and these boar's feet. Also to the lord of the hills he dedicated this quiver and the dog-collar, gifts of thanks for his success in boar-hunting. But, O Pan the highlander, do thou in the future, too, speed Polyaenus, the son of Semylas, in his hunting.

(W.R. Paton in the Loeb Classical Library, modified)

29–30. 'Sane deest *consecrat*, ut "Aeneas haec de Danais uictoribus arma"' (DServ.); similarly the Verona scholiast, also quoting *A*. 3. 288. The verb is often omitted in dedications.

29. saetosi ... apri: a new adjective with a brief history: here, Hor. *Serm*. 1. 5. 60–1 (Messius is compared to a mutilated unicorn) 'at illi foeda cicatrix / saetosam laeui frontem turpauerat oris', *Epod*. 17. 15–17 'saetosa duris exuere pellibus / laboriosi remiges Vlixei / uolente Circa membra', and Prop. 4. 1. 25 'uerbera pellitus saetosa mouebat arator'. But why did V. not write *saetigeri* instead of *saetosi*, especially as he was to write *ramosa* in the next line? Because *saetiger* was traditionally (Ennius?) attached to *sus*: Lucr. 5. 969 'saetigerisque ... subus', 6. 974 'saetigeris subus', *A*. 7. 17 'saetigerique sues', 11. 198 'saetigerosque sues', 12. 170 'saetigeri ... suis', Ov. *Met*. 8. 359 'sus', 376 'saetiger', 10. 549 'saetigerosque sues', *Fast*. 1. 352 'saetigerae ... suis', Silius 3. 23 'saetigeros ... sues', Stat. *Theb*. 1. 397 'saetigerumque suem', 8. 532–3 'qualis saetigeram Lucana cuspide frontem / strictus aper'. Cf. A. Ernout, *Les Adjectifs latins en* -osus *et en* -ulentus (Paris, 1949), 28: 'D'un coté, le dérivé *saetosus* et le mot courant *aper*; de l'autre, dans la langue épique, le composé *saetiger* et le nom ancien *sus*'.

Delia: although *Δήλιος* was a cult-title of Apollo, e.g. Callim. *Hymn* 4. 269 *Δήλιος Ἀπόλλων*, *Δηλία* was not a cult-title of Artemis: it is found in a few late dedications but nowhere in literature; see O. Jessen, *RE* iv. 2433. *Delia* seems therefore to be V.'s innovation, a convenient literary epithet of which poets from Ovid to Dracontius availed themselves; see *TLL* Onomast. s.v. *Delos* 90. 8.

paruus: an etymological adjective 'explaining' the Greek name Micon (*μικός, μικκός, μικρός*, see Gow on Theocr. 15. 12); no ancient etymologist was ever deterred by a false quantity. Thus the name of Virgil's fearsome monster Cacus was derived from *κακός*, an etymology underlying Ov. *Fast*. 1. 551–2 'Cacus ... / non leue finitimis hospitibusque malum'. Such adjectives are usually attached to place-names, e.g. *A*. 3. 703 'arduus inde Acragas'; see J. S. T. Hanssen, *SO* 26 (1948), 113–25, D. O. Ross, Jr., *Mnemosyne*, 4th ser., 26 (1973), 60–2. For the arrangement of adjective and noun in separate clauses, here perhaps calling attention to the etymology, cf. *G*. 3. 303–4, *A*. 3. 161–2, 7. 464–5, 9. 521–2, 11. 454, 659–60.

30. Micon: 3. 10 n.

ramosa: a Lucretian adjective, used of a tree in 5. 1096, but of atoms in 2. 446, and of clouds in 6. 133. Cf. *A.* 1. 190 'cornibus arboreis' (DServ.), Pliny, *NH* 8. 116 'indicia quoque aetatis in illis gerunt, singulos annis adicientibus ramos' (La Cerda).

uiuacis . . . cerui: 'Fables are raised concerning the vivacity of Deer' (Sir Thomas Browne, *Pseudodoxia Epidemica*, 3. 9); cf. Juv. 14. 251 'longa et ceruina senectus', with Courtney's note. *Viuax* is one of two adjectives in *-ax* in the *E.*, the other being *fallax* (4. 24). There are fifteen such adjectives in V.: *audax* first attested in Naevius; *edax ferax fugax loquax mendax procax rapax tenax* first in Plautus; *minax* first in Ennius; *fallax* first in Cato; *uiuax* first in Afranius 251 R.[3]; *sequax* first in Lucretius; and two apparently V.'s own: *pellax sternax*. V. tends to be sparing in his use of these adjectives, five of which he uses only once: *edax pellax procax rapax sternax*. For a detailed study see S. de Nigris Mores, *Acme*, 25 (1972), 263–313.

cornua cerui: an offering appropriate to Artemis (Diana), one of whose cult-titles was ἐλαφηβόλος, 'deer-slayer'; see Jebb on Soph. *OC* 1092.

31–2. If Micon gets what he hopes for—good success in hunting—he will make a splendid offering, a statue of the goddess herself all of marble.

31. si proprium hoc fuerit: cf. *A.* 6. 870–1 'nimium uobis Romana propago / uisa potens, superi, propria haec si dona fuissent', Hor. *Serm.* 2. 6. 4–5 'nil amplius oro, / Maia nate, nisi ut propria haec mihi munera faxis'.

31–2. leui de marmore tota / . . . stabis: cf. *G.* 3. 34 'stabunt et Parii lapides, spirantia signa', Hor. *Serm.* 2. 3. 183 'aeneus ut stes', Theocr. 10. 33 (Bucaeus and Bombyca) χρύσεοι ἀμφότεροί κ' ἀνεκείμεθα τᾷ Ἀφροδίτᾳ, 'Then should we both stand in gold as offerings to Aphrodite' (Gow), with Gow's note.

31. de marmore: cf. *G.* 3. 13 'templum de marmore ponam', *A.* 4. 457.

tota: emphatic, like *totis* in 1. 11; all of marble, no *acrolithon* with only the undraped and visible extremities—head, hands, and feet—of marble; see *RE* 1. 1198–9.

32. puniceo . . . suras euincta coturno: cf. Callim. *Hymn* 3. 15–16 (Artemis to Zeus) δὸς δέ μοι ἀμφιπόλους Ἀμνισίδας εἴκοσι νύμφας, | αἵ τε μοι ἐνδρομίδας . . ., 'Give me for handmaids twenty Nymphs of the Amnisus who (shall tend) my boots'. Her boots are a conspicuous feature of Artemis' iconography; cf. *LIMC* ii/2, especially pls. 204, 210, 224, 233, 239, 251, 264, 274, 307, 308, 319, 355, 356, 387, 392, 414, 416, 423, 435, 443, 496 (pp. 462–85); M. Bieber, *The Sculpture of the Hellenistic Age* (New York, 1955), 116 (the Artemis on the Altar of Pergamum): 'her shoes are a wonderwork of shoemakers' craft: high boots with elegant fur cuffs and finely drawn scallops and scrolls'. See also the 'Diana of the Springs' in G. M. A. Hanfmann, *Roman Art* (Greenwich, Conn., 1964), col. pl. xxiii: 'Her high hunter's boots are tied with red laces'. Servius compares, and misquotes, *A*. 1. 337 (Venus disguised as a huntress is speaking) 'purpureoque alte suras uincire coturno'. Dionysus also wears red boots, Nonnus, *Dionys*. 18. 200 ἐρευθιόωντι κοθόρνῳ (La Cerda).

euincta: with the exception of Tac. *Ann*. 6. 42.4 and 15. 2.4 (*euinxit*) only the perfect passive participle of this verb occurs, and that, with the exception of Cassius Hemina fr. 37 Peter, Tac. *Hist*. 4. 53.2, and Gellius 14. 1. 3, 20. 1. 54, only in poetry (*TLL* s.v.). Cf. *A*. 5. 269 'puniceis ibant euincti tempora taenis', 8. 286 'populeis adsunt euincti tempora ramis'. For the construction see 1. 54 n.

33. **sinum lactis:** *sinum* is defined by Varro, *LL* 5. 123, as 'uas uinarium grandius'; by Asper, quoted here by the Verona scholiast, as 'uas uinarium, ut Cicero significat, non, ut quidam, lactarium'. Cf. Valgius Rufus, *FPL* fr. 5 Büchner 'sed nos ante casam tepidi mulgaria lactis / et sinum bimi cessamus ponere Bacchi?' Milk instead of wine is a feature of the most ancient Roman cults and appropriate as a simple rustic offering (Wissowa 411 n. 7); but milk, sweet milk, as an offering to Priapus is unusual (H. Herter, *RE* xxii. 1934). Priapus was neither primitive nor indigenous; a Hellenistic import rather, never taken very seriously by the Romans, and with no official cult ('quotannis'; 1. 42 n.). Evidently Thyrsis is mocking Corydon's pretentiousness; hence the deliberate simplicity, emphasized by the rhythm (1. 70 n.), of this line, and the whimsical exaggeration of ll. 35–6.

liba: sacrificial cakes, for which Cato, *De agr*. 75, gives the recipe: brayed cheese, wheat flour, and an egg.

35. **marmoreum pro tempore:** marble for the time being, capping Corydon's 'de marmore' (31); 'haec festiva tantum ironia

dici in promptu est' (Forbiger). The Priapus in a poor man's garden would be a roughly carved image of figwood, painted red, and quite worthless; cf. Theocr. *Epigr.* 4. 2–3 Gow (= *AP* 9. 437) *σύκινον . . . ξόανον | ἀσκελὲς αὐτόφλοιον ἀνούατον*, 'an image of figwood, legless, earless, with the bark still on it', Hor. *Serm.* 1. 8. 1 'Olim truncus eram ficulnus, inutile lignum'. Martial mentions a marble Priapus, but 'ingenti . . . in horto' (6. 72. 3).

pro tempore: 'pro captu rei nostrae, pro uiribus, quae sunt hoc tempore' (DServ.).

37–40. Corydon now assumes the role of Polyphemus, imitating Theocr. 11. 19–21 *Ὦ λευκὰ Γαλάτεια . . . | λευκοτέρα πακτᾶς ποτιδεῖν, ἁπαλωτέρα ἀρνός, | μόσχω γαυροτέρα, φιαρωτέρα ὄμφακος ὠμᾶς*, 'O white Galatea . . . whiter than curd to look on, softer than the lamb, more skittish than the calf, sleeker than the unripe grape' (Gow). The ablative of comparison is rare in V. In the *E.* it is found, with the exception of 5. 53, only in this *Eclogue*, and only in imitation, direct or indirect, of Theocritus. See Löfstedt i. 315–20 and Nisbet–Hubbard on Hor. *Carm.* 1. 19. 6.

37. Nerine: this form is unique, the 'correct' form of the patronymic being *Nereine*, first attested in Catull. 64. 28 'tene Thetis tenuit pulcherrima Nereine?' (Haupt's emendation of *nectine*). No doubt Catullus had Hellenistic precedent, although *Νηρηίνη* is first attested in later Greek poetry: Opp. *Hal.* 1. 386, Quint. Smyrn. 3. 125, 596, 4. 128. Similarly, Ov. *Her.* 6. 103 'Phasias Aeetine' (as emended by Salmasius and Heinsius), but *Αἰητίνη* is first attested in Dionys. *Per.* 490. In early Greek poetry these patronymics ending in *-ίνη* are derived from names ending in *-os*: Hom. *Il.* 5. 412 *Ἀδρηστίνη*, 9. 557 *Εὐηνίνη*, Hes. *Theog.* 364, 389, 507, 956 *Ὠκεανίνη*. But Callimachus begins to experiment in the *Hecale*, fr. 103. 2 Hollis (= 302. 2 Pf.) *Δηωίνη* (by analogy with *ἡρωίνη*? Cf. Theocr. 13. 20, 26. 36, Callim. *Hymn* 4. 161) and fr. 140 Hollis (= 352 Pf.) *Νωνακρίνη* (see also *Suppl. Hell.* 250. 9). In the case of *Νωνακρίνη*, Callimachus removed what he took to be the ending (*Νωνακρ-ίς*) and added the suffix *-ίνη*. *Νηρίνη* from *Νηρ-εύς* is not unlike, nor is *Αἰητίνη* from *Αἰήτ-ης*. As Pfeiffer notes on fr. 352, these patronymics ordinarily stand at the end of the line. V. may have invented *Nerine*; more probably he found it in a Hellenistic poet and used it here to keep Galatea in her Theocritean place.

Hyblae: 1. 54 n.

38. hedera . . . alba: 3. 39 n. But V. was not so much concerned to distinguish a particular species of ivy as he was to frame his line with adjectives denoting the same colour; cf. *A*. 8.82 'candida per siluam cum fetu concolor albo'.

40. tui Corydonis: *E*. 4, Introduction, n. 11.

41–4. Thyrsis impersonates Galatea and returns a mocking answer, as in 3. 78–9 Menalcas replies to Damoetas by impersonating Iollas. V.'s model was Theocr. 6; see *E*. 5, Introduction, p. 154. See also R. Heinze, *Hermes*, 58 (1923), 112. Thyrsis' parody of Corydon, though on a sour note, is remarkably clever: 37 'thymo mihi dulcior Hyblae' ~ 41 'Sardoniis . . . tibi amarior herbis', 38 'candidior cycnis' ~ 42 'horridior rusco', 38 'formosior alba' ~ 42 'uilior alga', 39 'pasti . . . tauri' ~ 44 'pasti . . . iuuenci'. See Klingner 122.

41–2. For the harshness of the repeated *r* sound (contrasting with the smoothness of 37–8) see R. Heinze, *N. Jahrb.* 19 (1907), 166.

41. Sardoniis . . . herbis: *Ranunuculus sceleratus* L., 'the celery-leaved crowfoot, so acrid that its leaves applied externally produce inflammation. Those who ate it had their faces distorted into the proverbial Sardonic smile. Thyrsis contrasts it with the thyme of Hybla, as producing proverbially bitter honey, "Sardum mel," Hor. *A.P.* 375' (Conington). For Homer's 'sardonic smile' see Russo on *Od.* 20. 302.

42. rusco: Butcher's-broom, *Ruscus aculeatus* L., 'a low-growing rough shrub' (Mynors on *G*. 2. 413–15); see Abbe 65.

proiecta: cast up on the shore, sea-wrack.

uilior alga: apparently a proverbial expression; cf. Hor. *Serm.* 2. 5. 7–8 'atqui / et genus et uirtus nisi cum re uilior alga est', *Carm.* 3. 17. 10–12 'alga litus inutili / demissa tempestas ab Euro / sternet'.

43. si mihi non haec lux toto iam longior anno est: cf. Theocr. 12. 2 *οἱ δὲ ποθεῦντες ἐν ἤματι γηράσκουσιν*, 'but men with yearning grow old in a day'—'a line much cited in later antiquity' (G.–P. on Dioscorides 11. 3–4). Cf. Ov. *Her.* 11. 29 'et nox erat annua nobis', 18. 25 'septima nox agitur, spatium mihi longius anno'.

44. si quis pudor: 'Galatea' is impatient and annoyed; the steers are well fed and really ought to be going home. For the idiom, which Servius of course understood (not vulgar wit, as some commentators imagine), see Fedeli on Prop. 1. 9. 33, Housman, *JPhil* 22 (1894), 112 = *Class. Papers*, i. 335: 'The formula is employed not merely in serious objurgation ... but in mild or playful remonstrances, as at Prop. I ix 33 "quare, *si pudor est*, quam primum errata fatere", Ovid. am. III 2 23 sq. "tua contrahe crura, / *si pudor est*, rigido nec preme terga genu", Verg. buc. VII 44'.

45–60. These four quatrains are freely modelled on four quatrains in [Theocr.] 8. 33–48. Note that *herba* (45) and *umbra* (46) end the first two lines of the first quatrain as *herba* (57) and *umbras* (58) end the first two lines of the fourth. The contrast between summer and winter may have been suggested by [Theocr.] 9. 7–21; see below, 51 n.

45. muscosi: first attested in Catull. 68. 58 'riuus muscoso prosilit e lapide'; only here in V. Cf. Lucr. 5. 951 'umida saxa super uiridi stillantia musco', *G.* 4. 18 'at liquidi fontes et stagna uirentia musco'. See 1. 5 n.

somno mollior herba: Theocr. 5. 50–1 εἴρια ... | ... ὕπνω μαλακώτερα, 'fleeces softer than sleep', where see Gow. Homer uses the adjective of sleep, *Il.* 10. 2 μαλακῷ δεδμημένοι ὕπνῳ, 'overcome by soft sleep', and of soft grassy meadows, *Od.* 5. 72 λειμῶνες μαλακοί. Cf. Lucr. 5. 1392–3 'prostrati in gramine molli / propter aquae riuum sub ramis arboris altae', *E.* 3. 55 'in molli consedimus herba', *G.* 2. 470 'mollesque sub arbore somni'.

46. rara ... umbra: not entirely excluding the sun, sun-chequered shade.

arbutus: incorporated in the relative clause for metrical convenience; cf. e.g. *A.* 12. 181–2 'fontisque fluuiosque uoco, quaeque aetheris alti / religio', Ov. *Fast.* 1. 397–8, Mart. 1. 88. 5–6. See 3. 82 n.

47. solstitium: the summer solstice, the longest day in the year.

48. lento: 1. 4 n. The variant *laeto* may be owing to *G.* 2. 363–4 'dum se laetus ad auras / palmes agit', though the two words are easily confused, e.g. Ov. *Am.* 3. 6. 60, *Ars* 3. 452, Sen. *Tro.* 897.

turgent in palmite gemmae: Mynors on *G.* 2. 333: 'the woody vine-branches which remain on the stock throughout the winter, and are shortened by the pruner, are the *palmites*. Buds formed on these break in the spring (*E.* 7. 48, Ovid *fast.* 3. 238) and produce the rampant green shoots which bear leaves and grape-clusters, and these are the *pampini* (*E.* 7. 58, *G.* 2. 5)'. Swelling buds presage hot weather, and Corydon takes little pleasure in the thought (Wagner). See 8. 53 n.

49. **taedae pingues:** pine wood, resinous and smoky; cf. Lucr. 5. 296 'pingues multa caligine taedae'.

50. **adsidua postes fuligine nigri:** cf. Ov. *Fast.* 5. 505 'tecta senis subeunt nigro deformia fumo', Mart. 2. 90. 7–8 'me focus et nigros non indignantia fumos / tecta iuuant', Fraenkel on Aesch. *Ag.* 774: 'In the poor man's cottage kitchen and parlour are the same, so that the smoke and soot cover the whole place'.

51. Cf. [Theocr.] 9. 20–1 (Menalcas after extolling the comforts of his cave in winter) *ἔχω δέ τοι οὐδ' ὅσον ὤραν | χείματος*, 'I pay no heed at all to winter'.

52. **numerum:** 'subauditur ouium uel similium' (DServ.).

torrentia flumina ripas: 3. 94–7 n. Cf. Varro, *RR* 1. 12. 4 'nimbi repentini ac torrentes fluuii periculosi illis qui in humilibus ac cauis locis aedificia habent', and, for the noun, Ov. *Fast.* 2. 219–22 'torrens undis pluuialibus auctus / aut niue, quae Zephyro uicta tepente fluit, / per sata perque uias fertur nec, ut ante solebat, / riparum clausas margine finit aquas'.

53. **stant et iuniperi et castaneae hirsutae:** for the structure of this extraordinary line cf. *A.* 3. 74 'Nereidum matri et Neptuno Aegaeo', Ov. *Met.* 5. 312 'fonte Medusaeo et Hyantea Aganippe', 8. 310 'cumque Pheretiade et Hyanteo Iolao'. For the hiatus in the third foot see 3. 6 n.; for that in the fifth, 2. 24 n. It is of some importance perhaps that V., sparing of elision, and especially the elision of long vowels, in the *E.*, does not elide *-ae*; see F. Leo, *Plautinische Forschungen*[2] (Berlin, 1912), 357. If V. had a single Greek model in mind, it may have been Arat. *Phaen.* 942 *Πολλάκι λιμναῖαι ἢ εἰνάλιαι ὄρνιθες*, 'Often the birds of lake or sea', the opening line of a passage on weather-signs, which V. later imitated, after Varro of Atax, in *G.* 1. 373–89. For the metrical pattern *castaneae hirsutae* in subsequent Latin poetry see Fordyce on *A.* 7. 631, Bömer on Ov. *Fast.* 2. 43, *Met.* 3. 184.

stant et: this sequence—an initial monosyllabic verb followed by a long monosyllable—is all but unique in the *E.*, in 10. 50 'ibo et' elision is involved (Fantazzi–Querbach 364).

stant: opposed to *iacent* (54); cf. Lucr. 3. 887 'stansque iacentem'. Strength and rigidity are implied; cf. *G.* 3. 368, *A.* 2. 333, 6. 652.

iuniperi: 'in favourable positions . . . a shapely tree eighteen or twenty feet high. It is very common in Italy and attains this height in the lower country' (Sargeaunt 64).

hirsutae: not to be understood with *iuniperi*. Chestnuts have prickly husks; cf. Palladius, *De insit.* 131–2 'poma hirsuta uirentis / castaneae', and see O. Skutsch, *Gnomon*, 37 (1965), 168, Abbe 89–90. The adjective is first attested in Cicero, and in poetry first here (*TLL* s.v. 2824. 65).

54. sua quaeque: *sua quaque* (see app. crit.) is logical but unidiomatic, for in such proximity *quisque* was attracted to the case of *suus*; see Madvig on Cic. *De fin.* 5. 46, 'sua quaeque uis', where *cuiusque* might have been expected, Lachmann on Lucr. 2. 371 (quoting this line), Wackernagel i. 54.

55. omnia nunc rident: cf. Lucr. 5. 1395–6 'praesertim cum tempestas ridebat et anni / tempora pingebant uiridantis floribus herbas'. The metaphor of nature laughing or smiling is as old as Homer; see Livrea on Ap. Rhod. 4. 1171, Nisbet–Hubbard on Hor. *Carm.* 2. 6. 14. Cf. 4. 20.

formosus Alexis: Corydon 'remembers' his beloved Alexis (2. 1) and the love-tokens, chestnuts and fruit, with which he tried to win him (2. 51–3).

56. et flumina sicca: 'hyperbolicos: etiam iuges aquas, perpetuo fluentes, siccari conspicias' (Serv.).

57. aret ager . . .: cf. 10. 67, *G.* 1. 107 'exustus ager morientibus aestuat herbis'.

uitio . . . aëris: tainted air; cf. Lucr. 6. 1097 'morbidus aër', *G.* 3. 478 'morbo caeli', *A.* 3. 138 'corrupto caeli tractu'.

58. pampineas: first attested here. Following the lead of Lucretius (*fulmineus*) and Catullus (*aequoreus*), V. invented a number of new adjectives in *-eus* (but only *pampineus* in the *E.*); Norden on *A.* 6. 281 lists fourteen. See Ross (1969), 60–3.

collibus: vineyards are usually located on hillsides; cf. 9. 49 and see Mynors on *G.* 2. 113 'Bacchus amat collis'.

59–60. These lines correspond to 55–6; cf. [Theocr.] 8. 35–6 ~ 39–40 and 47–44 ~ 43–8. At the coming of Phyllis every tree will turn green and Jupiter will descend in a downpour of rain to fecundate and gladden the earth; cf. Lucr. 1. 250–3 'postremo pereunt imbres, ubi eos pater aether / in gremium matris terrai praecipitauit; / at nitidae surgunt fruges ramique uirescunt / arboribus', *G*. 2. 325–6 'tum pater omnipotens fecundis imbribus aether / coniugis in gremium laetae descendit' (Serv.).

80. plurimus: cf. *G*. 2. 166 (*Italia*) 'auro plurima fluxit' and see *TLL* s.v. *multus* 1609. 7.

61–2. Cf. Pliny, *NH* 12. 3 'arborum genera numinibus suis dicata perpetuo seruantur, ut Ioui aesculus, Apollini laurus, Mineruae olea, Veneri myrtus, Herculi populus'.

61. populus: white poplar, *Populus alba* L., called 'the holy plant of Heracles' in Theocr. 2. 121 Ἡρακλέος ἱερὸν ἔρνος; see Mynors on *G*. 2. 66, Abbe 73.

Alcidae: a patronymic derived from Alceus, the father of Amphitryon the putative father of Hercules; first attested in Callim. *Hymn* 3. 145 Ἀλκεΐδην (though [Probus] states that Pindar used it, fr. 291 Maehler), then in [Moschus], *Epitaph. Bion.* 117, in Meleager's *Garland*, Samius 1. 1 G.—P. (=*AP* 6. 116), and here.

Iaccho: 6. 15 n. *Alcidae* and *Iaccho* make a precious pair.

62. myrtus: above, 6 n.

laurea: an adjective used substantively, with *corona* or, as here, *arbos* understood; a usage as old as Plautus and Cato (*TLL* s.v. *laureus* 1058. 13, 70). Only here in V.

65. fraxinus: not a common poetic tree; apart from Accius, *FPL* fr. 4 Büchner, where it is used of a spear, *fraxinus* is found before V. only in Ennius in a line V. seems to have had in mind here, *Ann.* 177 Skutsch '*fraxinus* frangitur atque *abies* consternitur *alta*'.

pinus in hortis: the stone or umbrella pine, *Pinus pinea* L., 'a familiar object in the scenery of central and southern Italy, but not coming much north of the famous forest which it makes near Ravenna' (Sargeaunt 101). See Nisbet–Hubbard on Hor. *Carm.* 2. 3. 9–11 'quo pinus ingens albaque populus / umbram hospitalem consociare amant / ramis?'

66. in fluuiis: 'by streams'; a strange use of *in*, but Thyrsis is concerned to produce a series of symmetrical phrases; not therefore a Graecism (*LSJ* s.v. ἐν A. I. 4). Cf. Prop. 1. 3. 6 (*Edonis*) 'in herboso concidit Apidano', where, however, the adjective makes a difference. Abbe 73 quotes William Turner (London, 1548): 'Whyte Poplar. Thys kynde is commune about the bankes of the floude Padus'.

montibus altis: 1. 83 n.

67. at: postponed as in 10. 31; see Norden 403.

Lycida formose: Thyrsis and Corydon share an affection for Phyllis (59, 63), but Alexis—'formosus Alexis' (55)—is too closely associated with Corydon to be named as a lover by Thyrsis.

70. Corydon Corydon: on being repeated, not as in 2. 69 'a, Corydon, Corydon' but as a predicate, the name becomes metaphorical: not simply Corydon, therefore, but Corydon the ideal singer. Thus a singer of comparable quality might be termed 'a Corydon'; cf. Catull. 22. 18–20 'nimirum idem omnes fallimur, neque est quisquam / quem non in aliqua re uidere Suffenum / possis', with Fordyce's note; Shakespeare, *The Merchant of Venice*, IV. i. 221 'A Daniel come to judgment: yea, a Daniel!'

ECLOGUE 8

Introduction

A reader of the Eighth *Eclogue*, while admiring individual lines and passages, may become aware of a certain incoherence or forced unity in the poem as a whole.

The first word of the first line is noticeably insistent, 'Pastorum Musam Damonis et Alphesiboei'—*Pastorum*, as if to assert the pastoral character of the poem as a whole in anticipation of the reader's response to the unpastoral Muse of Alphesiboeus (64–109). Similarly, *certantis*, the first word of the third line, invites the reader to regard Damon and Alphesiboeus as engaged in a singing-match, although none of the preliminary formalities has been observed or even suggested.[1]

His pastoral décor thus provisionally in place, Virgil announces, with a graceful rephrasing of his first line, his intention—that of telling of the Muse of Damon and Alphesiboeus, 'Damonis Musam dicemus et Alphesiboei' (5). Instead of proceeding to do so, however, he addresses an unnamed patron.

> tu mihi, seu magni superas iam saxa Timaui
> siue oram Illyrici legis aequoris,—en erit umquam
> ille dies, mihi cum liceat tua dicere facta?
> en erit ut liceat totum mihi ferre per orbem
> sola Sophocleo tua carmina digna coturno?
> a te principium, tibi desinam: accipe iussis
> carmina coepta tuis, atque hanc sine tempora circum
> inter uictricis hederam tibi serpere lauros. (6–13)

These lines present the chief difficulty of the poem, and yet are, in a sense, extraneous to the poem; were they to be removed, their absence would not be felt.[2]

Who, then, is the patron so abruptly and cryptically addressed? There are, as there have been since antiquity, only two possible candidates: Octavian and Pollio;[3] and modern commentators,

[1] See *E.* 3, Introduction. See also E. Bethe, *RhM* 47 (1892), 590–6.

[2] P. Levi, *Hermes*, 94 (1966), 73–9, in fact argues that all or most of these lines should be removed.

[3] Serv. on l. 6: 'ubi ubi es, o Auguste, siue . . .'; DServ. on l. 10: 'alii ideo hoc de Pollione dictum uolunt, quod et ipse utriusque linguae tragoediarum scriptor fuit'. The name Augustus, conferred by the Senate in 27 BC, is no indication of date in an

MAP 1. The Roman Conquest of Dalmatia

without serious question, have preferred Pollio, largely because of his reputation as a tragic poet.[4] But in 1971 the question was reopened by Bowersock,[5] who argued (i) that the traditional dating of the Eclogues (*c.* 42–39 BC) is a scholiastic fabrication and, as such, worthless; and (ii) that the conqueror of the Parthini, a

ancient author. For convenience, modern scholars refer to Augustus as Octavian (Octavianus) before that date; but from the early 30s he called himself Imp. Caesar Diui f., Caesar being the potent name to which, Antony said, he owed everything. See R. Syme, *Historia*, 7 (1958), 172–88 = *Roman Papers*, i (Oxford, 1979), 361–77.

[4] Nothing survives of Pollio's tragedies; see 3. 86 n.

[5] Bowersock had been anticipated by H. W. Garrod, *CQ* 10 (1916), 216–17: 'Nor is it obvious what the conqueror of the Parthini was doing among "the rocks of the Timavus", i.e. in N. Istria . . . In fact, Virgil's language is less applicable to the circumstances of the year 39 than it would be to those of the years 35–33. In 35 BC Augustus first turned his attention to the subjugation of Dalmatia and Pannonia'.

people in the hinterland of Dyrrhachium, had no reason to be sailing past 'the rocks of the Timavus' some 400 miles to the north along the perilous Dalmatian coast.[6] No doubt Pollio returned to Italy the usual way in 39 BC, crossing over from Dyrrhachium to Brundisium, from which he had embarked in the previous year. After celebrating a triumph 'ex Parthineis'[7] he ostentatiously retired from public life to devote himself to literature. Since Pollio appears to be excluded on geographical (and other) grounds, Virgil's addressee—unnamed here as he is in the First *Eclogue*—must be Octavian, who in 35 BC initiated a series of campaigns in northern Dalmatia in the general region of the Timavus.[8]

But can this identification be reconciled with Virgil's reference to Sophoclean tragedy? Suetonius reports, though without indicating a date, that Augustus began to compose a tragedy, an *Ajax*, with great energy, but, his style proving inadequate, deleted it, and when friends inquired after his *Ajax*, replied—no doubt a much-appreciated witticism—that Ajax had fallen on his sponge.[9] Obviously, Virgil's enthusiasm is out of proportion to a single, abortive effort,[10] as Octavian's *Ajax* must appear in retrospect. But

[6] And in an extreme angle of the Adriatic; see map, p. 234. The Timavus flows into the Adriatic between Aquileia and Tergeste. Since the identity of V.'s patron depends on his association with the Timavus, the Timavus cannot be dismissed as a geographical imprecision of the sort occasionally found in poetry.

[7] See T. R. S. Broughton, *The Magistrates of the Roman Republic*, ii (New York, 1952), 387–8. Horace's 'Delmatico . . . triumpho' (*Carm.* 2. 1. 16) is not, nor was it intended to be, geographically exact; J. J. Wilkes, *Dalmatia* (Cambridge, Mass., 1969), 45: 'Horace was employing a well-known triumphal title, in preference to the technically correct but hardly flattering *Parthinicus*'.

[8] Cf. Wilkes, ibid. 50: 'Octavianus probably began his advance in 35 BC against the Iapodes from the Liburnian port Senia . . .'. It was a difficult campaign, and Octavian himself was wounded in the assault on Metulum, the chief stronghold of the Iapodes. Note that the Timavus is called Iapydian in *G.* 3. 475 'Iapydis arua Timaui'. For the spelling of their name—Iapodes, Iapudes, Iapydes—see *RE* ix. 724.

[9] Suet. *Aug.* 85. 2 'nam tragoediam magno impetu exorsus, non succedenti stilo, aboleuit quaerentibusque amicis quidnam Aiax ageret respondit Aiacem suum in spongiam incubuisse'; cf. Macrob. *Sat.* 2. 4. 2 'Aiacem tragoediam scripserat eandemque quod sibi displicuisset deleuerat. postea L. Varius tragoediarum scriptor interrogabat eum quid ageret Aiax suus. et ille, "in spongiam", inquit, "incubuit"'. For Varius see 9. 35 n.

[10] Hence the desperate expedient of Beroaldus, who interpreted 'tua carmina' as 'songs about you'; see Van Sickle 21. Three parallel phrases: 'tua . . . facta', 'tua carmina', 'iussis . . . tuis'—why should the second phrase be different? And, if so,

suppose that Octavian, with an enthusiasm communicated to his friends, had only begun his *Ajax* when Virgil wrote these lines.[11] And if Octavian was too little known as a tragic poet, Pollio was too well known, it might be argued, to be so praised: high hopes for the future are better suited to a poet as yet little known, of whom great things may be expected.

Virgil implies a relation between his patron's poetry (10 'tua carmina') and his own, begun at his patron's command (11–12 'iussis / carmina coepta tuis'). Octavian could, as a patron, give such an order and can, as a poet himself, judge the result (11 'accipe'). If Pollio ever was Virgil's patron,[12] he has been superseded,[13] since the First *Eclogue* effectively dedicates the Book of *Eclogues* to Octavian; and here the poet, by an understood fiction, ascribes to his patron's command the poetry he would have written anyhow.[14]

This argument is confirmed, finally, by Virgil's declaration that he will begin with his patron and end with his patron, 11 'a te principium, tibi desinam'—a formula applied to Zeus,[15] to Agamemnon,[16] and, by inference, to Ptolemy Philadelphus, Theocr. 17. 1–4 *Ἐκ Διὸς ἀρχώμεσθα καὶ ἐς Δία λήγετε Μοῖσαι,* | *ἀθανάτων τὸν ἄριστον* ... | *ἀνδρῶν δ' αὖ Πτολεμαῖος ἐνὶ πρώτοισι λεγέσθω* | *καὶ πύματος καὶ μέσσος· ὃ γὰρ προφερέστατος ἀνδρῶν*, 'From Zeus let us

why the reference to Sophocles? Epic deeds demand epic praise. Köhnken n. 43 asserts that 'Sophocleo ... coturno' stands by metonymy for the high style in general (*genus grande*); an arbitrary interpretation unsupported by evidence, for in Prop. 2. 34. 41, which he cites, 'Aeschyleo ... coturno' refers to the high style of tragedy.

[11] In any case, excessive praise of a ruler's poetic achievement (10 'sola ... tua carmina') should occasion no surprise. Cf. e.g. Ben Jonson, *Epigram* 4, *To King James* 1–2 'How, best of kings, dost thou a sceptre bear! / How, best of poets, dost thou laurel wear!', with Samuel Johnson's opinion of poetic veracity: 'as much veracity as can be properly exacted from a poet professedly encomiastick' (*Life of Prior*). See below for Quintilian's praise of Domitian as 'the greatest of poets'.

[12] Not a relationship easy to define. Nisbet and Hubbard, however, on Hor. *Carm.* 2. 1. 12, define Pollio as Virgil's 'old patron' and find it hard to believe that Virgil would offer him 'so unnecessary an insult'. See also Tarrant 197–8.

[13] Though not entirely: 3. 84–91, 4. 11–14. Perhaps Pollio may be regarded as a secondary dedicatee, as he is in Horace's *Carmina* (2.1), which are primarily dedicated to Maecenas (1. 1).

[14] Cf. *G.* 3. 41 'tua, Maecenas, haud mollia iussa'.

[15] Notably by Aratus, *Phaen.* 1. *Ἐκ Διὸς ἀρχώμεσθα*. See Gow on Theocr. 17. 1 f.; also M. Fantuzzi, *MD* 5 (1980), 163–72.

[16] *Il.* 9. 97, a passage Virgil had in mind; see ad loc.

begin, and with Zeus in our poems, Muses, let us make end, for of immortals he is best; but of men let Ptolemy be named, first, last, and in the midst, for of men he is most excellent' (Gow). Such intimations of sublimity are unsuited, as Quintilian seems to have recognized, to a mere proconsul, for in congratulating Domitian on his accession to the throne, and extolling his literary genius, he quotes the last line of Virgil's dedication.

Germanicum Augustum ab institutis studiis deflexit cura terrarum, parumque dis uisum est esse eum maximum poetarum. quid tamen his ipsis eius operibus in quae donato imperio iuuenis secesserat sublimius, doctius, omnibus denique numeris praestantius? quis enim caneret bella melius quam qui sic gerit? quem praesidentes studiis deae propius audirent? cui magis suas artis aperiret familiare numen Minerua? dicent haec plenius futura saecula, nunc enim ceterarum fulgore uirtutum laus ista praestringitur. nos tamen sacra litterarum colentis feres, Caesar, si non tacitum hoc praeterimus et Vergiliano certe uersu testamur

inter uictrices hederam tibi serpere laurus.[17]

Apart from the opening lines (1–5), which introduce the singers and set the scene, and the dedication (6–13), the Eighth *Eclogue* is composed almost entirely of two songs, or two performances: the first by Damon (17–61), of which the idea and several details are borrowed from Theocritus' Third *Idyll*; and the second by Alphesiboeus (64–109), which is modelled on the incantation of Simaetha in Theocritus' Second *Idyll* (17–63).

The second song must be considered first, along with Simaetha's incantation, if the structure of the *Eclogue* is to be understood. Her incantation consists of nine quatrains—an appropriate number as being a multiple of three, the magic number[18]—each of which is accompanied by a refrain 'Draw to my house that man of mine'.[19]

[17] Quintil. 10. 1. 91–2, cited by Bowersock (1971), 79. Domitian assumed the title Germanicus after his campaign against the Chatti in AD 84. For his literary studies cf. Tac. *Hist.* 4. 86. 2 'Domitianus sperni a senioribus iuuentam suam cernens modica quoque et usurpata antea munia imperii omittebat, simplicitatis ac modestiae imagine in altitudinem conditus studiumque litterarum et amorem carminum simulans', Suet. *Dom.* 20 'numquam tamen aut historiae carminibusque noscendis operam ullam aut stilo uel necessario dedit. praeter commentarios et acta Tiberi Caesaris nihil lectitabat'.

[18] See note on ll. 73–4.

[19] For the arrangement of the refrain see Gow on Theocr. 2. 17–63.

Similarly, the song of Alphesiboeus consists of nine stanzas[20]—deliberately varied in length, however, but, like Simaetha's incantation, amounting to thirty-six lines—each of which is followed by a refrain 'Bring Daphnis home from town'. Virgil, it seems, conceived of his stanzas in groups of three, for the same numbers of lines recur in a triadic pattern, thus: 4, 3, 5; 4, 5, 3; 5, 3, 4.[21] The number four is prominent and may recall Simaetha's quatrains.

Since Damon and Alphesiboeus are engaged, at least nominally, in a singing-match, the reader will assume that the second song conforms to the first, that Alphesiboeus is following the lead of Damon; and had Virgil chosen to elaborate his fiction, the incongruity of the last three stanzas in the second song with the last three in the first might have seemed a reason for adjudging the prize to Damon. In fact, the second song must be primary and the first song secondary, modelled, that is, on the second song, because the refrain, while necessary to Alphesiboeus' song, is unnecessary, indeed inappropriate, to Damon's;[22] hence Virgil's concern to justify it (22–4). Let the reader try the experiment of reading first Theocritus' Third *Idyll* and then, omitting ll. 22–4, Damon's song without the refrain.

In a study of the frequency and character of elision in the *Eclogues*, N.-O. Nilsson[23] found, somewhat to his dismay, that the metrical technique of Alphesiboeus differs from Damon's and resembles that of the Second and Third *Eclogues*, which are generally accepted as being Virgil's earliest. Nilsson rejected the obvious chronological explanation, however, and attributed the difference to a difference of tone in the two songs.[24] But a subjective interpretation of metrical evidence will always be

[20] On the assumption that ll. 26–30 and 73–8 are single stanzas; otherwise Virgil's imitation of Simaetha's incantation would have a stanza too many. See note on l. 28ª.

[21] In the song of Alphesiboeus; in Damon's song the arrangement of the last three stanzas differs (4, 5, 3)—a further indication that Virgil was not seriously committed to the fiction of a singing-match.

[22] So Bethe (above, n. 1), 595–6. See note on ll. 18–20.

[23] 'Verschiedenheiten im Gebrauch der Elision in Vergils Eklogen', *Eranos*, 58 (1960), 80–91.

[24] P. 90: 'Die Verschiedenheit der Elisionfrequenz innerhalb der achten Ekloge wird vielmehr mit der Verschiedenheit der Stimmung zusammenhängen: tragischer Ernst bei Damon, hoffnungsvoller Eifer nebst abgespanntem, malerischem Realismus bei Alphesiboeus'; an interpretation accepted by Schmidt 32 n. 5.

dubious, and here a much more plausible explanation presents itself: Pliny, *NH* 28. 19 'hinc Theocriti apud Graecos, Catulli apud nos proximeque Vergilii incantamentorum amatoria imitatio'. Catullus practiced the art of translation, and it is easy to see why Simaetha's incantation, so beautifully composed and pathetic, would have appealed to him.[25] Prompted by the example of Catullus, apparently, and by his own deeper instinct, Virgil imitated Theocritus, or rather, Theocritus and Catullus;[26] and some years later (Virgil likes to reuse his own poetry) reused his imitation—modified in certain particulars[27] no doubt, but not so modified as to efface the impression of his earlier metrical technique—for the song of Alphesiboeus. If so, then Damon's song may be among the latest of Virgil's pastoral compositions,[28] and the Eighth *Eclogue*, as a whole, contemporary with the First, which, like the Eighth, honours Octavian.

Bibliography
(ll. 6–13)

G. W. Bowersock, 'A Date in the *Eighth Eclogue*', *HSCP* 75 (1971), 73–80.

W. Clausen, 'On the Date of the *First Eclogue*', *HSCP* 76 (1972), 201–5.

E. A. Schmidt, *Zur Chronologie der Eklogen Vergils* (SB Heidelberg, 1974/6).

[25] See Wilamowitz, *Die Textgeschichte der griechischen Bukoliker* (Philol. Untersuch. 18, Berlin, 1906), 112, R. Reitzenstein, *RE* vi (1907), 110, G. Jachmann, *Gnomon* 1 (1925), 203, S. Timpanaro, *Contributi di filologia e di storia della lingua latina* (Roma, 1978), 276–7. Catullus' imitation, like several of his poems, has been lost; see the OCT edition of Mynors, p. 106.

[26] Cf. Clausen 20: 'Almost as a matter of principle, a Hellenistic poet—and by now it should be abundantly clear that Virgil is a Hellenistic poet, writing in that tradition—will choose, where conveniently possible, to imitate two, or even more, poets simultaneously, or to add to his imitation of one poet from another'. It may not be absurd to suggest that Virgil's interest in Theocritus began with Simaetha's incantation.

[27] Quite possibly the happy ending of Alphesiboeus' song (Simaetha's magic is ineffective), which contrasts with the sad ending of Damon's song.

[28] And perhaps produced in some haste; so Bethe (above, n. 1), 594. It contains the affecting lines admired by Voltaire, André Chénier, and Thackeray (37–41) as well as the frigid lines which have given so much offence (47–50).

R. J. Tarrant, 'The Addressee of Virgil's Eighth Eclogue', *HSCP* 82 (1978), 197–9.

G. W. Bowersock, 'The Addressee of Virgil's Eighth Eclogue: A Response', *HSCP* 82 (1978), 201–2.

J. Van Sickle, '*Commentaria in Maronem Commenticia*: A Case History of Bucolics Misread', *Arethusa*, 14 (1981), 17–34.

A. Köhnken, 'Sola . . . Tua Carmina', *WJA* 10 (1984), 77–90.

J. E. G. Zetzel, 'Servius and Triumviral History in the *Eclogues*', *CP* 79 (1984), 139–42.

D. Mankin, 'The Addressee of Virgil's Eighth Eclogue: A Reconsideration', *Hermes*, 116 (1988), 63–76.

M. Pavan, *Enc. Virg.* s.v. *Timavo.*

J. Farrell, 'Asinius Pollio in Vergil, *Eclogue* 8', *CP* 86 (1991), 204–11.

(Damon's song, ll. 17–61)

B. Otis, *Virgil: A Study in Civilized Poetry* (Oxford, 1963), 105–20.

(the poem as a whole)

Klingner 134–46.

A. Richter, *Virgile: La huitième bucolique*, (Paris, 1970).

V. Tandoi, 'Lettura dell'ottava bucolica', in Gigante 265–317.

1. Musam: 'la dolcissima musa di Damone e di Alfesibeo' (Sannazaro, *Arcadia*, prosa 10). See 1. 2 n.

Damonis: the name of a herdsman in 3. 17, 23; non-Theocritean.

Alphesiboei: 5. 73 n.

2–4. An amazed heifer, enchanted lynxes, and rivers stayed in their course—no singer in Theocritus possesses such Orphean power over the natural world. Cf. 6. 27–30.

2. immemor herbarum: cf. *G.* 3. 498 (*equus*) 'immemor herbae', Hor. *Carm.* 1. 15. 30 (*ceruus*) 'graminis immemor', where Nisbet–Hubbard remark: 'one suspects a common source, perhaps the passage from the *Eclogues*, but conceivably the *Io* of Calvus'; see below, 4 n.

2–3. mirata ... / ... stupefactae: cf. *G.* 4. 363 (Aristaeus) 'mirans', 365 'stupefactus', *A.* 10. 445–6 (Pallas) 'iussa superba / miratus stupet in Turno'.

3. certantis: the merest indication of a singing-match; see below, 62 n.; also 3. 31 n.

stupefactae carmine lynces: exotic beasts, here consorting with the domestic heifer in an imaginary landscape. Lynxes are nowhere to be found in Theocritus, but Callimachus has one, on Mt. Maenalus, *Hymn* 3. 87–9 (Artemis) *ἵκεο δ' αὖλιν | Ἀρκαδικὴν ἔπι Πανός. ὁ δὲ κρέα λυγκὸς ἔταμνε | Μαιναλίης*, 'you came to the Arcadian fold of Pan. And he was cutting up the flesh of a lynx of Maenalus'; see below, 22 n. The only lynx known to V. and his contemporaries was the African caracal (*RE* xiii. 2476–7); V. therefore describes a huntress in the woods near Carthage as wearing a spotted lynx-hide (*A.* 1. 323). Lynxes were reputedly fond of music: Eur. *Alc.* 579 (to Apollo) *σὺν δ' ἐποιμαίνοντο χαρᾷ μελέων βαλιαί τε λύγκες*, 'and with the flocks roamed spotted lynxes rejoicing in your songs'.

4. et mutata suos requierunt flumina cursus: DServius quotes the *Io* of Calvus, *FPL* fr. 13 Büchner 'sol quoque perpetuos meminit requiescere cursus'. For inchoative verbs used transitively see Munro on Lucr. 4. 1282, Löfstedt i. 239–40, and for *requiesco* transitive, A. Taliercio, *RCCM* 28 (1986), 117–29. V. seems to have found the construction harsh, however, and modified it with *mutata* (La Cerda); cf. *A.* 1. 658 'faciem mutatus et ora Cupido'. Contrast *Ciris* 233 'quo rapidos etiam requiescunt flumina cursus'. The rivers are 'changed' in their course, that is, they stopped to listen to the Orphean singing of Damon and Alphesiboeus; cf. Ap. Rhod. 1. 26–7 (Orpheus) *τόν γ' ἐνέπουσιν ἀτειρέας οὔρεσι πέτρας | θέλξαι ἀοιδάων ἐνοπῇ ποταμῶν τε ῥέεθρα*, 'men say that he charmed hard rocks upon the mountains and the flow of rivers by the music of his songs', Hor. *Carm.* 1. 12. 9–10 (*Orphea*) 'rapidos morantem / fluminum lapsus' (cf. 3. 11. 14), Prop. 3. 2. 3–4 'Orphea delenisse feras et concita dicunt / flumina Threicia sustinuisse lyra', Ov. *Fast.* 2. 84 (Arion) 'carmine currentes ille tenebat aquas', *Culex* 117–18 'Oeagrius Hebrum / restantem tenuit ripis siluasque canendo', Sen. *Herc.* 573 'ars, quae praebuerat fluminibus moras', *Med.* 626–7 'cuius ad chordas modulante plectro / restitit torrens', [Sen.] *Herc.* [*Oet.*] 1036–9 'Illius stetit ad modos / torrentis rapidi fragor, / oblitusque sequi fugam / amisit liquor impetum', Claudian, *De Rapt. Pros. prol.* 2. 18 'pigrior astrictis torpuit Hebrus aquis'. See K. Ziegler, *RE* xviii. 1250, Housman, *JPhil* 16 (1888), 27–8 = *Class. Papers*, i. 48–9, Pease on *A.* 4. 489.

5. Damonis Musam ... et Alphesiboei: 'bene repetit, ne longum hyperbaton sensum confunderet' (DServ.); but of course that was not the only reason for the repetition. A similar repetition occurs in Alphesiboeus' song, below, 85–9.

6. tu mihi: the reader expects *dexter ades*, which often occurs in prayers and invocations, e.g. Ov. *Fast.* 1. 65–7 'Iane biceps ... dexter ades ducibus', or something of the sort ('subaudiendum *faue*', La Cerda); but the poet breaks off under the stress of strong excitement as in *G.* 1. 384, 4. 67, 252. Since the address has some of the attributes of an address to a deity (naming the alternative places where the addressee may be found, e.g. Theocr. 1. 123–6) the reader hears *tu* with some vocative force. Cf. Stat. *Silu.* 2. 7. 107–20 'At tu, seu ..., seu ..., adsis', 5. 3. 19–28 'At tu, seu ..., seu ..., da'.

magni ... saxa Timaui: 'the Timavus, so poetically (in every sense of the word) described by Virgil' (Gibbon, *Decline and Fall of the Roman Empire*, ch. 7 n. 27). J. Henry, who explored the Timavus in 1865, describes it as follows, *Aeneidea*, i (London, 1873), 523: 'At the foot of Monte Albio (Schneeberg), the last of the Julian Alps eastward, rises a river, which at San Canziano, sixteen miles from its source, becomes subterranean, and (having flowed from San Canziano eighteen miles underground) emerges from under the mountain at San Giovanni di Tuba, in numerous so-called springs or *sorgenti* coalescing almost immediately again into a single deep and broad stream, which, after a slow, smooth, and noiseless course of scarcely more than an Italian mile through the flat and marshy litoral, discharges itself into the Adriatic by a single mouth'. The Latin poets permit themselves a good deal of licence in describing rivers (see Tarrant on Sen. *Ag.* 318 ff.); and V. here accommodates the size of the river to the importance of his patron. The rocks at the river's mouth, to some extent conventional, are also of the poet's making. V. may have been thinking of the notoriously dangerous Illyrian coast; cf., noting the emphasis on *tutus*, *A.* 1. 242–6 (Venus to Jupiter) 'Antenor potuit mediis elapsus Achiuis / Illyricos penetrare sinus atque intima tutus / regna Liburnorum et fontem superare Timaui, / unde per ora nouem uasto cum murmure montis / it mare proruptum et pelago premit arua sonanti'. (Venus devotes two lines of her impassioned speech to the *mirabilia* of the Timavus; see R. Heinze, *Virgils epische Technik*[3] (Stuttgart, 1957), 480 n. 1.) V.'s reference to the Timavus here, while serving to establish his patron's whereabouts, is hardly so simple: the Timavus was strange and portentous, a river of legend, down which Jason and the Argonauts had sailed on their improbable return; see H. Philipp, *RE*, 2nd ser., vi. 1244.

superas: 'sail past'; 'nauticus sermo est', DServ. on *A.* 1. 244, quoting Lucil. fr. 125 M. 'promontorium remis superamus Mineruae'; cf. Livy 26. 26. 1 'nauibus superato Leucata promunturio'. Octavian did not, so far as is known, sail past the Timavus; he probably began his campaign in 35 BC from the Liburnian port Senia (now Senj in Croatia), which lies to the south of the Timavus; see Introduction, n. 8, and map, p. 234.

7. **oram . . . legis:** the first extant occurrence of *lego* in this sense (*TLL* s.v. 1127. 54); the phrase is used metaphorically in *G.* 2. 44.

en . . . umquam: 1. 67 n.

8. **tua dicere facta:** 4. 54.

10. **Sophocleo . . . coturno:** the cothurnus (κόθορνος) was the high boot or buskin worn by actors in Greek tragedy, and hence stands for tragedy, here Sophoclean tragedy; cf. Hor. *Carm.* 2. 1. 11–12 (Pollio) 'grande munus / Cecropio repetes coturno'.

digna: 9. 35–6 n.

11. **a te principium, tibi desinam:** an allusion to the formula (as it had become) Ἐκ Διὸς ἀρχώμεσθα; see Introduction, p. 236. More especially, V. had in mind two passages: Theocr. 17. 1 Ἐκ Διὸς ἀρχώμεσθα καὶ ἐς Δία λήγετε Μοῖσαι, 'From Zeus let us begin, and with Zeus make end, Muses', and *Il.* 9. 96–8 (Nestor speaking) Ἀτρεΐδη κύδιστε, ἄναξ ἀνδρῶν Ἀγάμεμνον, | ἐν σοὶ μὲν λήξω, σέο δ' ἄρξομαι, οὕνεκα πολλῶν | λαῶν ἐσσι ἄναξ, 'Most glorious son of Atreus, king of men Agamemnon, with you will I end, with you begin, for you are king of many peoples'. Note the metrical correspondence of *a te* to ἐν σοὶ, *tibi* to σέο, and ἄρξομαι (with the final syllable shortened in hiatus before the bucolic diaeresis) to *desinam*.

desinam: a prosodic hiatus unparalleled in V.; cf. the hiatus, also unparalleled in V., at *A.* 1. 405 'et uera incessu patuit dea. ille ubi matrem'. Hiatus occurs at the bucolic diaeresis several times in Theocritus; see Gow on 2. 83 ('usually accompanied, as here, by a sense-pause'). See also 2. 53 n. *Desinet* (see app. crit.) is evidently an attempt to remove the offence.

13. uictricis: cf. *A.* 3. 54 'uictricia . . . arma' and, for such adjectives, see Wackernagel ii. 54.

hederam . . . lauros: 'nam uictores imperatores lauro, hedera coronantur poetae' (Serv.). See 7. 25 n.

serpere: Cicero, who likes to use this verb metaphorically, uses it once of the vine, *De sen.* 52 'serpentem multiplici lapsu et erratico'. The intertwining word-order is no doubt intentional.

14–16. Pastoral order is re-established after the interruption of 6–13. Damon and Alphesiboeus had driven afield several hours before sunrise, 'cum sole nondum orto iam lucet' (Censorinus 24. 2), as V., reusing l. 15, urges real shepherds to do, *G.* 3. 324–6 'Luciferi primo cum sidere frigida rura / carpamus, dum mane nouum, dum gramina canent, / et ros in tenera pecori gratissimus herba'. Cf. Varro, *RR* 2. 2. 10 (flocks) 'aestate, quod cum prima luce exeunt pastum, propterea quod tunc herba roscida meridianam, quae est aridior, iucunditate praestat'.

16. incumbens tereti . . . oliuae: Servius, followed by La Cerda and some older commentators, thought a tree was meant. Other ancient readers disagreed: 'alii *tereti oliuae* baculum de oliua accipiunt' (DServ.). Shepherds have made their staves of olive-wood ever since Polyphemus: *Od.* 9. 319–21 *μέγα ῥόπαλον . . . | χλωρὸν ἐλαΐνεον τὸ μὲν ἔκταμεν, ὄφρα φοροίη | αὐανθέν*, 'a great staff of green olive-wood, which he cut to carry when dried'. Damon's posture is, however, unusual, for pastoral singers are usually seated while performing; see 3. 55 n. Perhaps V. was thinking of the solitary, lovelorn singer in Theocr. 3. 38 *ᾀσεῦμαι ποτὶ τὰν πίτυν ὧδ' ἀποκλινθείς*, 'I will step aside under the pine here and sing' (Gow), where *ἀποκλινθείς* has generally been understood to mean 'leaning against', and may have been so understood by V.

17. nascere praeque diem ueniens age, Lucifer, almum: 'ordo est *nascere, Lucifer, praeueniensque age diem clarissimum*' (DServ., Serv.). Cf. *A.* 2. 801–2 'iamque iugis summae surgebat Lucifer Idae / ducebatque diem', Ov. *Fast.* 5. 547–8 'quid solito citius liquido iubar aequore tollit / candida, Lucifero praeueniente, dies?'. For the tmesis see 6. 6 n., L. Mueller, *De re metrica*² (Leipzig, 1894), 458–60, Housman on Manil. 1. 355.

nascere: used also by V. of the rising sun (*G.* 1. 441), and by Horace of the new moon (*Carm.* 3. 23. 2).

Lucifer: the Morning Star, that is, the planet Venus, when visible in the east before sunrise; Cic. *De nat. deor.* 2. 53 'stella Veneris, quae *Φωσφόρος* Graece, Lucifer Latine dicitur cum antegreditur solem, cum subsequitur autem *Ἕσπερος*'. That the Morning and Evening Star were one and the same was so well known in antiquity as to have become a poetic conceit, e.g. Catull. 62. 34–5 'nocte latent fures, quos idem saepe reuertens, / Hespere, mutato comprendis nomine Eous', where see Fordyce and Ellis. See also Mynors on *G.* 3. 324–6, Nisbet–Hubbard on Hor. *Carm.* 2. 9. 10.

almum: V. seems to have been the first to apply this adjective to *dies* and *lux* (*TLL* s.v. 1704. 41); cf. *A.* 5. 64 'diem . . . almum', 1. 306 'lux alma'. Here as bringing some relief from the ghastly night that Damon has suffered through.

17–60. Damon delivers a dramatic monologue (not unlike that of the unnamed goatherd in Theocr. 3 or Corydon in *E.* 2), impersonating a rejected suitor (32–5) who has lain awake during the night that precedes his girl's wedding to someone else. He prays for the interminable night to end; the Morning Star is either just rising or has not yet risen. He imagines the ceremony that will take place in the evening (29–30), passionately recalls his childhood memories of the girl (37–41), and, finally, after expressing his grief and bewilderment in a series of *adynata*, decides to commit suicide (52–60).

18–20. Nysa had sworn to be true for ever, their understanding ratified, to Damon's satisfaction, by vows of fidelity which she has now broken. Damon's reference to Nysa as his wife (18 'coniugis') embarrasses commentators ('it was as his wife that Damon loved her', Conington; 'an "Arcadian" wife, with whom vows had been exchanged (19–20) but no formal union solemnized', Coleman). But unlike Galatea (1. 30–2), Nysa seems a respectable country girl, if proud and false (but why believe a complaining lover?), who is now about to be married in style (29–30). The word *coniunx* occurs only in this *E.* (again in l. 66 'coniugis') and owes its presence here not to the requirements of the immediate context, which it apparently disrupts, but to V.'s desire to relate Damon's song to the song of Alphesiboeus; see Introduction, p. 238. Under the circumstances, Servius' interpretation is probably as good as can be invented: 'non quae erat sed quae fore sperabatur'.

18. indigno . . . amore: so Damon feels it to be; unrequited love, as in 10. 10.

Nysae: a curious name, before V. attested only as the name of the Nymph who nursed Dionysus on Mt. Nysa; see Roscher's *Lexikon* s.v. 567.

19–20. Cf. Catull. 64. 188–91 (the abandoned Ariadne) 'non tamen ante mihi languescent lumina morte, / . . . / quam iustam a diuis exposcam prodita multam / caelestumque fidem postrema comprecer hora'.

19. quamquam nil testibus illis: 'sic hoc dicit tamquam iusiurandum inter eum et Nysam intercesserit' (DServ.). The impunity of perjured lovers is a commonplace of ancient poetry, e.g. Dioscorides 6. 3–4 G.–P. (= *AP* 5. 52) *κενὰ δ' ὅρκια· τῷ δ' ἐφυλάχθη | ἵμερος· ἡ δὲ θεῶν οὐ φανερὴ δύναμις*, 'vain were her oaths, yet his love persisted; the power of the gods was not manifest', *A.* 4. 520–1 (Dido) 'tum, si quod non aequo foedere amantis / curae numen habet iustumque memorque, precatur', Ov. *Am.* 3. 3. 1 'Esse deos, i, crede: fidem iurata fefellit', *Ars* 1. 633–4 'Iuppiter ex alto periuria ridet amantum / et iubet Aeolios inrita ferre Notos'. See 3. 72–3 n.

20. extrema moriens: extremely pathetic; echoed in his final line (60).

21. The refrain is reminiscent of that in Theocr. 1. 64 *ἄρχετε βουκολικᾶς, Μοῖσαι φίλαι, ἄρχετ' ἀοιδᾶς*, 'Begin, dear Muses, begin the pastoral song'. See below, 61 n., 68 n., 109 n., and Introduction, pp. 237–8.

mea tibia: 1. 2 n.

22. Maenalus: to explain 'Maenalios . . . uersus' in the refrain. (*Carmina* occurs at the same place in Alphesiboeus' song (69), although 'mea carmina' (68) requires no explanation.) Maenalus, a mountain sacred to Pan in Arcadia, is mentioned only once by Theocritus, in a prayer to Pan (1. 124). The adjective occurs twice in Callimachus, *Hymn* 3. 89 (quoted above, 3 n.) and 224. See *E.* 10, Introduction, p. 289.

argutum: cf. 7. 1.

pinusque loquentis: cf. Catull. 4. 12 ('comata silua') 'loquente saepe sibilum edidit coma', Manil. 3. 655–6 'totumque canora / uoce nemus loquitur', where Housman cites Petron. 120. 72–3 'non uerno persona cantu / mollia discordi strepitu uirgulta loquuntur', *A.* 11. 458 'dant sonitum rauci per stagna loquacia cycni'.

23. semper ... semper: repetition with a shift of the ictus; see 6. 9 n.

amores: 'cantica de amoribus' (DServ.).

24. Panaque, qui primus: 2. 32–3 n.

calamos: 2. 37 n.

26–8. Nysa is given in marriage to Mopsus—in such a topsy-turvy world what may we lovers not expect? Unnatural unions are a well-attested form of *adynaton*; cf. Hor. *Epod.* 16. 30–2 'nouaque monstra iunxerit libidine / mirus amor, iuuet ut tigris subsidere ceruis, / adulteretur et columba miluo' and see *E.* 4, Appendix, pp. 147–8.

26. Mopso: the name of the younger singer in *E.* 5.

datur: old and common in this sense (*TLL* s.v. 1693. 51).

27. iam: 'soon', in contrast with what will happen later, 'aeuoque sequenti'.

grypes: griffins, fabulous animals, with the head and wings of an eagle and the body of a lion, who inhabit Scythia and guard its gold; between them and the one-eyed Arimaspian horsemen, who purloin the gold, there is constant warfare; see Aesch. *Prom.* 803–6, with Griffith's note. Grypomachies with horses are represented in Greek art; see K. Ziegler, *RE* vii. 1927, J. D. P. Bolton, *Aristeas of Proconnesus* (Oxford, 1962), 36–7.

28. canibus ... dammae: cf. Theocr. 1. 135 *τὰς κύνας ὤλαφος ἕλκοι*, 'let the stag worry the hounds' (Gow), one of a series of *adynata* imitated below, 52–6.

timidi ... dammae: 'ne homoeoteleuton faceret dicendo *timidae dammae*' (DServ.); see *TLL* s.v. *damma* 14, Norden 406. Again in *G.* 3. 539; cf. *G.* 1. 183 'capti ... talpae', with Mynors's note.

pocula: drinking-water, with no reference to cups, as in Columella 7. 10. 7 (for pigs) 'puteis extracta et large canalibus immissa praebenda sunt pocula'.

28[a]. This line, which does not appear in any of the capital MSS (see app. crit.), should be deleted along with the corresponding l. 76. See Introduction, p. 238; also O. Skutsch, *BICS* 18 (1971), 26, *RPh* 51 (1977), 366.

29–30. The impending ceremony is described, although sarcastically and with much bitterness, in terms appropriate to a legal Roman marriage.

29. nouas . . . faces: torches for the wedding procession in the evening, the *deductio*, when the bride was conducted to her new home; cf. Catull. 61. 114–15 'tollite, o pueri, faces: / flammeum uideo uenire'. And *nouas*, 'as the occasion would doubtless seem to require new torches' (Conington).

incide faces: cut (wood for) torches; for similarly compressed expressions cf. *A.* 1. 552 'stringere remos', strip (boughs for) oars, Livy 33. 5. 4 'uallum caedere', cut (stakes for) a palisade, *TLL* s.v. *caedo* 57. 3.

30. nuces: walnuts were scattered among the crowd during the *deductio*; see Fordyce on Catull. 61. 121.

tibi: 'Haec repetitio, *tibi ducitur uxor, tibi deserit Hesperus Oetam*, iucunda est' (La Cerda).

Hesperus: this seems to be V.'s spelling (MV), though the Palatinus (P) has *-os*, for in 10. 77 all three capital MSS (MPR) have *-us*. Cf., however, Censorinus 24. 4 'stellae quam Plautus Vesperuginem, Ennius Vesperum, Vergilius Hesperon appellat', with Skutsch's note on Enn. *Op. Inc.* 29.

Oetam: a mountain in southernmost Thessaly, over the ridge of which the Evening Star rises in Latin poetry; cf. Catull. 62. 7 'nimirum Oetaeos ostendit noctifer ignes', with Fordyce's note. See also *Suppl. Hell.* 666. The Veronensis (V), a highly respectable witness, and the Verona scholiast have *Oetan*, which Sabbadini, Geymonat, and Coleman accept. It is possible that V. wrote *Hesperus* (or *Hesperos*) *Oetan*, but not probable, since this Greek ending is not securely attested before Ovid: *Andromedan*, *Ars* 1. 53, *Met.* 4. 671, 757 'Andromedan et tanti'; *Electran*, *Fast.* 4. 32, 174, *Trist.* 2. 395 'Electran et egentem'.

32–3. dum . . . / dumque . . . dumque: 'The repetition of *dum*, with or without asyndeton, is a characteristic Virgilian touch'—so Pease on *A*. 4. 53, where he cites, in addition to these lines, *G*. 1. 214, 2. 362–3, 3. 165, 325, 428, *A*. 1. 453–4, 494–5, 4. 336, 8. 580–1, 11. 671–2.

34. hirsutumque supercilium promissaque barba: the shaggy brow belongs to Polyphemus in Theocr. 11. 31 *λασία . . . ὀφρύς*, the jutting beard to the goatherd in Theocr. 3. 9 *προγένειος*, where see Gow. An iambic word is expected after *hirsutumque*, e.g. *A*. 1. 487 'tendentemque manus Priamum conspexit inermis'; see J. Hellegouarc'h, *Le Monosyllabe dans l'hexamètre latin* (Paris, 1964), 287–9. The uncouth rhythm may be taken as suggesting Damon's rough, unkempt appearance; cf. Hor. *Epist*. 1. 1. 94 'si curatus inaequali tonsore capillos'.

37–8. These lines are modelled on Theocr. 11. 25–7 (Polyphemus to Galatea) *ἠράσθην μὲν ἔγωγε τεοῦς, κόρα, ἁνίκα πρᾶτον | ἦνθες ἐμᾷ σὺν ματρὶ θέλοισ' ὑακίνθινα φύλλα | ἐξ ὄρεος δρέψασθαι, ἐγὼ δ' ὁδὸν ἁγεμόνευον*, 'I fell in love with you, maiden, when first you came with my mother to gather hyacinths on the hill, and I led the way'. See *E*. 2, Introduction, p. 63.

37. saepibus in nostris: Columella 5. 10. 1 'modum pomarii, priusquam semina seras, circummunire maceriis uel saepe uel fossa praecipio'.

roscida mala: cf. 4. 30 'roscida mella'. Ursinus compares Theocr. *Epigr*. 1. 1 Gow *Τὰ ῥόδα τὰ δροσόεντα*, 'The dew-drenched roses'.

38. cum matre: 'deest pronomen, et ideo uel huius uel puellae matrem intellegere possumus' (DServ.). La Cerda, Conington, and T. E. Page rather against his better judgment understand 'my mother'. 'The reference of "matre" is fixed by the passage in Theocr. *ἐμᾷ σὺν ματρί*' (Conington)—as if V. could not, as he repeatedly does, modify Theocritus to suit his own purpose. Heyne, Wagner, and Forbiger remain prudently silent; Klingner and Coleman understand 'your mother'. Polyphemus' mother was the sea-nymph Thoösa (see Gow ad loc.), and when the Nereid Galatea first emerged from the sea with her to pick flowers on dry land, Polyphemus was their guide; but Damon's mother would hardly need to be shown around the family enclosure. V. identifies *matre* by placing the parenthetic *dux ego uester eram* before it (compare the position of ἐγὼ δ' ὁδὸν ἁγεμόνευον), for the reference of *uester* is normally plural; see Housman, '*Vester = Tuus*', *CQ* 3 (1909), 244–8 = *Class. Papers*, ii. 790–4. Nysa was very young then, too young and small to come without her mother, and Damon was only twelve. An overwhelming first experience of sexual passion, every detail of which remains vivid in memory—like Dante's passion for Beatrice (*Vita nuova* 2) or Byron's for Mary Duff (*Journal*, 26 Nov. 1813). But Damon's passion is only a fiction, a 'memory' of Theocritus.

39. **alter ab undecimo:** next after the eleventh, that is, the twelfth. Cf. 5. 49 'tu nunc eris alter ab illo' and see Housman on Manil. 4. 445.

41. **ut uidi, ut perii, ut me malus abstulit error:** 'I saw, and I was lost, and madness carried me away'. As so often, V. had two passages in mind, one from each of the *Idylls* he was chiefly imitating in this *Eclogue*: Theocr. 2. 82 χὠς ἴδον, ὣς ἐμάνην, ὥς μοι πυρὶ θυμὸς ἰάφθη, 'I saw, and madness seized me, and my hapless heart was aflame' (Gow), and 3. 41–2 ἁ δ' Ἀταλάντα | ὡς ἴδεν, ὣς ἐμάνη, ὣς ἐς βαθὺν ἅλατ' ἔρωτα, 'and Atalanta saw, and frenzy seized her and deep in love she plunged' (Gow). The hiatus after ἐμάνη may have prompted that after *perii*, but the same hiatus occurs several times elsewhere in the *E.*; see 3. 6 n. S. Timpanaro, *Contributi di filologia e di storia della lingua latina* (Rome, 1978), 219–87, demonstrates, in impressive detail, that the ancient, and still generally accepted, interpretation of this line is mistaken: 'unum *ut* (*ut* DServ.) est temporis, aliud quantitatis; nam hoc dicit: mox uidi, quemadmodum perii' (Serv.); 'uel primum *ut* postquam, duo sequentia pro admirandi significatione posita sunt' (DServ.). Cf. T. E. Page: 'It is usually rendered "*when* I saw, *how* I fell in love!"' which saves the grammar but destroys the charm of the phrase, for this consists in the parallelism of *ut* . . . *ut* and the exact balance thus established between the ideas of "seeing" and "loving"' ('il modesto ma stilisticamente fine commento di T. E. Page', Timpanaro 226). The three clauses, Timpanaro argues, must be correlative, as Heyne understood, 'cum vidi, tum statim amore exarsi'. Timpanaro points out, after La Cerda and others, that Ovid imitates V.'s line (and Theocr. 2. 82), *Her*. 12. 33 'et uidi et perii nec notis ignibus arsi'; 'Nel nostro caso, mentre Virgilio

ha voluto mantenersi fedele al modello ellenistico a prezzo di un grecismo alquanto audace, Ovidio ha voluto rendere lo stesso concetto con un'espressione più facile e più latina' (278 n. 87). See Clausen 132 n. 15.

malus . . . error: 'definitio amoris' (Serv.); cf. Prop. 1. 13. 35 'qui tibi sit felix, quoniam nouus incidit, error', Ov. *Am.* 1. 10. 9–10 'animique resanuit error, / nec facies oculos iam capit ista meos'.

43. nunc scio quid sit Amor: cf. Theocr. 3. 15–16 *νῦν ἔγνων τὸν Ἔρωτα· βαρὺς θεός· ἦ ῥα λεαίνας | μαζὸν ἐθήλαζεν, δρυμῷ τέ νιν ἔτραφε μάτηρ*, 'Now I know Love, and a cruel god he is. Surely he sucked the dug of a lioness, and in the wildwood his mother reared him'.

scio: iambic shortening, again in *A.* 3. 602; in *A.* 10. 904 *-o* is elided before a short vowel, 'scio acerba'. Cf. below, 107 'nescio quid', 3. 103 'nescio quis', and see Austin on *A.* 2. 735.

43–5. Cf. *A.* 4. 365–7 (Dido to Aeneas) 'nec tibi diua parens, generis nec Dardanus auctor, / perfide, sed duris genuit te cautibus horrens / Caucasus Hyrcanaeque admorunt ubera tigres', with Pease's note. The pedigree of Love was notoriously uncertain; see Gow on Theocr. 13. 2, D. Page, *Sappho and Alcaeus* (Oxford, 1955), 271.

44. aut Tmaros aut Rhodope aut extremi Garamantes: cf. Theocr. 7. 77 *ἢ Ἄθω ἢ Ῥοδόπαν ἢ Καύκασον ἐσχατόωντα*, 'or Athos or Rhodope or remotest Caucasus', *G.* 1. 332 'aut Atho aut Rhodopen aut alta Ceraunia telo'.

Tmaros: or Tomaros (see Schwyzer, *Griech. Gramm.* i. 278), a mountain range dominating the western side of the valley in which the oracle of Dodona was situated; see E. Polaschek, *RE*, 2nd ser., vi. 1697–8, D. M. Nicol, *G&R*, NS 5 (1958), 128–43, N. G. L. Hammond, *Epirus* (Oxford, 1967), 168–9. The name is first attested in Pindar, *Paean* 6. 109 *Τομάρου*, but for V. it no doubt had a Callimachean colouring, *Aet.* 23. 3 Pf. *ἐνὶ Τμαρίοις οὔρεσιν*, *Hymn* 6. 51 *ὤρεσιν ἐν Τμαρίοισιν*, 'in the mountains of Tmaros'.

Rhodope: 6. 30 n.

extremi Garamantes: tribesmen of the eastern Sahara, here introduced into poetry; see Pease on *A.* 4. 198 and cf. Catull. 11. 2 'extremos . . . Indos', *A.* 6. 794 'super et Garamantas et Indos'; also *Od.* 1. 23 (*Αἰθίοπας*) *ἔσχατοι ἀνδρῶν*, '(the Ethiopians) remote from men', *A.* 8. 727 'extremique hominum Morini'.

45. nec generis nostri puerum nec sanguinis: 'no boy of our breed or blood' (Lee); inhuman therefore ('nihil habentem in se humanitatis', Serv.) and savage. See McKeown on Ov. *Am.* 1. 1. 25–6.

edunt: the present tense, as often in Greek and Latin poetry (though more often in Greek and not attested before V. in Latin) of birth and upbringing; cf. *G.* 1. 279 *creat*, *A.* 8. 141 *generat*, 10.518 *educat*, Prop. 4. 4. 54 *nutrit*, and see Tränkle 73–4, Schwyzer–Debrunner, *Griech. Gramm.* ii. 272.

47–50. 'Pitiless Love once taught a mother to pollute / Her hands with blood of sons; you too were cruel, mother. / Who was more cruel, the mother or that wicked boy? / That wicked boy was; yet you too were cruel, mother' (Lee). This is the most straightforward way of taking the Latin and agrees with Servius: 'utitur autem optima moderatione; nam nec totum Amori imputat, ne defendat parricidam, nec totum matri, ne Amorem eximat culpa, sed et illam quae paruit et illum qui coegit incusat'. Perhaps no two lines of V., certainly no two in the *E.*, have been so persistently or so violently misconstrued as 49–50; for discussion see Conington, and Coleman, and especially J. Vahlen, *Opuscula Academica*, ii (Leipzig, 1908), 526–44. Ancient scholars, too, were perplexed, many of whom, says DServius, attempted to extenuate the mother's guilt with this punctuation: 'ille, improbus ille puer crudelis; tu quoque, mater'. But why so much trouble with these apparently simple lines? Two reasons may be offered: (i) such deliberate frivolity seems to strike a wrong note in Damon's song ('Inest enim ieiuni et inepti lusus nescio quid', Heyne); and (ii) the style or manner offends: minimal reference to the story ('et bene fabulam omnibus notam per transitum tetigit', Serv.), sympathetic apostrophe (cf. 6. 47, 52), displaced emphasis (see 6. 79–81 n.), formal repetition (see 4. 58–9 n.)—in a word, Callimachean; and how else was a New Poet to tell an old story?

47. saeuus Amor: famously, Medea had been wounded by savage love, Enn. *Medea exul* 254 V.[2] = 216 J. 'Medea animo aegro amore saeuo saucia'. Cf. Prop. 3. 19. 17–18 'nam quid Medeae referam, quo tempore matris / iram natorum caede piauit amor?', Ov. *Trist.* 2. 387–8 (Medea is not named) 'tingeret ut ferrum natorum sanguine mater, / concitus a laeso fecit amore dolor'. See Clausen 40–1.

48. commaculare: in this non-metaphorical sense occurs again in classical Latin only in Tac. *Ann.* 1. 39. 4 'legatus populi Romani Romanis in castris sanguine suo altaria deum commaculauisset' (*TLL* s.v. 1818. 75). The simple form, however, is fairly common, e.g. *A.* 3. 28–9 'huic atro liquuntur sanguine guttae / et terram tabo maculant'.

49. improbus: cf. *A.* 4. 412 'improbe Amor', with Pease's note; also see Mynors on *G.* 1. 145–6.

52–6. Damon has no desire to go on living in such an impossible world (cf. above, 27–8). His *adynata* are modelled on those of another victim of love, Daphnis in Theocr. 1. 132–6 *νῦν ἴα μὲν φορέοιτε βάτοι, φορέοιτε δ' ἄκανθαι, | ἁ δὲ καλὰ νάρκισσος ἐπ' ἀρκεύθοισι κομάσαι, | πάντα δ' ἄναλλα γένοιτο, καὶ ἁ πίτυς ὄχνας ἐνείκαι, | Δάφνις ἐπεὶ θνάσκει, καὶ τὰς κύνας ὥλαφος ἕλκοι, | κὴξ ὀρέων τοὶ σκῶπες ἀηδόσι γαρύσαιντο,* 'Now violets bear, ye brambles, and, ye thorns, bear violets, and let the fair narcissus bloom on the juniper. Let all be changed, and let the pine bear pears since Daphnis is dying. Let the stag worry the hounds, and from the mountains let the owls cry to nightingales' (Gow). V.'s imitation is accurate but subtle; see L. Braun, *Philologus*, 113 (1969), 292–3. On the rarity of botanical *adynata* see D. O. Ross, Jr., *Virgil's Elements: Physics and Poetry in the Georgics* (Princeton, 1987), 107.

52. ultro: emphasizing the reversal of their natural roles; cf. 3. 66.

52–3. aurea durae / mala ferant quercus: impossibilities in the real world will be realized in the Golden Age, 4. 30 'durae quercus sudabunt roscida mella'. Cf. 3. 71 'aurea mala'.

53. narcisso floreat alnus: V. has transferred Theocritus' narcissus to the alder. Cf. Lucr. 5. 911–12 (the first extant Latin *adynaton*) 'aurea tum dicat per terras flumina uulgo / fluxisse et gemmis florere arbusta suesse'; *gemmis*, 'precious stones', the metaphorical (and prevalent) sense is here especially apt; see 7. 48 n. See also *TLL* s.v. *floreo* 917. 36–7.

54. pinguia corticibus sudent electra myricae: cf. Catull. 64. 106 'conigeram sudanti cortice pinum' and see Clausen, *BICS*, *Suppl.* 51 (1988), 17 n. 17. The tree that exudes thick amber is not the tamarisk but the poplar (6. 63 n.).

55. certent et cycnis ululae: cf. Theocr. 5. 136–7 *οὐ θεμιτόν . . . ποτ' ἀηδόνα κίσσας ἐρίσδειν, | οὐδ' ἔποπας κύκνοισι,* 'it is not right for jays to contend with a nightingale, nor hoopoes with swans', Lucr. 3. 6–7 'quid enim contendat hirundo / cycnis?'

certent: 5. 8 n.

cycnis: 9. 29 n.

55–6. sit Tityrus Orpheus, / Orpheus in siluis, inter delphinas Arion: 'uilissimus rusticus Orpheus putetur in siluis, Arion uero inter delphinas' (Serv.). Arion, here a sort of marine Orpheus, was thrown overboard by sailors but carried ashore on the back of a dolphin who had been charmed by his music; for the story see Herod. 1. 23–4, Ov. *Fast.* 2. 79–118. L. P. Wilkinson, *CR* 50 (1936), 120–1, conjectures that V.'s dolphins (there are none in Theocritus) come from Archilochus, fr. 122. 6–9 W., the first extant *adynaton.*

Nonnus has a line oddly similar to l. 56 in shape, *Dionys.* 16. 135 Ἄρτεμις ἐν σκοπέλοισι καὶ ἐν θαλάμοις Ἀφροδίτη, (that you may appear) 'Artemis among the rocks and in the bedroom Aphrodite'.

58. omnia uel medium fiat mare: for the singular verb cf. *Dirae* 46 'cinis omnia fiat', Ov. *Met.* 1. 292 'omnia pontus erat', and see Löfstedt ii. 119. It is unnecessary to assume, with Conington and T. E. Page, that V. misread the variant ἔναλλα, 'changed', in Theocr. 1. 134 as ἐνάλια, 'in the sea'; see Gow ad loc.

medium: adds a slight emphasis; cf. Plaut. *Truc.* 527–8 'si hercle me ex medio mari / sauium petere tuom iubeas', Lucr. 4. 1100 'in medioque sitit torrenti flumine potans', *A.* 10. 305 (*puppis*) 'soluitur atque uiros mediis exponit in undis' (*TLL* s.v. 585. 44).

uiuite siluae: 'farewell the woods', 'ualete' (Serv.); perhaps suggested by Theocr. 1. 115–18. Cf. Hor. *Serm.* 2. 5. 110 'uiue ualeque', *Epist.* 1. 6. 67 'uiue, uale'.

59–60. praeceps aërii specula de montis in undas / deferar: cf. Theocr. 3. 25–6 τὰν βαίταν ἀποδὺς ἐς κύματα τηνῶ ἀλεῦμαι, | ὧπερ τὼς θύννως σκοπιάζεται Ὄλπις ὁ γριπεύς, 'I will strip off my cloak and leap into the waves from the cliff whence Olpis, the fisherman, watches for the tunny' (Gow). And in Hermesianax, *Leontion* fr. 3 Powell, Menalcas, rejected by Euhippe, leaps to his death from a cliff.

59. aërii specula de montis: cf. Catull. 68. 57 'in aerii . . . uertice montis' and see *TLL* s.v. *aerius* 1063. 18. For the anastrophe of the preposition cf. Lucr. 3. 1088 'tempore de mortis' and see Mynors on *G.* 4. 333 'thalamo sub fluminis'; see also Munro on Lucr. 3. 140, Housman on Manil. 1. 245, and, for the anastrophe of prepositions in general, Wackernagel ii. 196–200.

60. extremum . . . morientis: echoing l. 20, 'extrema moriens'.

munus: probably not Damon's song (Heyne, Coleman) but his death, which he imagines will give pleasure to Nysa; see above, 59–60 n., and cf. Theocr. 3. 27 *καἴ κα δὴ 'ποθάνω, τό γε μὲν τεὸν ἁδὺ τέτυκται*, 'And if I die, at least your pleasure will have been done', 54 (the singer, finally, will lie down and die and be eaten by wolves) *ὡς μέλι τοι γλυκὺ τοῦτο κατὰ βρόχθοιο γένοιτο*, 'And sweet as honey in the throat may this be to you'. Cf. also *A.* 4. 429 'extremum hoc miserae det munus amanti'.

61. The refrain is modelled on that in Theocr. 1. 127 *λήγετε βουκολικᾶς, Μοῖσαι, ἴτε λήγετ' ἀοιδᾶς*, 'Cease, Muses, come cease the pastoral song', which replaces the earlier refrain *ἄρχετε βουκολικᾶς, Μοῖσαι φίλαι, ἄρχετ' ἀοιδᾶς*, 'Begin, dear Muses, begin the pastoral song', as Thyrsis' song draws to a close; see above, 21 n. So here *desine* replaces *incipe*.

62. responderit: maintains the fiction of a singing-match; see above, 3 n., and cf. 7. 5 'respondere parati'.

63. dicite, Pierides: cf. 6. 13 'Pergite, Pierides'. The poet unexpectedly appeals to the Muses for help, as though recalling the second song were beyond his unaided powers; see 7. 19 n. In much the same way—'si parua licet componere magnis'—*G.* 3 breaks in the middle and the second half begins with an invocation, or rather, a digression (284–94), in which V. displays great art and artfulness while questioning his ability to render poetic the care of sheep and goats.

non omnia possumus omnes: cf. Lucil. 218 M. 'non omnia possumus omnes', a proverb 'in the typical form of the paroemiac' (Fraenkel on Aesch. *Ag.* 1527). Cf. e.g. Theocr. 15. 62 *πείρᾳ θην πάντα τελεῖται*, 'everything's done by trying' (Gow), and see Gow on Theocr. 5. 38, McLennan on Callim. *Hymn* 1. 9. For the sentiment, which is as old as Homer, see Otto, no. 1288. Cf. 7. 23.

64–109. Alphesiboeus impersonates an unnamed countrywoman—and, surprisingly, her assistant (105–6)—as she performs a magic ceremony in the hope of compelling her absent 'husband' to return home from town. Alphesiboeus' song is modelled more or less closely on Theocr. 2. 1–63: the preparations of Simaetha, seduced and abandoned, for a magic ceremony, and her incantation. There is nothing pastoral about Theocritus' *Idyll*, however; Simaetha lives in a town near the sea, and her faithless lover, Delphis, frequents the local wrestling-school. V. necessarily adapted his imitation to the pastoral mode: hence the rural setting, with a hint of the opposition between town and country; the names Amaryllis and Daphnis; the elaborate, pathetic simile of the weary heifer (V. is more self-consciously 'poetic' than Theocritus); and, finally, though perhaps more rustic than pastoral, mention of a werewolf and crops spirited away. For the structure of Alphesiboeus' song and its relation to Damon's song and Simaetha's incantation see Introduction, pp. 237–8.

64–82. Cf. Apul. *Apol.* 30 'at si Vergilium legisses, profecto scisses alia quaeri ad hanc rem solere; ille enim, quantum scio, enumerat uittas mollis et uerbenas pinguis et tura mascula et licia discolora, praeterea laurum fragilem, limum durabilem, ceram liquabilem'.

64. effer aquam: the water, it seems, is to be brought out into the *atrium*, where the altar stands in the open not far from the outer door of the house (101, 107). The command is addressed to her assistant, whose name, the reader presently learns, is Amaryllis (77); see 2. 10 n. So Simaetha bids Thestylis bring her the bay-leaves and love-charms, Theocr. 2. 1 *Πᾶ μοι ταὶ δάφναι; φέρε, Θεστυλί. πᾶ δὲ τὰ φίλτρα;* Two is the usual number—a woman and her assistant or accomplice—in such a scene; cf., besides Theocritus 2, the fragment of Sophron's mime (D. L. Page, *Select Papyri* (Loeb Classical Library, iii. 328–31), Hor. *Epod.* 5, *Serm.* 1. 8.

molli cinge haec altaria uitta: cf. Theocr. 2. 2 *στέψον τὰν κελέβαν φοινικέῳ οἰὸς ἀώτῳ*, 'Wreathe the bowl with fine crimson wool' (Gow). Both wool and crimson, as Gow observes ad loc., have apotropaic power. Cf. e.g. Prop. 4. 6. 6 'terque focum circa laneus orbis eat', Tac. *Hist.* 4. 53. 2 'spatium omne, quod templo dicabatur, euinctum uittis coronisque'.

molli: 'lanea scilicet' (DServ.).

65. uerbenas: boughs or plants used for decorating the altar; see Nisbet–Hubbard on Hor. *Carm.* 1. 19. 14. Boughs of olive, bay, and myrtle are mentioned in the magical papyri; see A. Abt, *Die*

Apologie des Apuleius von Madaura und die antike Zauberei (Giessen, 1908), 145–6.

adole: an old ritual word meaning 'to set on fire'; cf. Lucr. 4. 1236–7 'multo sanguine ... / conspergunt aras adolentque altaria donis', *A.* 3. 546–7 'rite / Iunoni Argiuae iussos adolemus honores'. See A. Ernout, *Philologica* (Paris, 1946), 53–8, and Mynors on *G.* 4. 379.

pinguis: full of sap, good for burning; cf. 7.49.

mascula tura: the choicest frankincense, white globules greasy within when crushed and quick-burning (Dioscorides, *De mat. med.* 1. 68. 1 λίβανος ... πρωτεύει δὲ ὁ ἄρρην ...); Pliny, *NH* 12. 61 'masculum aliqui putant a specie testium dictum', where Ernout remarks: 'L'expression est demeurée en français: oliban ou encens mâle'.

66. coniugis: in her view, though hardly in a legal sense; see above, 18–20 n.

magicis ... sacris: the first extant occurrence of the adjective in Latin; cf. *A.* 4. 493 'magicas ... artis'.

66–7. sanos auertere ... / ... sensus: she wishes Daphnis again to be madly in love with her, not, as he now is, heart-whole and indifferent; cf. Catull. 83. 3–4 'si nostri oblita taceret, / sana esset'. So Simaetha wishes Delphis to return to her 'like a madman' (Theocr. 2. 51 μαινομένῳ ἴκελος) from the wrestling-school.

67. nihil hic nisi carmina desunt: for the attraction of the verb cf. Ov. *Trist.* 1. 2. 1 'quid enim nisi uota supersunt?' and see Löfstedt ii. 117.

68. ducite ab urbe domum, mea carmina, ducite Daphnin: see above, 21 n., and cf. the refrain of Simaetha's incantation, Theocr. 2. 17 ἶυγξ, ἕλκε τὺ τῆνον ἐμὸν ποτὶ δῶμα τὸν ἄνδρα, 'My magic wheel, draw to my house that man of mine'.

69–71. carmina ... / carminibus ... / ... cantando: elegantly imitated by Tibullus, 1. 8. 19–21 'cantus uicinis fruges traducit ab agris, / cantus et iratae detinet anguis iter, / cantus et e curru lunam deducere temptat'. The omnipotence of charms, *carmina*, is a commonplace of later Latin poetry; see Pease on *A.* 4. 487.

69. deducere lunam: drawing down the moon, 'the most famous and picturesque charm in all antiquity' (K. F. Smith on Tib. 1. 2. 43). It was always a love charm, as Smith notes, associated with Thessaly, a land of magic, and, in the later tradition, often attributed to Medea. See D. E. Hill, 'The Thessalian Trick', *RhM* 116 (1973), 221–38, Fedeli on Prop. 1. 1. 19.

70. Circe: Simaetha prays that her drugs may be no less potent than those of Circe or Medea (Theocr. 2. 15–16).

Vlixi: for the form of the genitive see M. Leumann, *MH* 2 (1945), 245–7, 251–2=*Kleine Schriften* (Zurich, 1959), 116–18, 122–4.

71. frigidus ... anguis: 3. 93 n.

cantando rumpitur anguis: by a charm or incantation that causes the snake to swell up until it bursts; cf. Lucil. 575–6 M. 'iam disrumpetur, medius iam, ut Marsus colubras / disrumpit cantu, uenas cum extenderit omnis', Pomponius 118 R.[3] 'mirum ni haec Marsa est, in colubras callet cantiunculam', Ov. *Med. fac.* 39 'nec mediae Marsis finduntur cantibus angues', and see K. F. Smith on Tib. 1. 8. 20, Abt (above, 65 n.), 127–8. The Marsi, who still retain their ancient fame as snake-charmers, 'serpari' (see H. V. Morton, *A Traveller in Southern Italy* (London, 1969), 36–42), were believed to be descended, as V., with his interest in primitive Italy, would have known, from a son of Circe; so Pliny, *NH* 7. 15 'in Italia Marsorum genus durat, quos a Circae filio ortos ferunt', 25. 11, Gellius 16. 11. 1.

For the construction *cantando rumpitur*, with the gerund implying a subject other than that of the verb to which it is attached, cf. *G.* 2. 239 'mansuescit arando', 250 'lentescit habendo', 3. 215 'uritque uidendo', 454 'uiuitque tegendo', *A.* 12. 46 'aegrescitque medendo', and see Munro on Lucr. 1. 312 'anulus in digito subter tenuatur habendo'.

73–4. terna tibi haec primum triplici diuersa colore / licia circumdo: 'First with these triple threads in separate colours three / I bind you' (Lee)—'id est tria alba, tria rosea, et tria nigra' (DServ.), the colours of Hecate; see Abt (above, 65 n.), 148–50, 156, S. Eitrem, *Gnomon*, 2 (1926), 97. 'Fairies red, black, white' were believed to exist in Ireland; see Samuel Johnson on Shakespeare, *Macbeth*, IV. i. 1. The number three and its multiples play a large part in Roman and Greek ritual and magic; see Pease on *A*. 4. 510, Gow on Theocr. 2. 43.

73. tibi: explained by *effigiem* (75). 'In all enchantments that which is done to the image of a person is supposed to affect the person himself: the threads which bind the image will also bind Daphnis' (T. E. Page). Such images were often made of wax, e.g. Hor. *Epod.* 17. 76 'cereas imagines'. For the use of images or puppets in witchcraft see Pease on *A*. 4. 508, Gow on Theor. 2. 28.

74. licia: an old word used in legal formulae (*TLL* s.v. 1373. 64); here first in a magic ritual (*TLL* s.v. 1374. 12), but probably in common use.

altaria circum: *A*. 2. 515, 4. 145, where see Pease, 8. 285. V. likes to place this preposition at the end of the line with a neuter plural preceding—a pattern found twice in Lucretius, 1. 937 (=4. 12) 'pocula circum', 4. 220 'litora circum'. And here 'circum / . . . duco' echoes 'circumdo'.

75. deus: 'aut quicumque superorum, aut Hecaten dicit' (Serv.). Cf. Theocr. 2. 28 *ὡς τοῦτον τὸν κηρὸν ἐγὼ σὺν δαίμονι τάκω*, 'As, with the goddess's aid, I melt this wax' (Gow). The goddess is Hecate.

impare: '*impare* autem propter metrum ait' (Serv.); see *TLL* s.v. 516. 73.

76. See above, 28[a] n. In Catull. 64 the refrain was interpolated after l. 377, causing the Oxford MS to omit ll. 379–81.

77–8. Cf. Theocr. 2. 18–21 *ἄλφιτά τοι πρᾶτον πυρὶ τάκεται. ἀλλ' ἐπίπασσε, | Θεστυλί. δειλαία, πᾷ τὰς φρένας ἐκπεπότασαι; | ἦ ῥά γέ θην, μυσαρά, καὶ τὶν ἐπίχαρμα τέτυγμαι; | πάσσ' ἅμα καὶ λέγε ταῦτα· "τὰ Δέλφιδος ὀστία πάσσω"*, 'First barley groats smoulder on the fire. Nay, strew them on, Thestylis. Poor fool, whither have thy wits taken wing? Am I become a mock, then, even to thee, wretch? Strew them on, and say the while, "I strew the bones of Delphis"' (Gow).

77. necte tribus nodis ternos, Amarylli, colores: 'It seems clear from the use of the distributive *ternos* and *necte* "twine" that each knot is to be twined with three colours . . . the use of *terna*, *triplici* line 73 and *ternos necte* here certainly suggests that Virgil was not thinking of single threads but of threads each twined with three differently-coloured strands' (T. E. Page). Cf. Petron. 131. 4 'illa de sinu licium protulit uarii coloris filis intortum ceruicemque uinxit meam'.

necte . . . nodis: for the connection of these words, which may in fact be connected (Ernout–Meillet, *Dict. étym.* s.v.), cf. Cic. *Arat.* fr. 32. 4 Soubiran 'conectere nodum', *A.* 12. 603 'nodum . . . nectit', and see T. E. V. Pearce, *CQ*, NS 20 (1970), 154–5.

78. modo: with 'a colouring of impatience' (Mynors on *G.* 3. 73).

80–1. limus ut hic durescit, et haec ut cera liquescit / uno eodemque igni, sic nostro Daphnis amore: cf. Theocr. 2. 28–9 ὡς τοῦτον τὸν κηρὸν ἐγὼ σὺν δαίμονι τάκω, | ὣς τάκοιθ' ὑπ' ἔρωτος ὁ Μύνδιος αὐτίκα Δέλφις, 'As I melt this wax by the goddess's aid, so may Delphis the Myndian at once melt with love', where Gow comments: 'It is however possible that the wax is, in Simaetha's rite, not an image at all but a symbol, like the bay and the barley-groats'. True, κηρός is unqualified, as is *cera*, but Abt (above, 65 n.), 156–7, remarks, with reference to Theocritus and V., that the wax used in such a ceremony was traditionally in the shape of a human being ('Aber die gesamte sonstige Überlieferung spricht von geformtem Wachs, von einem Wachsbilde'). La Cerda, observing that Canidia uses two images in Hor. *Serm.* 1. 8. 30–3, one of wool representing herself, and the other of wax representing the faithless lover whom she means to torture to death, supposes that Daphnis' 'wife' also uses two images, but both representing Daphnis, one of wax, which she first binds and carries around the altar (75 *effigiem*) and then melts, and the other of clay. As fire hardens clay and melts wax, such, she prays, 'let the soul of cruel Daphnis be— / Hard to the rest of women, soft to me' (Dryden). This, the generally accepted interpretation, is as old as the tenth century (and no doubt much older), for it is found in the Vaticanus Reginensis 1495 (R) of Servius; see Thilo's app. crit. ad loc.

There is another interpretation, however, that of DServ.: 'se de limo facit, Daphnidem de cera', which was adopted by H. J. Rose, *The Eclogues of Vergil* (Berkeley, 1942), 157, and has recently been defended by C. A. Faraone, *CP* 84 (1989), 294–300. Faraone 'can find no parallels . . . for a spell that attempts simultaneously to change a victim into diametrically opposed states, such as hard and soft' (295); even so, it is easier to imagine V. manipulating the practice of magic for his own purpose—hence perhaps the emphasis 'uno eodemque igni'—than it is to intrude Daphnis' abandoned 'wife' into a passage that has no room for her.

80. limus ut hic durescit, et haec ut cera liquescit: this artful line, with its parallel clauses and rhyming words, in which ictus and accent coincide (see 1. 70 n.), has been designed to suggest the assonantal, accentual character of primitive spells or charms (*carmina*), e.g. 'terra pestem teneto, salus hic maneto' (Varro, *RR* 1. 2. 27), 'nouum uetus uinum bibo, nouo ueteri morbo medeor' (Varro, *LL* 6. 21). For a collection of examples from Latin, Greek, and German see Norden, *Die antike Kunstprosa* (Leipzig, 1898), ii. 819–24. V. may also be indebted here to Lucr. 1. 305–6 'suspensae in litore uestes / uuescunt, eaedem dispansae in sole serescunt'.

81. uno eodemque igni: cf. *A*. 10.487 'una eademque uia', 12. 847 'uno eodemque . . . partu'.

82. molam: salt mixed with spelt, *mola salsa*; see Pease on *A*. 4. 517.

fragilis . . . lauros: 'brittle', as in Lucr. 6. 112 'fragilis <sonitus> chartarum', Prop. 4. 7. 12 (Cynthia's ghost) 'pollicibus fragiles increpuere manus'. Cf. Theocr. 2. 24 χὠς αὕτα λακεῖ μέγα καππυρίσασα, 'And as this (the bay leaf) crackles catching fire'. 'The extent to which laurel crackles when it burns is proverbial', K. F. Smith on Tib. 2. 5. 81 'et succensa sacris crepitet bene laurea flammis', quoting Lucr. 6. 154–5 'nec res ulla magis quam Phoebi Delphica laurus / terribili sonitu flamma crepitante crematur'. Cf. also Ov. *Fast*. 1. 344 'et non exiguo laurus adusta sono' (La Cerda).

bitumine: cf. Hor. *Epod*. 5. 81–2 'quam non amore sic meo flagres uti / bitumen atris ignibus' and see G. Tabarroni, *Enc. Virg.* s.v. *bitumen*.

83. Daphnis me malus urit, ego hanc in Daphnide laurum: a line modelled on Theocr. 2. 23 Δέλφις ἔμ' ἀνίασεν· ἐγὼ δ' ἐπὶ Δέλφιδι δάφναν | αἴθω, 'Delphis brought trouble on me, and I for Delphis burn this bay' (Gow). Gow on Theocr. 2. 1: 'Bay however is not otherwise associated with love charms except at Virg. *E*. 8. 82, Prop. 2. 28. 36, and both passages seem dependent on T.'; cf. Serv.: 'aut intellegamus supra Daphnidis effigiem eam laurum incendere propter nominis similitudinem'.

85–8. No doubt, as La Cerda notices, V. was thinking of Lucretius' pathetic description of a cow looking for her lost calf

(2. 352–66), and incorporates, as if to alert his reader, a Lucretian phrase, 87 'propter aquae riuum' (2. 30, 5. 1393).

85. qualis cum fessa . . .: the construction is *qualis amor buculam tenet cum . . . .*

86. bucula: first attested here and again in *G.* 1. 375, where Mynors remarks that words for domesticated animals tend to be diminutive in form. See also Axelson 40.

87–8. uiridi procumbit in ulua / perdita, nec serae meminit decedere nocti: cf. *G.* 3. 466–7 (a sheep) 'medio procumbere campo / pascentem et serae solam decedere nocti'.

88. perdita, nec serae meminit decedere nocti: a line borrowed, but not the pathos, which is V.'s own, from Varius' *De morte* (written in 44 or the first part of 43 BC; see A. S. Hollis, *CQ*, NS 27 (1977), 187–8) describing the pursuit of an aged doe by a Cretan hound, *FPL* fr. 4. 5–6 Büchner 'non amnes illam medii, non ardua tardant, / perdita nec serae meminit decedere nocti'. Varius' hound has become a heifer enamoured of a young bull whom she pursues through the groves and clearings until finally, late at night, she sinks down exhausted, 'amore consumpta' (Serv.). For Varius see 9. 35 n.; for lovesick animals, 3. 100 n.

nec meminit: for 'not remembering' to perform a habitual action where no question of memory is involved see Mynors on *G.* 1. 399–400 'non ore solutos / immundi meminere sues iactare maniplos'.

89. talis amor: see above, 5 n.

mederi: cf. 10. 60.

91. exuuias: pieces of clothing left behind by Daphnis; cf. *A.* 4. 496, with Pease's note. In Theocr. 2. 53–4, Simaetha shreds the fringe of Delphis' cloak and throws it into the fire.

perfidus: cf. *A.* 4. 305, with Pease's note, and see Clausen 47.

92. pignora cara sui: 'dear pledges of himself' (she loves him still, despite his treachery) which she buries under the threshold. In Theocr. 2. 59–60, Simaetha orders Thestylis to knead magic herbs over the threshold of Delphis. The door or any part of it was efficacious in ancient magic; see M. B. Ogle, 'The House-Door in Greek and Roman Religion and Folk-Lore', *AJP* 32 (1911), 251–71.

93. debent: cf. *A.* 12. 317 'Turnum debent haec iam mihi sacra'.

95. has herbas atque haec ... uenena: 'these poisonous plants'; described by Mynors on *G.* 2. 192 as 'a hendiadyoin of closer definition (epexegesis)'. So 2. 8, *G.* 1. 106, 4. 56, 388–9, *A.* 1. 61. Cf. Hor. *Epod.* 5. 21–2 'herbasque, quas Iolcos atque Hiberia / mittit uenenorum ferax', *A.* 4. 514 'pubentes herbae nigri cum lacte ueneni', Tib. 2. 4. 55–6 'quidquid habet Circe, quidquid Medea ueneni, / quidquid et herbarum Thessala terra gerit', where K. F. Smith remarks: 'The distinction between a drug, a poison, and a magic philtre tends to disappear as we approach the primitive stage of popular belief'.

96. ipse ... Moeris: 'a noted country wizard' (Conington); non-Theocritean, the name of the dispossessed farmer in *E.* 9.

Ponto: Pontus was known for its poisons because of Mithridates, and aconite, the deadliest of poisons, grew there (Pliny, *NH* 27. 4), but V. probably means Colchis, the country of Medea; cf. Cic. *De imp. Pomp.* 22 'ex suo regno sic Mithridates profugit ut ex eodem Ponto Medea illa quondam fugisse dicitur', Juv. 14. 114 'Hesperidum serpens aut Ponticus' (Forbiger).

97. his: Moeris changed himself into a wolf with drugs (95 'uenena'), but the ghosts of the dead were raised and crops conveyed elsewhere by incantation; cf. Tib. 1. 2. 45–6 'haec cantu finditque solum manesque sepulcris / elicit' and 8. 19 'cantus uicinis fruges traducit ab agris', with K. F. Smith's notes. Cf. also the Twelve Tables 8. 8a 'Qui fruges excantassit' (C. G. Bruns, *Fontes Iuris Romani Antiqui* (Leipzig, 1893), 30).

Belief in werewolves is ancient and universal; see J. A. MacCulloch, *Hastings' Encyclopaedia of Religion and Ethics*, s.v. *lycanthropy*. For classical antiquity see W. Kroll, *RE Suppl.* vii. 423–6, *TLL* s.v. *lupus* 1853. 25.

97–9. ego saepe ... / ... saepe ... / ... uidi: very emphatic, as if anticipating disbelief. Cf. 9. 51 'saepe ego', *G.* 1. 316–18 'saepe ego ... uidi'.

98. excire: first attested here in this sense (*TLL* s.v. 1246. 83); the usual verb is *elicio* (*TLL* s.v. 366. 56). See Pease on Cic. *De diu* 1. 132 (*psychomantia*).

101–2. In Theocr. 24. 93–6, one of the serving-women is to gather up the ashes of the fire at dawn, carry them across the river to the rocks and throw them beyond the boundary, then return without looking back. Sinister things, the remains of witchcraft, and to be disposed of as expeditiously as possible; here Amaryllis is ordered to throw the ashes into the river, which will bear them away to the sea; cf. Ov. *Fast.* 6. 227–8 'donec ab Iliaca placidus purgamina Vesta / detulerit flauis in mare Thybris aquis' (La Cerda).

101. cineres: the poetic plural is first attested here; see Maas 519=560. Cf. 106 *cinis.*

101–2. riuoque fluenti / transque caput iace, nec respexeris: her instructions are exact and particular (and supported by the rhythm? Note the diaereses in the second and fourth feet of l. 102). For the dative cf. *A.* 12. 256 'proiecit fluuio' and see 6. 85 n.

102. nec respexeris: looking back could be dangerous and was commonly forbidden in Greek and Roman ritual; cf. Plaut. *Most.* 523 'caue respexis', Ov. *Fast.* 6. 163–4 'sic ubi libauit, prosecta sub aethere ponit, / quique adsint sacris, respicere illa uetat', and see Gow on Theocr. 24. 96, Bömer on Ov. *Fast.* 5. 439.

his ego: repeating 'his ego' (97) with a certain emphasis and the same reference (so Klingner and Coleman); not the ashes (so T. E. Page), which would have no magical potency. Cf. Vahlen (above, 47–50 n.), i. 397 (with regard to 101–2 'fer—respexeris'): 'quod medium interiectum est inter duas partes unius sententiae . . . nec obfuit perspicuitati et hanc moratam orationem decuit'. She now intends to employ more drastic means—the drugs given her by Moeris, the potency of which she has often observed (97–8)—and is only prevented by the sudden and unexpected arrival of Daphnis.

105 aspice: 'hoc ab alia dici debet' (DServ.). The vocative *Amarylli* (101) prepares for her speech. Direct speech, surprising in a song, would be appropriate in a mime, and Alphesiboeus' song may originally have been conceived as a mime; see Introduction, p. 239, E. Bethe, *RhM* 47 (1892), 591, and Gow's Preface to the Second *Idyll*, pp. 33–5.

tremulis . . . flammis: cf. Cic. *Arat.* fr. 22. 3 Soubiran 'tremulam . . . flammam', Lucr. 4. 404 'tremulis . . . ignibus'.

106. bonum sit!: she hopes that the sudden blaze may be a good omen.

107. nescio quid certe est: colloquial; cf. Catull. 80. 5, Pers. 5. 51.

Hylax: the correction appears to have been made by A. Mancinellus (1490); see J. Van Sickle, *RIFC* 102 (1974), 311–13. A very suitable name; cf. Ov. *Met.* 3. 224 'acutae uocis Hylactor' and see 3. 18 n.

108. credimus? an, qui amant, ipsi sibi somnia fingunt?: modelled on Lucr. 1. 104–5 'quippe etenim quam multa tibi iam fingere possunt / somnia'; but Lucretius is speaking of seers who invent terrifying fantasies for others, V. of lovers who invent their own fantasies of happiness. La Cerda compares Publ. Syr. A 16 Meyer 'amans, quod suspicatur, uigilans somniat'.

qui amant: prosodic hiatus, that is, the shortening of a long syllable in hiatus; cf. e.g. Plaut. *Merc.* 744 'nam qui amat', *Amph.* 597 'ita me di ament', Catull. 97. 1 'ita me di ament', Hor. *Serm.* 1. 9. 38 'si me amas', and see Munro on Lucr. 2. 404, Leumann, *Lat. Laut- und Formenlehre*², 105.

109. parcite, ab urbe uenit, iam parcite carmina, Daphnis: an ingenious reworking of the refrain 'ducite ab urbe domum, mea carmina, ducite Daphnin'; see above, 61 n. Like the concluding line of *E.* 10, 'ite domum saturae, *uenit Hesperus*, ite capellae', the concluding line of *E.* 8 contains an 'explanation' of its imperatives, 'parcite, *ab urbe uenit*, iam parcite carmina, *Daphnis*'.

ECLOGUE 9

Introduction

The Ninth *Eclogue* is usually read in conjunction with the First, rarely as a poem in its own right. Virgil's arrangement of the two *Eclogue*s in his Book and the temporal sequence suggested by his fiction invite, it must be admitted, such a reading. In the Ninth *Eclogue*, the land-confiscations are over. There is no turmoil, as in the First, no hopeless flight; only, for Moeris, the dreary routine of a menial existence embittered by memory. Peace has returned to the countryside, a desolate peace.

The Ninth is, however, the earlier *Eclogue*;[1] and while writing it Virgil could hardly have supposed that one day he would be writing another *Eclogue* on the same subject, for how could he, then, have imagined the experience differently? His later *Eclogue*, the First in his Book, is owing to changed circumstances, or rather, to Virgil's changed perception of circumstances that had once seemed to him irredeemably sad.

'Going where the road leads, Moeris? to town?' So begins the Ninth *Eclogue*, with a casual yet carefully placed question. Given the opening, Moeris immediately unburdens himself of his troubles—'O Lycidas, we have finally come to this, that a stranger should take possession of my farm, order me off. And now, defeated, disheartened, I am taking these kids (may no good come of it) to him'. From this dramatic outburst the reader learns all that he needs to know, for the moment.

'Veteres migrate coloni' (4). But Moeris has been allowed to remain on his old farm as a tenant paying rent in kind, like Horace's philosophical Ofellus.[2] Evidently the truculent soldier, unused to tilling the soil, relented and now lives in town. Such

[1] See *CHCL* ii. 314–15.

[2] *Serm.* 2. 2. 114–15 'uideas metato in agello / cum pecore et gnatis fortem mercede colonum'; W. E. Heitland, *Agricola* (Cambridge, 1921), 234: 'Ofellus, one of the yeomen of the old school. He had been a working farmer on his own land, but in the times of trouble his farm had been confiscated and made over to a discharged soldier. But this veteran wisely left him in occupation as cultivator on terms'. Unlike Moeris, Ofellus feels no bitterness, 133–5 'nunc ager Vmbreni sub nomine, nuper Ofelli / dictus, erit nulli proprius, sed cedet in usum / nunc mihi, nunc alii'.

details Virgil leaves to the reader's imagination; his poetic interest lies elsewhere.

It was rumoured, says Lycidas, that Menalcas had saved the district, all of it, with his songs, 'omnia carminibus' (10).

> qua se subducere colles
> incipiunt mollique iugum demittere cliuo,
> usque ad aquam et ueteres, iam fracta cacumina, fagos. (7–9)

Virgil is quite willing to sacrifice dramatic probability (for Moeris would know the lie of the land in his own neighbourhood) to the creation of an ideal landscape—a landscape here defined as by a surveyor's eye and yet, with its literary beeches,[3] curiously vague and indeterminate.

There was such a rumour, Moeris replies, but songs count for little in a violent world, as little as doves when the eagle comes, and discloses that 'your Moeris here' (16) and Menalcas had almost been killed in a quarrel.[4] Appalled, Lycidas apostrophizes the absent Menalcas—'Oh, that you and the solace of your songs were nearly lost to us! Who would sing of the Nymphs, of the ground sprinkled with flowers, of green-shadowed springs?' The tone being thus elevated above the conversational level, Lycidas sings three lines of a song by Menalcas, lines translated from Theocritus, which he had overheard Moeris singing (23–5). And Moeris responds with three lines of another song, begun but still unfinished, which Menalcas had sung to Varus in the vain hope of saving Mantua (27–9).

Lycidas now urges the older and more accomplished singer to sing a song all the way through. 'The Muses have made me a poet too', he blandly observes, 'and the shepherds call me a bard, but I mistrust them' (32–4). Virgil's indebtedness to Theocritus' Seventh *Idyll*, indicated elsewhere by the name Lycidas, the

[3] See l. 9 n.

[4] Presumably with the soldier to whom a section including Moeris' farm had been allotted in the centuriation or land-division. Because of the rigidity of centuriation—an *ager centuriatus* was divided into square sections all of the same size—a section could hardly coincide with a previous holding. See J. Bradford, *Ancient Landscapes* (London, 1957), 145–216. But how was Menalcas involved? And why was he compelled to leave? The pursuit of Virgil's farm, once so popular and not quite relinquished yet, is a sterile exercise. It is not Virgil's experience of the confiscations, but Virgil's experience transmuted into poetry, the poem itself, that matters—except to a reader for whom a connection with 'real life' enhances a work of art.

journey to town, and the wayside tomb of Bianor, is here apparent as he makes the opportunity of complimenting, in the person of Lycidas, Varius and Cinna.[5] In the Seventh *Idyll*, a younger and an older singer meet, Simichidas (whose words Virgil's Lycidas here echoes[6]) and Lycidas, and exchange songs; whence, it may be, the idea for the intricate pattern of exchanges in the Ninth *Eclogue*.[7]

While Lycidas talks complacently of his poetic aspirations, Moeris reminisces and manages to produce five lines of a song by Menalcas, again Theocritean, describing the amenity of Polyphemus' enclosed garden, 'Come hither, Galatea' (39–43). Possibly a contrast is implied with the plain Moeris and Lycidas are crossing, in which there seem to be no vineyards, no animals, no human beings[8]—a bleak and empty landscape brightened intermittently by passages of song.

Lycidas, it now appears, had been eavesdropping on another occasion too, and heard Moeris singing, alone, beneath a clear night sky. He recalls the music but is unsure of the words; he collects his thoughts, however, and sings five lines of a song, like the abortive song to Varus, a Roman song, 'Caesar's Star' (46–50).

Despite Lycidas' insistence, Moeris, pleading loss of memory, declines to sing any further; a sickness of the soul, which the younger man cannot comprehend, oppresses him.[9] Menalcas, he says, will often sing those songs for you when he comes.

Poetry fails in the end. Moeris has forgotten so many songs; it will soon be dark and rain threatens; when will Menalcas come?

Bibliography

F. Leo, 'Vergils erste und neunte Ekloge', *Hermes*, 38 (1903), 1–18 = *Ausgewählte kleine Schriften* (Rome, 1960), ii. 11–28.

Klingner 148–56.

G. Williams, *Tradition and Originality in Roman Poetry* (Oxford, 1968), 313–27.

[5] See l. 35 n.

[6] See ll. 32–6 n.

[7] See ll. 39–50 n.

[8] Lycidas (60–1) points out the place where farmers strip the trees of their boughs for fodder, but no farmers are present.

[9] Lycidas' offer to relieve Moeris of the kids, his physical burden, is curtly rejected (65–6).

G. Jachmann, 'Die dichterische Technik in Vergils Bukolika', *N. Jahrb.* 49 (1922), 110–19 = *Ausgewählte Schriften* (Königstein im Taurus, 1981), 312–22.

1. Quo te, Moeri, pedes?: a casual question, apparently suggested by Theocr. 7. 21 *Σιμιχίδα, πᾶ δὴ τὺ μεσαμέριον πόδας ἕλκεις;*, 'Where are you going, Simichidas, with such energy at midday?' Cf. Hor. *Serm.* 2. 4. 1 'Vnde et quo Catius?' But the gambit of a conversation during a walk is as old as Plato, *Phaedr.* 227 A *Ὦ φίλε Φαῖδρε, ποῖ δὴ καὶ πόθεν;*, 'My dear Phaedrus, whither (going) and whence (come)?', *Symp.* 173 B *πάντως δὲ ἡ ὁδὸς ἡ εἰς ἄστυ ἐπιτηδεία πορευομένοις καὶ λέγειν καὶ ἀκούειν*, 'and in fact the road to town is suitable for conversation as we walk along'. Cf. Izaak Walton, *The Compleat Angler*, ch. 1 'You are well overtaken, Gentlemen! A good morning to you both! I have stretched my legs up Tottenham Hill to overtake you'.

Moeri: 8. 98 n.

pedes: if a verb is to be understood (which is doubtful), then *ducunt*; 'subaudis *ducunt*, et est zeugma a posterioribus' (Serv.). Cf. Hor. *Epod.* 16. 21 'ire pedes quocumque ferent', and see Gow on 13. 70 *ᾇ πόδες ἆγον*, 'wherever his footsteps led him'; also Vian on Ap. Rhod. 4. 66 *πόδες φέρον*: 'l'expression se dit le plus souvent d'une personne qui marche machinalement'.

2–3. The artfully disordered word-order produces an effect of agitated, vivid speech.

2. uiui: cf. Lucr. 3. 1046 'mortua cui uita est prope iam uiuo atque uidenti', Cic. *Pro Sest.* 59 'uiuus, ut aiunt, est et uidens cum uictu ac uestitu suo publicatus', Otto, no. 1933. V. has apparently shortened a familiar proverbial expression to heighten its effect by reducing the element of cliché.

2–3. nostri / . . . agelli: cf. E. Fraenkel, *JRS* 51 (1961), 47 = *Kl. Beiträge* (Rome, 1964), ii. 116–17: '. . . the note of burning indignation in the words of the wronged farmer . . . "I, I am the owner of this piece of land" he cries—the strong emphasis is brought out by the wide hyperbaton *nostri . . . agelli*'. Cf. Plaut. *Bacch.* 842–3 '*meamne* hic Mnesilochus, Nicobuli filius, / per uim ut retineat *mulierem*?', *A.* 4. 318–19 '*istam*, / oro, si quis adhuc precibus locus, exue *mentem*', Hor. *Carm.* 1. 20. 10–12 '*mea* nec Falernae /

temperant uites neque Formiani / *pocula* colles'; and see Fraenkel on Aesch. *Ag.* 13–14, *Iktus und Akzent* (Berlin, 1928), 114–15, *Horace*, 241 n. 1. Note the collocation 'aduena nostri'.

3. **possessor:** a legal term; cf. Macer, *Dig.* 2. 8. 15. 1 'possessor ... is accipiendus est qui in agro uel ciuitate rem soli possidet' and see *TLL* s.v. 102. 30.

agelli: a truly affectionate diminutive, as in Lucr. 5. 1367–8 (primitive men) 'inde aliam atque aliam culturam dulcis agelli / temptabant'. In any case, the holding of a *colonus* would be small; cf. Tib. 1. 1. 22 'nunc agna exigui est hostia parua soli', Ov. *Fast.* 5. 499 'senex Hyrieus, angusti cultor agelli', *Moretum* 3 'Simulus exigui cultor ... rusticus agri'. The diminutive *agellus* does not occur elsewhere in Lucretius or V.

4. **haec mea sunt:** a quasi-legal phrase, with a hint of violence.

ueteres ... coloni: not necessarily of long standing, though that would increase the pathos. As La Cerda notes, *ueteres* is opposed to *aduena*; cf. *A.* 12. 27 'ueterum ... procorum' of the suitors for Lavinia's hand who preceded Aeneas, an *aduena* (12. 261).

migrate: a harsh and unfeeling order. The soldier has taken over the land of several *coloni*; see Introduction, n. 4.

5. A sad, slow line.

6. **(quod nec uertat bene):** a parenthetical curse, as in Ter. *Ad.* 191 'minis uiginti tu illam emisti (quae res tibi uortat male)'.

nec: an archaism for *non*, here embedded in an old formula; see Löfstedt i. 338–40.

uertat bene: *bene uertat*, the *lectio facilior* (*bene* or *male* ordinarily precedes the verb in Plautus), is attested by Donatus, Servius, and Nonius (see app. crit.) but is rhythmically unparalleled, for in only ten lines of the *E.* is the fourth foot a spondee consisting of a single word: 1. 80, 81, 2. 27, 3. 51, 56, 4. 24, 6. 10, 11, 39, 7. 45 (in 9. 12 *inter Martia* form a rhythmical unit, as do *in longum* in 9. 56), and in none of these is there the slightest pause at the diaeresis. For this 'antiquated rhythm' see Housman, *JPhil* 21 (1893), 150–1 = *Class. papers*, i. 269.

mittimus: he carries the kids in a basket (65), an uncomfortable position for singing. V. occasionally uses the first person plural of a singular subject, e.g. 5. 50–2, 85. See K. F. Smith on Tib. 1. 2. 11, Kühner–Stegmann i. 88–9.

7. equidem: 1. 11 n.

se subducere: 'to draw themselves up from the plain—the slope being regarded from below, as in "iugum demittere" it is regarded from above' (Conington).

8. molli . . . cliuo: *G.* 3. 293; cf. Pliny, *NH* 3. 147 'qua mitescentia Alpium iuga . . . molli . . . deuexitate considunt'.

iugum demittere: cf. Curtius 5. 4. 23 'qua se montium iugum paulatim ad planiora demittit'.

9. ueteres, iam fracta cacumina, fagos: cf. 2. 3 'densas, umbrosa cacumina, fagos'; the broken tree-tops are in harmony with the general mood. The beech does not grow near water (if that is what V. means here), nor is it mentioned elsewhere as a boundary tree; cf. C. Thulin, *Corpus Agrimensorum Romanorum* (Leipzig, 1913), 107 (Siculus Flaccus on *arbores finales*) 'alicubi enim pinos inuenimus, alicubi cypressos, alibi fraxinos aut ulmos aut populos quaeque alia ipsis possessoribus placuerunt', Varro, *RR* 1. 15 'Praeterea sine saeptis fines praedi satione arborum tutiores fiunt, ne familiae rixent cum uicinis ac limites ex litibus iudicem quaerant. serunt alii circum pinos, ut habet uxor in Sabinis, alii cupressos, ut ego habui in Vesuuio, alii ulmos, ut multi habent in Crustumino', Hor. *Epist.* 2. 2. 170–1 'qua populus adsita certis / limitibus uicina refugit iurgia'.

10. omnia carminibus uestrum seruasse Menalcan: not merely the *agellus* of Moeris but the whole district just described. Hence the plural reference of *uestrum* (8. 38 n.); Menalcas is a general benefactor.

11. audieras: repeating 'audieram' (7). 'Latin has no "yes", and often expresses assent by repeating the word used by the first speaker' (T. E. Page). Cf. e.g. Plaut. *Rud.* 1267 'PL. Repperit patrem Palaestra suom atque matrem? TR. Repperit'.

et: adds a certain emphasis, 'and in fact' (*TLL* s.v. 892. 77).

13. Chaonias . . . columbas: an exquisitely pointless epithet (see, however, G. Zanker, *CQ*, NS 35 (1985), 235–7), apparently borrowed from Euphorion, fr. 48 Powell Ζηνὸς Χαονίοιο προμάντιες ηὐδάξαντο, 'the priestesses of Chaonian Zeus spoke', *G.* 2. 67 'Chaoniique patris glandes'. The reference is to the sanctuary of

Dodona, the most venerable in the Greek world. In the course of its long and varied history, several methods of divination were practised at Dodona, but it was chiefly famous for its sacred oak and oracular doves or dove-priestesses, πέλειαι. Cf. Nonnus, *Dionys.* 3. 293 Χαονίῃ ... πελειάδι, 'the Chaonian dove', and see O. Kern, *RE* v. 1257–65, H. W. Parke, *The Oracles of Zeus* (Cambridge, Mass., 1967), 1–163. Dodona was not, however, situated in Chaonia, a part of northern Epirus, but in Thesprotia to the south; cf. Aesch. *PV* 830–1 Δωδώνην, ἵνα | μαντεῖα θᾶκός τ' ἐστὶ Θεσπρωτοῦ Διός, 'Dodona, where is the prophetic seat of Thesprotian Zeus', N. G. L. Hammond, *Epirus* (Oxford, 1967), 464 (map). V., following Euphorion (or possibly Callimachus?), uses the metrically convenient name of a part of Epirus for the whole, as, in *G.* 1. 492, he places the battle of Philippi in Emathia, a part of Roman Macedonia that did not include Philippi; see Gow–Page, *The Garland of Philip* (Cambridge, 1968), on Adaeus 4. 1 (= *AP* 7. 238). Synecdoche of this sort is Hellenistic: thus Callimachus, *Hymn* 3. 178, uses Στυμφαιίδες, the name of a town and a mountain in Epirus, for cattle from Epirus. See 10. 59–60 n.

dicunt: implies a proverbial saying, but this is the first reference in Latin to eagles molesting doves (*TLL* s.v. *columba* 1731. 69).

aquila ueniente columbas: Ursinus compares Lucr. 3. 751–2 (an *adynaton*) 'tremeretque per auras / aeris accipiter fugiens ueniente columba'. The dove's natural enemy is the hawk: Hom. *Il.* 15. 237–8, 21. 493–4, 22. 139–40, *Od.* 15. 525–7 (the eagle in *Od.* 20. 242–3 has a special significance, as does the eagle in Quint. Smyrn. 1. 198–200), Aesch. *Suppl.* 223–4, *PV* 857, Eur. *Andr.* 1140–1, Ap. Rhod. 1. 1049–50, 3. 541, 4. 485–6, Quint. Smyrn. 1. 572, 12. 12, Nonnus, *Dionys.* 42. 535–6, V. *A.* 11. 721–2, Hor. *Carm.* 1. 37. 17–18, Ov. *Ars* 2. 363, *Met.* 5. 605–6, *Fast.* 2. 90, *Trist.* 1. 1. 75–6, *Ex Pont.* 2. 2. 35, Silius 4. 105–9, 5. 282, Varro, *RR.* 3. 7. 6, Columella 8. 8. 4, Pliny, *NH* 10. 108; but the fierce and predatory eagle, the *aquila* of the legions, suits better here, 'tela inter Martia'.

ueniente: with hostile intent, as in Lucr. 3. 752; cf. *A.* 10. 456 'Turni uenientis imago', 12. 299, 510, 540, 595.

14. quacumque: 'quacumque ratione' (Serv.); *quacumque* nearly always has a local sense, as elsewhere in V., *G.* 1. 406, *A.* 10. 49, 12. 368, 913.

incidere: 'pro *decidere*' (DServ.). This metaphorical sense first occurs in Cicero (*TLL* s.v. 909. 53).

15. sinistra . . . cornix: birds on the left were considered favourable in Roman augury, although something seems to have depended on the kind of bird. See Pease on Cic. *De diu.* 1. 12 (*cornicem*): 'its omens, though occasionally good, are usually unfavorable'. See 1. 16 n. As Hollis observes on Ov. *Met.* 8. 237, a bird talking or watching from a tree is a motif of Hellenistic poetry: Callim. fr. 194. 61–3 Pf., *Hec.* fr. 74. 11 Hollis, Ap. Rhod. 3. 927–31, Ov. *Met.* 2. 557.

16. hic: probably not the adverb but the pronoun (cf. Cic. *De sen.* 4 'cum hoc C. Laelio') and reminiscent of the colloquial *hic homo*, as in Ter. *Heaut.* 356 'tibi erunt parata uerba, huic homini (= *mihi*) uerbera', Hor. *Serm.* 1. 9. 47 'hunc hominem' (= *me*); see *TLL* s.v. *hic* 2703. 38. By having Moeris speak of himself in the third person V. secures the balance of the line, 'nec . . . Moeris nec . . . Menalcas'.

uiueret: effectively juxtaposed with 'ipse Menalcas'.

17. heu . . . heu: a highly pathetic repetition, not found before V., and found again in V. only in *A.* 6. 878 'heu pietas, heu prisca fides' (*TLL* s.v. 2674. 46). See 2. 58 n.

cadit in quemquam: a Ciceronian idiom, e.g. *Pro Sulla* 27 'non . . . cadit . . . in hunc hominem ista suspicio'; see *TLL* s.v. *cado* 30. 64.

17–18. tua . . . / . . . solacia: 'tua carmina, quibus consolamur' (DServ.).

19. quis caneret Nymphas? quis . . .: cf. Longus 1. 14. 4 (Chloe to the Nymphs) *τίς ὑμᾶς στεφανώσει μετ' ἐμέ; τίς τοὺς ἀθλίους ἄρνας ἀναθρέψει; τίς τὴν λάλον ἀκρίδα θεραπεύσει . . .;*, 'Who will put garlands on you when I'm gone? Who will rear the poor lambs? Who will look after my chattering locust?' (Turner).

19–20. Cf. 5. 40.

quis humum florentibus herbis / spargeret?: 'id est aspersam floribus caneret' (DServ.); for the construction, perhaps made easier here by *caneret*, see 6. 46 n. Cf. Lucr. 2. 32–3 'cum tempestas arridet, et anni / tempora conspergunt uiridantis floribus herbas' and see Mynors on *G.* 3. 126 'florentis . . . herbas'.

20. uiridi . . . umbra: a striking phrase, though less so in Latin than it is in English; cf. Marvell, *The Garden* 47–8 'Annihilating all that's made / To a green Thought in a green Shade'.

fontis induceret umbra: reads like an elegant variation of 5. 40 'inducite fontibus umbras'. Cf. *A.* 8. 457 'tunicaque inducitur artus' (a Virgilian development of the construction described in 1. 54 n.).

21. uel . . .: having apostrophized Menalcas, Lycidas turns to Moeris, 'or (who would sing) the song I recently stole from you?' *Quis caneret* is easily supplied from l. 19.

sublegi: cf. Plaut. *Mil.* 1090 'clam nostrum hunc sermonem sublegerunt' (misquoted by DServ. here). As Moeris went off to serenade Amaryllis he was rehearsing his song, a translation of Theocr. 3. 3–5 by Menalcas (below, 23–5 n.), and Lycidas was eavesdropping, picking up words and music; he now quotes, or rather sings, the first three lines of the song.

tibi: not Menalcas (so Servius, La Cerda, Heyne, Conington, Leo, Klingner, Coleman) but Moeris (so Voss and T. E. Page). It would be out of character for Lycidas to address the master-singer of the neighbourhood so familiarly.

22. cum te ad: an elision found only once again in V., *A.* 6. 900 'tum se ad'; see J. Hellegouarc'h, *Le Monosyllabe dans l'hexamètre latin* (Paris, 1964), 43 n. 2.

nostras: opposed to *te*, '*my* sweetheart'; slightly aggrieved, as if to justify his own behaviour. Not therefore the sweetheart of Lycidas and Moeris ('communem amicam', Serv.) and certainly not 'everyone's favourite' (Coleman)—'such that the swains desired her' (Conington).

23–5. 'Theocriti sunt uersus, uerbum ad uerbum translati' (Serv.). Cf. Theocr. 3. 3–5 *Τίτυρ', ἐμὶν τὸ καλὸν πεφιλημένε, βόσκε τὰς αἶγας, | καὶ ποτὶ τὰν κράναν ἄγε, Τίτυρε· καὶ τὸν ἐνόρχαν, | τὸν Λιβυκὸν κνάκωνα, φυλάσσεο μή τυ κορύψῃ*, 'Tityrus, dear friend, feed the goats and drive them to the spring, Tityrus; and mind the he-goat, the tawny Libyan, lest he butt you' (after Gow). An unnamed goatherd on his way to serenade Amaryllis asks Tityrus to tend his goats; his speech is plain, as plain as may be in Theocritus. V.'s translation, while in part studiously accurate, is more artful, for he wished to improve upon Theocritus. Note the repetitions (a native resource of Latin rhetoric) *pasce pastas, age agendum*, and the parentheses, each of three words, following the caesura in the first and third lines (see 1. 31 n.). Gellius 9. 9. 7–11 offers a brief criticism of V.'s translation; see L. A. Holford-Strevens, *Aulus Gellius* (London, 1988), 149.

23–4. Tityre . . . Tityre: cf. 1. 1–4.

23. dum redeo: 'until I return'; for this construction, colloquial in tone, see *TLL* s.v. 2216. 81. Cf. e.g. Plaut. *Rud.* 879 'manete dum ego huc redeo', Ter. *Ad.* 196 'delibera hoc dum ego redeo'.

24. potum pastas age: cf. 7. 11, Varro, *RR* 2. 2. 12 'pascunt diem totum ac meridiano tempore semel agere potum satis habent'.

inter agendum: DServius compares 'inter loquendum' from Afranius (422 R.[3]) and 'inter ponendum' from Ennius (*Inc.* 2 V.[2]); see Hofmann–Szantyr 233.

25. occursare capro . . . caueto: the infinitive with *caueo* is first attested in Catull. 50. 21 'laedere hanc caueto', preceded by the usual construction, 19 'caue despuas'; cf. *A.* 11. 293 'ast armis concurrant arma cauete'. See *TLL* s.v. 635. 68 and Shackleton Bailey on Cic. *Ad Att.* 3. 17. 3.

26. immo haec: tired and depressed, Moeris no longer takes any pleasure in elegant imitations of Theocritus. He replies with three lines from the unfinished song which Menalcas addressed to Varus.

necdum perfecta: not an inert detail but added to give these lines the appearance of a fragment; the imperfect *canebat* also contributes to the effect.

27. Vare: corresponding to *Tityre* (23); see 1. 1 n. For his identity see 6. 6–7 n.

superet: *supersit;* cf. *G.* 1. 189, 3. 286.

27–8. Mantua nobis, / Mantua: 6. 21 n.

28. uae miserae . . . Cremonae: there are three reasons for thinking repeated *-ae* expressive: (i) *uae* is found only here in V.; (ii) V. ordinarily takes pains to avoid similar case endings (Norden 405–7); and (iii) the existence of parallels elsewhere in V., whose ear was so finely attuned to the sound of the sense: *A.* 2. 282 'qu*ae* tant*ae* tenuere mor*ae*? quibus Hector ab oris . . .?', 5. 80–1 'saluete, recepti / nequiquam cineres anim*ae*que umbr*ae*que patern*ae*',

6. 426–7 (emphasized by the caesura) 'continuo audit*ae* uoces uagitus et ingens / infantumque anim*ae* flentes'. Cf. also [Moschus], *Epitaph. Bion.* 99 αἰαῖ ταὶ μαλάχαι μέν, ἐπὰν κατὰ κᾶπον ὄλωνται, 'Ah me, the mallows, when they wither in the garden'.

uae: colloquial in tone; see Nisbet–Hubbard on Hor. *Carm.* 1. 13. 3. In early Latin *uae* is used with the dative, e.g. Ter. *Hec.* 605 'uae misero mihi', but here, as first in Catull. 64. 196 'uae misera', parenthetically.

nimium uicina: some forty miles away, but emotionally much closer, 'Mantua . . . Cremonae'.

29. cantantes . . . cycni: proverbially the most musical of birds, ἀοιδότατοι πετεηνῶν (Callim. *Hymn* 4. 252); see 8. 55 n. But swans, as V. knew from his own experience, make a harsh, unmelodious sound, *A.* 11. 457–8 'piscosoue amne Padusae / dant sonitum rauci per stagna loquacia cycni'. There were swans on the river near Mantua, *G.* 2. 198–9, where again V. regrets Mantua's lost fields. See D'A. W. Thompson, *A Glossary of Greek Birds* (Oxford, 1936), 180–3.

30–2. sic tua Cyrneas fugiant examina taxos, / sic . . ., / incipe: 'so may your bees . . ., so may your cows . . .'; that is, so may you prosper in the future, begin your song. A form of oath or asseveration frequent in Augustan poetry, probably colloquial in origin and first attested in Catull. 17. 5–7 'sic tibi bonus ex tua pons libidine fiat, / . . . / munus hoc mihi . . . da', where see Fordyce; see also Tränkle 166, Nisbet–Hubbard on Hor. *Carm.* 1. 3. 1. Cf. 10. 4–6.

30. Cyrneas: an adjective derived from Κύρνος, the Greek name of Corsica; like *Hyblaeus*, not attested in Latin before V. (1. 54 n.). Unlike *Hyblaeus*, however, it was to have scant success, reappearing finally in Claudian, *De cons. Stilich.* 3. 314, and in Rutil. Namat. 1.437, but in neither connected with honey (*TLL* Onomast. s.v. *Cyrnos*).

taxos: Corsican honey was notoriously bad, Ov. *Am.* 1. 12. 7–10 'ite hinc, difficiles, funebria ligna, tabellae, / tuque, negaturis cera referta notis, / quam, puto, de longae collectam flore cicutae / melle sub infami Corsica misit apis', Pliny, *NH* 30. 28 'melle Corsico, quod asperrimum habetur'; and since the yew is bad for bees (*G.* 4. 47, where see Mynors), V. blames the yew for the badness of the honey. But no one else mentions yews growing in Corsica—no one, that is, except Servius, who would have them growing there in abundance, 'taxus uenenata arbor est, quae abundat in Corsica'. The tree responsible, according to Theophrastus, *HP* 3. 15. 5, was the box; so also Diod. Sic. 5. 14. 3 and Pliny, *NH* 16. 71. Corsica produced large amounts of honey and wax (Diod. Sic. 5. 13. 4); in 181 BC, after a military victory, the Romans exacted 100,000 lb. of wax (Livy 40. 34. 12).

31. cytiso: 1. 78 n.
distendant: 7. 3 n.

32. incipe, si quid habes: something by Menalcas; so DServ.: 'non dixit tuum sed Menalcae'. In 3. 52 'quin age, si quid habes' and 5. 10–11 'incipe . . . si quos . . . ignis / . . . habes' the singer is invited to sing something of his own.

So far two 'fragments' of song have been heard (the dramatic fiction being that 23–5 and 27–9 are out of context: they were, of course, composed precisely for this context). Lycidas now urges Moeris to begin a song, that is, to sing a song through from the beginning.

32–6. et me fecere poetam . . .: two passages are necessary for an understanding of these lines: Theocr. 7. 37–41 (Simichidas to Lycidas) *καὶ γὰρ ἐγὼ Μοισᾶν καπυρὸν στόμα, κἠμὲ λέγοντι | πάντες ἀοιδὸν ἄριστον· ἐγὼ δέ τις οὐ ταχυπειθής, | οὐ Δᾶν· οὐ γάρ πω κατ' ἐμὸν νόον οὔτε τὸν ἐσθλόν | Σικελίδαν νίκημι τὸν ἐκ Σάμω οὔτε Φιλίταν | ἀείδων, βάτραχος δὲ ποτ' ἀκρίδας ὥς τις ἐρίσδω*, 'For I too am a clear voice of the Muses, and all call me the best of singers; but I am slow to credit them, faith I am. For in my own esteem I am as yet no match in song either for the great Sicelidas from Samos or for Philetas, but vie with them like a frog against grasshoppers' (Gow); and Varro, *LL* 7. 36 '*Versibus quos olim Fauni uatesque canebant* [=Enn. *Ann.* 207 Skutsch]: Fauni dei Latinorum, ita ut et Faunus et Fauna sit; hos uersibus quos uocant Saturnios in siluestribus locis traditum est solitos fari futura, a quo fando Faunos dictos. antiqui poetas uates appellabant a uersibus uiendis, ut de poematis cum scribam ostendam'. Varro may have known that Ennius was attacking Naevius and the old Saturnian poetry, but if so he gives no hint; he simply states that the Ancients called poets *uates* and supplies an etymology (what he later wrote 'de poematis' cannot, unfortunately, be known). V. was thus provided with a word of antique dignity which exactly served his purpose.

Much, too much perhaps, has been made of the distinction between *poeta* and *uates*: H. Dahlmann, 'Vates', *Philologus*, 97 (1948), 337–53 = *Kleine Schriften* (Hildesheim, 1970), 35–51, on which see O. Skutsch, *Enniana* (London, 1968), 28 n. 9; J. K. Newman, *The Concept of Vates in Augustan Poetry* (Coll. Latomus, 89; Brussels, 1967), 99–206; P. R. Hardie, *Virgil's Aeneid: Cosmos and Imperium* (Oxford, 1986), 16–22; M. Massenzio, *Enc. Virg.* s.v. *vates*. Certainly too much has been made of it as far as V. is concerned, since for him the distinction was hardly more than an expedient, the solution to a particular literary problem, that is, his need for another, more elevated word meaning 'poet'. He repeats the distinction in 7. 25–8—a passage secondary, for obvious reasons, to this—and then gives it up. *Vates* occurs four times in the *G.* and thirty-six times in the *A.*, but always in its old-fashioned sense of seer or prophet. Once, it is true, V. applies it to himself, *A.* 7. 41 'tu uatem, tu, diua, mone', where he assumes the traditional role of divinely inspired interpreter; cf. *A.* 7. 645–6 'et meministis enim, diuae, et memorare potestis; / ad nos uix tenuis famae perlabitur aura', Ov. *Met.* 15. 622–3 'pandite nunc, Musae, praesentia numina uatum / (scitis enim, nec uos fallit spatiosa uetustas)', Callim. *Hymn* 3. 186 *εἰπέ, θεή, σὺ μὲν ἄμμιν, ἐγὼ δ' ἑτέροισιν ἀείσω*, 'Speak, goddess, you to me, and I shall sing to others', Livrea on Ap. Rhod. 4. 1381.

33. **Pierides:** 3. 85 n.

34. **credulus:** here first with the dative (*TLL* s.v. 1152. 33).

35. The pastoral illusion is momentarily shattered, as it is in Theocr. 7. 40 by the introduction of Sicelidas (Asclepiades) and Philitas. The reader expects, unless he has Theocritus clearly in mind, not Varius and Cinna but Menalcas.

Vario: L. Varius Rufus, a close friend of V., and, like V., an early member of Maecenas' circle (the two of them sponsored Horace, *Serm.* 1. 6. 54–5). When Maecenas was travelling to Brundisium in the spring of 37 BC, Varius, Plotius Tucca, and Virgil joined his party in Campania, Hor. *Serm.* 1. 5. 40–2 'Plotius et Varius Sinuessae Vergiliusque / occurrunt, animae qualis neque candidiores / terra tulit neque quis me sit deuinctior alter'; cf. *Serm.* 1. 10. 81 'Plotius et Varius, Maecenas Vergiliusque'. Years later, Varius and Plotius would be V.'s literary executors. Varius was admired as an epic poet (*Serm.* 1.10.43–4) and for his tragedies, especially a *Thyestes* produced on the occasion of Octavian's triple triumph in 29 BC, and was not, apparently, inferior to V. in Horace's judgment, *Ars* 55 'Vergilio Varioque', *Epist.* 2. 1. 247 'Vergilius Variusque poetae'. See 8. 88 n., R. Syme, *AClass* 28 (1985), 43–4 = *Roman Papers*, v (Oxford, 1988), 637–8, C. Murgia, *CQ*, NS 41 (1991), 190–3, 212 n. 73.

Cinna: Varius, a contemporary and a friend—but why Cinna? Why not rather Calvus, to whose *Io* V. alludes in 6. 47? Possibly V. had known Cinna and felt he could make a personal reference to him (that the reference is personal is indicated by the parity of the names). Calvus died at about the same time as Catullus or not much later (G. W. Bowersock, *Fondation Hardt, Entretiens*, 25 (Geneva, 1979), 60–1), but Cinna lived on until March 44 BC, when he was mistaken for one of Caesar's assassins and killed by the angry mob; see F. Skutsch, *RE* viii. 226, Housman, *JPhil* 12 (1883), 167 = *Class. Papers*, i. 9, *JPhil* 35 (1920), 315–16 = *Class. Papers*, iii. 1040, T. P. Wiseman, *Cinna the Poet* (Leicester, 1974), 44–6, J. D. Morgan, *CQ*, NS 40 (1990), 558–9. After completing his studies at Cremona, and a short stay in Milan, V. moved to Rome; *Vita Donati* 7 'Vergilius a Cremona Mediolanum et inde paulo post transiit in Vrbem'; see Clausen 3–5. Would not a young and ambitious poet have made himself known to the last important survivor of the New Poets? The subtlety of V.'s imitation here was noticed by Wiseman 56–7. Theocritus names two poets: Asclepiades, whom he calls Sicelidas, a contemporary (and V. may have imagined that he was also a friend), and Philitas, an older poet, revered as a founder of the Alexandrian school; see R. Pfeiffer, *History of Classical Scholarship* (Oxford, 1968), 88–9. It is of some interest that the New Gallus 8–9 names two critics: Viscus, a contemporary (Hor. *Serm.* 1. 9. 22, 10. 83, 2. 8. 20), and Valerius Cato, an older critic and poet, whose *Dictynna* was admired by Cinna (Suet. *De gramm.* 11. 3).

35–6. nam neque adhuc Vario uideor nec dicere Cinna / digna: cf. the New Gallus 6–7 'tandem fecerunt carmina Musae / quae possem domina deicere digna mea', *E.* 4. 3, 8. 10, and see S. Hinds, 'Carmina Digna', *Papers of the Liverpool Latin Seminar, Fourth Volume* (ARCA 11; 1983), 43–7.

36. argutos inter . . . anser olores: 'et alludit ad Anserem quendam, Antonii poetam, qui eius laudes scribebat, quem ob hoc per transitum carpsit' (Serv.); a poet whose existence might be doubted were it not for Ov. *Trist.* 2. 435 'Cinnaque procacior Anser'; cf. Cic. *Phil.* 13. 11 (cited here by Servius), with Shackleton Bailey's note. Does Ovid's pairing of Cinna and Anser indicate that he was aware of V.'s pun? That Propertius was aware of it is indicated by the adjective *indocto* in his reference to this line, 2. 34. 83–4 'nec minor hic animis, ut sit minor ore, canorus / anseris indocto carmine cessit olor'. The wit is V.'s own, for Theocritus' frog is only a frog.

olores: the native Latin word, surviving mainly in poetry (*TLL* s.v. 571. 69) and ultimately displaced by the loanword *cycnus*.

37. id quidem ago: this elision of a dactyl ending in *-m* is unique in the *E.*; see Soubiran 222–3. Cf. Lucr. 3. 904 'tu quidem ut', with Munro's note.

tacitus . . . mecum ipse uoluto: cf. Livy 9. 17. 2 (of himself) 'quibus saepe tacitus cogitationibus uolutaui animum'.

38. si ualeam: 'in the hope that . . .'; cf. 6. 57, *A.* 1. 181–2 (Aeneas) 'prospectum late pelago petit, Anthea si quem / iactatum uento uideat'.

39–50. Lines 23–5 are Theocritean and sung by Lycidas, 27–9 Roman and sung by Moeris; but now their roles are reversed: ll. 39–43 are Theocritean and sung by Moeris, 46–50 Roman and sung by Lycidas. The structure of 39–43 and 46–50 is nearly parallel: each passage consists of five lines, with the phrase *huc ades* enclosing the first (39, 43), and the vocative *Daphni*, a little less obviously, the second (46, 50); intervening in the first passage is a tricolon crescendo with anaphora (40–2), in the second a tricolon with epanalepsis (47–9; a comma should be added after *astrum* and *frugibus*); and each passage is introduced with two lines concerning memory and song: 38 *meminisse . . . carmen*; 44 *canentem*, 45 *memini*.

There is a textual problem here. Lines 46–50 are assigned to Lycidas in the Mediceus and Palatinus, but to Moeris in the Carolingian MSS, by an ancient corrector of the Palatinus, by DServius, and by most modern editors and commentators, Ribbeck, Forbiger, Klingner, and Mynors excepted. The vulgate arrangement is apparently easier, and therefore suspect: Lycidas having forgotten the words, Moeris again sings five lines and then stops, complaining of his memory. The chief objection to it is that it destroys the symmetry of V.'s 'quotations'. Lycidas sings the five lines he remembers, and Moeris sadly remarks, for he has no wish to continue, that he has forgotten so many of the songs he knew as a boy.

39. huc ades, o Galatea: cf. Theocr. 11. 19 *Ὦ λευκὰ Γαλάτεια*, 'O white Galatea', 42 *ἀλλ' ἀφίκευσο ποθ' ἀμέ*, 'Do come to me'. Polyphemus then proceeds to enumerate the various delights of his garden, 45–9: there are bays in it, there are slender cypresses, there is dark ivy, there is the vine with its sweet fruit, there is water cold from Etna's snowfields, a drink divine! Who would choose instead of these the sea and its waves? See 7. 9 n.

quis ... nam: *nam* is loosely attached to *quis* as occasionally in Plautus, e.g. *Aul.* 136 'quis ea est nam optuma?' See Mynors on *G.* 4. 445–6 *nam quis*.

40–1. hic ... hic ... / ... hic: cf. 10. 42–3.

40. uer purpureum: the 'purple spring' of the poets; A. Johnston, '"The Purple Year" in Pope and Gray', *Review of English Studies*, NS 14 (1963), 389–93. The Italian spring discloses itself in a profusion of flowers, many of them red or red-tinted; cf. *G.* 2. 319 'uere rubenti', 4. 53–4 (bees in springtime) 'saltus siluasque peragrant / purpureosque metunt flores', 306 (in early spring) 'ante nouis rubeant quam prata coloribus', *Dirae* 21 (*serta*) 'purpureo campos quae pingunt uerna colore'. For this difficult adjective in general see J. André, *Étude sur les termes de couleur dans la langue latine* (Paris, 1949), 90–102.

40–1. uarios ... / ... flores: various bright-coloured flowers; cf. *A.* 6. 708 'floribus ... uariis', 4. 202 'uariis florentia limina sertis'.

41. candida populus: the white poplar, properly *alba*; see 7. 61 n., *TLL* s.v. *albus* 1504. 29, s.v. *candidus* 242. 29. Polyphemus' cave in Homer is shaded by bays (*Od.* 9. 183), but here the poplar seems a picturesque and humanizing detail; cf. Hor. *Carm.* 3. 22. 5 'imminens uillae tua pinus esto'.

antro: 1. 75 n. Especially appropriate here, Theocr. 11. 44 *ἐν τὤντρῳ*, 'in the cave'.

42. lentae: 1. 4 n.

texunt umbracula: V.'s metaphor, as Heinsius recognized, was prompted by Theocritus' description of the spring Burina,

7. 7–8 *ταὶ δὲ παρ' αὐτάν | αἴγειροι πτελέαι τε εὔσκιον ἄλσος ὕφαινον* (Heinsius: *ἔφαινον* codd.), 'and hard by the spring, poplars and elms . . . wove a shady precinct' (Gow). Cf. Mart. 12. 31. 1–2 'Hoc nemus, hi fontes, haec textilis umbra supini / palmitis'.

umbracula: Macrob. *Sat.* 6. 4. 8 'sunt qui aestiment hoc uerbum *umbracula* Vergilio auctore compositum, cum Varro rerum diuinarum libro decimo dixerit "non nullis magistratibus in oppido id genus umbraculi concessum"; et Cicero in quinto de legibus "uisne igitur . . . descendatur ad Lirim eaque quae restant in illis alnorum umbraculis persequamur?"; similiter in Bruto (37)'. Cf. also Varro, *RR* 1. 51. 2, and DServ. 'Cicero "umbraculisque siluestribus"'. Here first in poetry, then in Tib. 2. 5. 97.

43. **insani . . . fluctus:** a metaphor as old as Semonides fr. 7. 39–40 W. and found in Moschus fr. 1. 5 Gow; cf. Catull. 25. 12–13 'uelut minuta magno / deprensa nauis in mari uesaniente uento', Prop. 1. 8. 5 'uesani murmura ponti', 3. 7. 6 'insano . . . mari', Hor. *Carm.* 3. 4. 30 'insanientem . . . Bosphorum'.

44. **quid, quae . . . ?:** 'what of the song . . . ?'; cf. *G.* 3. 265 'quid, quae . . . ?' and see Mynors on *G.* 3. 258.

pura . . . sub nocte: Arat. *Phaen.* 323 *καθαρῇ ἐνὶ νυκτί*, which Cicero translates as 'nocte serena' (*Arat.* 104). See K. F. Smith on Tib. 1. 9. 36.

46. **antiquos:** 'pro *antiquorum*' (DServ.), the only instance of transferred epithet or hypallage in the *E.*; appropriate to the sudden elevation of tone here and metrically convenient. See Mynors on *G.* 2. 143, Austin on *A.* 6. 2.

signorum . . . ortus: *G.* 1. 257.

47. **Dionaei:** as being descended from Venus (Aphrodite); cf. Theocr. 15. 106 *Κύπρι Διωναία*, 'O Cypris, daughter of Dione', *Orph. Arg.* 1323 *Διωναίῃ Ἀφροδίτῃ*. Dione was the mother of Aphrodite (*Il.* 5. 370–1), and until the third century BC mother and daughter remained distinct, Theocr. 17. 36 *Κύπρον ἔχοισα Διώνας πότνια κούρα*, 'the Queen of Cyprus, Dione's august daughter' (Gow). But in Theocr. 7. 116 *ξανθᾶς ἕδος αἰπὺ Διώνας*, 'the steep high abode of golden-haired Dione', Dione 'seems to be Aphrodite herself . . .; it should however be observed that this use of the name, common enough in Latin poets, is now usually removed from Bion 1. 93, the only other place in Greek where it occurred'

(Gow). Gow's 'common enough' is misleading: with the exception of Claudian, *De rapt. Pros.* 3. 433, Dione always means Venus in Latin poetry, beginning with Catull. 56. 6 (*TLL* Onomast. s.v.); and P. Mass, *CR*, NS 4 (1954), 11 = *Kleine Schriften* (Munich, 1973), 97, has shown that Dione should not be removed from Bion.

processit: 6. 86 n.

Caesaris astrum: a comet appeared during the games held in his honour by Octavian in July 44 BC; Suet. *Iul.* 88 'stella crinita per septem continuos dies fulsit exoriens circa undecimam horam, creditumque est animam esse Caesaris in caelum recepti'. See S. Weinstock, *Divus Julius* (Oxford, 1971), 370–84. Caesar's star is portrayed for the first time on an issue of coinage produced by Agrippa in 38 BC; see M. H. Crawford, *Roman Republican Coinage* (Cambridge, 1974), 744. Propertius has Caesar looking down approvingly from his star on the battle of Actium, 4. 6. 59 'pater Idalio miratur Caesar ab astro'. For references to Caesar in Augustan poetry see P. White, *Phoenix*, 42 (1988), 345–53.

47–8. astrum, / astrum, quo segetes: cf. Catull. 64. 285–6 'Tempe, / Tempe, quae siluae', *E.* 10. 72–3 'Gallo, / Gallo, cuius amor'. See 6. 21 n.

48. segetes: 'fields'; cf. Varro, *RR* 1. 29. 1 'seges dicitur quod aratum satum est', *G.* 1. 47, 4. 129.

48–9. gauderent ... duceret: subjunctives expressing purpose; Caesar's star rose to shed its stellar virtue on field and vineyard.

49. duceret ... colorem: the idiom seems to have been *colorem trahere*, *A.* 4. 701, Ov. *Met.* 2. 236, 14. 393, Pliny, *NH* 19. 134 'ut in lactucis, cum coeperint colorem trahere', Quintil. 12. 10. 44 'lacertos . . . colorem trahere naturale est'; cf., however, Ov. *Met.* 3. 484–5. For a similar variation cf. Ov. *Am.* 2. 2. 33 'uir traxit uultum', *Ex Pont.* 4. 1. 5 'trahis uultus', 3. 7 'contraxit uultum Fortuna', 8. 13–14 'si lectis uultum tu uersibus istis / ducis'.

apricis in collibus: cf. *G.* 2. 522 'mitis in apricis coquitur uindemia saxis'.

50. insere, Daphni, piros: carpent tua poma nepotes: cf. 1. 73. The pear-tree lives and bears fruit to a very great age. There are, or rather were, in Salem, Mass., two ancient trees, the Endicott Pear and the Orange Pear. The Endicott Pear is said to have been brought over from England in June 1630 on the Arabella with Governor Winthrop. 'As early as 1763 the tree was very old and decayed. It was much injured in the gale of 1804. In the gale of 1815 it was so much shattered that its recovery was considered doubtful. It was injured again in a gale about 1843. . . . In 1837 it was eighty feet high by measurement and fifty-five feet in the circumference of its branches, and does not probably vary much from these dimensions now' (R. Manning, Jr., *Proceedings of the American Pomological Society* (1875), quoted by U. P. Hedrick, *The Pears of New York*, State of New York, Dept. of Agriculture, Twenty-ninth Annual Report 2, part 2 (1921), 41). The Orange Pear seems to have been about a decade younger. 'The Rev. Dr. Bentley, who died about 1820, investigated the history of this tree and found it to be then 180 years old, which would make it now 235 years old . . . . In the very favorable pear season of 1862 it bore thirteen and a half bushels of pears' (ibid. 42). (I am indebted to D. Hull for this reference.) The Orange Pear has disappeared entirely; of the Endicott Pear, when Roger Mynors and I sought it out in the spring of 1977, only a few small branches were visible growing from the roots.

51. omnia fert aetas: 'id est aufert' (Serv.); see 1. 3 n. La Cerda has a long and learned note on this commonplace. Wilamowitz, on Eur. *Her.* 669, thinks that V. misunderstood [Plato], *AP* 9. 51. 1 *Αἰὼν πάντα φέρει*, 'Time brings everything' as 'takes away'.

animum quoque: 'etiam memoriam' (Serv.).

saepe ego: 8. 97–9 n.

51–2. Ursinus noticed the allusion to Callimachus 34. 2–3 G.–P. (=*AP* 7. 80). *ἐμνήσθην δ' ὁσσάκις ἀμφότεροι | ἠέλιον λέσχῃ κατεδύσαμεν*, 'I remembered, how often you and I / Had tired the sun with talking and sent him down the sky' (Cory). *Longos* (cf. above, 2–3 'nostri / . . . agelli') and *puerum* are emphatic: as a boy Moeris could sing all day long, but now he is old and tired.

52. condere soles: *sol* in this sense is first attested in Lucr. 6. 1219, then here, where V. has varied the phrase *diem condere* because of Callimachus, and again in A. 3. 203–4 'tris adeo incertos caeca caligine soles / erramus pelago, totidem sine sidere noctes'. Cf. *G.* 1. 458 (*sol*) 'cum referetque diem condetque relatum', Hor. *Carm.* 4. 5. 29 'condit quisque diem collibus in suis', Pliny, *Epist.* 9. 36. 4 'ita uariis sermonibus uespera extenditur, et quamquam longissimus dies bene conditur'.

53. oblita: passive, but elsewhere in V. (11 times) active or reflexive; see *TLL* s.v. 111. 79.

53–4. Moerim / . . . Moerim: Moeris repeats his name, as if recalling the boy and man he had once been.

54. lupi Moerim uidere priores: Pliny, *NH* 8. 80 'in Italia quoque creditur luporum uisus esse noxius uocemque homini quem priores contemplentur adimere ad praesens'. This belief was widespread among the Ancients; see Gow on Theocr. 14. 22 *οὐ φθεγξῇ; λύκον εἶδες;*, 'Won't you speak? Seen a wolf?' See also Sir Thomas Browne, *Pseudodoxia Epidemica*, 3. 8.

55. satis . . . saepe: 'ordo est *satis saepe*' (DServ.).

56. causando: 'causas nectendo differs nostra desideria' (Serv.). Cf. Lucr. 1. 398 'quamuis causando multa moreris'; the only occurrence of this verb in either poet.

57–8. et nunc omne tibi stratum silet aequor, et omnes, / aspice, uentosi ceciderunt murmuris aurae: cf. Theocr. 2. 38 *ἠνίδε σιγῇ μὲν πόντος, σιγῶντι δ' ἀῆται*, 'Look, quiet is the sea, quiet the winds', *A*. 5. 763 'placidi strauerunt aequora uenti'. Since Mantua is far from the sea, this passage was misunderstood in antiquity, '*aequor* spatium campi' (DServ.); but, as T. E. Page remarks, *stratum* prevents this interpretation. The landscape here is no more 'real' than that described above, 7–8; there are no high hills or mountains near Mantua. See above, p. xxx.

58. uentosi: 1. 5 n.

uentosi . . . murmuris aurea: the poets use *murmur* more often of the noise of water, of the sea especially, and the sky than of the sound of wind, e.g. Lucr. 1. 276 'saeuitque minaci murmure uentus', 5. 1220–1 'fulminis horribili cum plaga torrida tellus / contremit et magnum percurrunt murmura caelum', Prop. 1. 8. 5 'uesani murmura ponti', Ov. *Met*. 9. 40–1 'moles, magno quam murmure fluctus / oppugnant', Silius 4. 640–1 'furit unda sonoris / uerticibus, sequiturque nouus cum murmure torrens'. *Murmur* seems in general to have denoted a sound far harsher than English 'murmur', and V. evidently felt that *murmuris*, not *aurae* ('the more natural expression', Conington), required qualification. See Mynors on *G*. 4. 484.

ceciderunt: of the wind, *G.* 1. 354, *TLL* s.v. 27. 43.

59–60. hinc adeo media est nobis uia; namque sepulcrum / incipit apparere Bianoris: cf. Theocr. 7. 10–11 *κοὔπω τὰν μεσάταν ὁδὸν ἄνυμες, οὐδὲ τὸ σᾶμα | ἁμῖν τὸ Βρασίλα κατεφαίνετο*, 'We had not yet completed half our journey, nor was the tomb of Brasilas in sight'. Going from town to the Haleis for the harvest-festival, Simichidas and his friends passed the tomb of Brasilas, a name of some local repute but unknown to wider fame (Gow ad loc.). A tomb bearing a name that had no further identity, a venerable landmark by the side of the road—so V. imagined it, for he creates just such an impression here. L. Herrmann, *CRAI* 1930, 35, and independently S. Tugwell, *CR*, NS 13 (1963), 132–3, suggest that V. found the tomb of Bianor in Diotimus 4. 3 G.–P. (=*AP* 7. 261) *ἠιθέῳ γὰρ σῆμα Βιάνορι χεύατο μήτηρ*, 'so his mother built a tomb for young Bianor'; an epigram included in Meleager's *Garland* and therefore, in all probability, known to V. But V.'s ancient commentators did not (or could not) appreciate his fiction; desiderating a real tomb in an imaginary landscape, they arbitrarily identified Bianor with Ocnus, the legendary founder of Mantua, 'hic est qui et Ocnus dictus est' (Serv.). See Williams 320–1; also F. E. Brenk, *AJP* 102 (1981), 427–30, S. V. Tracy, *CP* 77 (1982), 328–30. For the rhythm of l. 60 see 2. 6 n.

61. stringunt frondes: cf. Cato, *De agr.* 54. 2 'si fenum non erit, frondem iligneam et hederaciam dato', *G.* 1. 305 (Serv.), Hor. *Epist.* 1. 14. 28, Columella 6. 3. 6–7; C. T. Ramage, *The Nooks and By-Ways of Italy* (Liverpool, 1868), 3–4 (Paestum): 'in the low warm country they find it difficult to procure green food for their cattle and horses, and are obliged to strip even the trees of their leaves for this purpose'.

64. Cf. Theocr. 7. 35–6 *ἀλλ' ἄγε δή, ξυνὰ γὰρ ὁδὸς ξυνὰ δὲ καὶ ἀώς, | βουκολιασδώμεσθα*, 'But come, for the way and the day are yours and mine alike, let us make country music'—though the tone here is very different.

65. hoc te fasce leuabo: Lycidas offers to relieve Moeris of the basket or crate in which he is carrying the kids so that he can sing more easily. Cf. *G.* 3. 347 (the Roman soldier) 'iniusto sub fasce uiam cum carpit' (Serv.), *Moretum* 79 'uenalis umero fasces portabat in urbem'.

66. desine plura, puer: cf. 5. 19.
puer: for the prosody see 1. 38 n.

67. ipse: Menalcas, the master-singer himself.

ECLOGUE 10

Introduction

The last poem of his book, Virgil's 'last labour', will be a poem for Gallus, a small poem for Gallus, but such as Lycoris herself may read. Virgil's concern is not with the erudite poet of the Sixth *Eclogue* but, as the reference to Lycoris indicates, with Gallus the love-poet, poet of the *Amores*[1] and lover of Lycoris, the mistress or *domina* of his poetry.[2] But how was this urbane, personal poet and soldier to be accommodated in Virgil's pastoral landscape? Virgil dealt with this difficulty—if Virgil was aware of a difficulty—by relating Gallus to Daphnis, in a close and sustained imitation of Theocritus' First *Idyll*, and by transferring him to Arcadia.

Gallus seems to be on leave in Arcadia, where he has assumed, this curiously inert warrior, the posture of a disconsolate elegiac lover, prostrate beneath a lonely crag.[3] Trees and mountains weep over him, while countryfolk and rural deities attempt to solace him. Shepherds come, swineherds slowly come, and Menalcas comes wet from the winter mast. All ask about his love. Apollo comes, rebukes Gallus for his mad passion, and tells him that

[1] The title, apparently, of the four books of love-elegies which he composed; 'amorum suorum de Cytheride scripsit libros quattuor' (Serv. on 10. 1); for Cytheris see note on line 2. Before the discovery of the New Gallus only one line was known: 'uno tellures diuidit amne duas' (*FPL* fr. 1 Büchner). Note the recurrence of the word *amores* in this *Eclogue*: 6 'sollicitos Galli . . . amores', 34 'meos . . . amores', 53 'meos . . . amores', 54 'amores'. Elsewhere in the *E.*, *amores* is found only in 3. 109, 8. 23, and, in a non-erotic sense, 9. 56. See Ross (1975) 73. Ovid called his five books (later reduced to three) of love-elegies *Amores*, presumably out of deference to Gallus.

[2] See the New Gallus 6–7 'tandem fecerunt carmina Musae / quae possem domina deicere digna mea', with the note on *domina* (p. 144); also *TLL* s.v. *dominus* 1938. 1. Cf. Prop. 2. 34. 91–2 (written after the disgrace and suicide of Gallus in 27 BC) 'et modo formosa quam multa Lycoride Gallus / mortuus inferna uulnera lauit aqua', Ov. *Am.* 1. 15. 29–30 'Gallus et Hesperiis et Gallus notus Eois, / et sua cum Gallo nota Lycoris erit'.

[3] 14 'sola sub rupe iacentem'; cf. Prop. 1. 9. 3 'ecce iaces supplexque uenis ad iura puellae', 1. 16. 23–4 'me mediae noctes, me sidera prona iacentem, / frigidaque Eoo me dolet aura gelu', 2. 14. 32 'uestibulum iaceam mortuus ante tuum', Tib. 2. 4. 22 'ne iaceam clausam flebilis ante domum', 2. 5. 109–11 'iaceo cum saucius annum / et faueo morbo . . . / usque cano Nemesim', Ov. *Am.* 2. 10. 15 'sed tamen hoc melius, quam si sine amore iacerem', *Ars* 2. 238 'frigidus et nuda saepe iacebis humo'. The posture of Daphnis is not indicated.

Lycoris has followed another through the snow and the rough camps. To this distressing news Gallus' response is a pathetic wish, deferred (48–9), that the cold may not injure her and oh, may the sharp ice not cut her tender feet! Silvanus also comes, crowned with a towering garland of fennel and lilies, and, last of all, comes Pan himself, Arcadia's god, whose doleful observation, beautifully expressed—cruel Love will never be satiated with tears, never . . . —introduces the lament of Gallus.

> tristis at ille 'tamen cantabitis, Arcades', inquit
> 'montibus haec uestris, soli cantare periti
> Arcades'. (31–3)

So Gallus begins, indulging a melancholy pleasure in the thought that Arcadian herdsmen, the unrivalled masters of pastoral song, will sing to their hills, even when he is dead, of his unrequited love for Lycoris. But why does Virgil place Gallus in Arcadia, a landscape unknown to Theocritus and merely alluded to elsewhere in the *Eclogues*?[4] Why not, rather, in Sicily, the scene of the First *Idyll*, or in Virgil's own Po valley? These landscapes were too familiar, it would seem, too accessible to suit Virgil's purpose—the Po valley with its bitter memories of the land-confiscations, in which Gallus may have played a part, and the Sicilian landscape with its burden of history. Virgil required a different landscape, one remote from experience and therefore permitting the utmost freedom of invention. Arcadia offered, obscure and wild, with its old poetic mountains, Maenalus, where Pan was worshipped, Erymanthus, Cyllene, Stymphalus, Parthenius, and, the birthplace of Pan, Lycaeus.

Rarely, if ever, can a poetic act be explained satisfactorily, and Virgil's appropriation of Arcadia was an intensely poetic act.[5] Perhaps Virgil had read the chapter in which Polybius, a native of

[4] See 4. 58–9, 7. 4, 26, 8. 21, 22. Far too much has been made of Virgil's Arcadia. Arcadia conceived as an ideal or symbolic landscape, 'la pastorale Arcadia', is the invention of Jacopo Sannazaro and Sir Philip Sidney—that is, a feature of the pastoral tradition—and should not be imposed retroactively on Virgil, as e.g. by E. R. Curtius, *European Literature and the Latin Middle Ages*, tr. W. R. Trask (Princeton, NJ, 1953), 190: 'Sicily, long since become a Roman province, was no longer a dreamland. In most of his eclogues Virgil replaces it by romantically faraway Arcadia, which he himself had never visited'. See R. Jenkyns, 'Virgil and Arcadia', *JRS* 79 (1989), 26–39.

[5] Cf. Wallace Stevens, *Opus Posthumous*, ed. M. J. Bates (New York, 1989), 255–6.

Arcadia, describes the musical character of the Arcadians;[6] perhaps not. But even if he had, would it have made, with no predisposing cause, such an impression on him? The Arcadian poetic tradition, tenuous as it now appears, is probably sufficient to account for Virgil's choice of Arcadia.[7] Here, for example, is an epigram by Anyte,[8] an Arcadian poet contemporary with Theocritus or a little older:

—Τίπτε κατ' οἰόβατον, Πὰν ἀγρότα, δάσκιον ὕλαν
ἥμενος ἁδυβόᾳ τῷδε κρέκεις δόνακι;
—Ὄφρα μοι ἑρσήεντα κατ' οὔρεα ταῦτα νέμοιντο
πόρτιες ἠυκόμων δρεπτόμεναι σταχύων.

—Why, rustic Pan, do you sit in the lonesome, shadowy wood playing on this sweet-voiced reed?
—So that the heifers may range over these dewy hills cropping the luxuriant herbage.

Significant, too, is the fact that Virgil appeals, at the beginning of his poem, to Arethusa as to a Muse. Virgil of course knew that Moschus[9] had identified Arethusa as the source of pastoral song, and himself plays, in ll. 4–5, with the conceit of her derivation from Arcadia.

Gallus' lament has nothing in common with the lament of Daphnis. In fact, apart from ll. 37–43, there is little or no imitation of Theocritus in Gallus' lament; only towards the end, in ll. 65–8, as Virgil prepares for the elaborate cadence to his poem and to his book, is Theocritus, but not the First *Idyll*, again imitated. Virgil ends the relationship between Gallus and Daphnis because he wishes to reproduce, as far as the hexameter allows, the tone and movement of an elegy, and overt reference to Theocritus would impair the effect.

Gallus now speaks in his own person, imitating himself, as it were; his self-dramatization, the shifting attitudes he adopts—

[6] Polybius 4.20. A connection with Virgil was posited by B. Snell, *Die Entdeckung des Geistes*[2] (Hamburg, 1948), translated by T. G. Rosenmeyer, *The Discovery of the Mind* (Cambridge, Mass., 1953), 281–309 'Arcadia: The Discovery of a Spiritual Landscape'. As Walbank notes on Polybius ad loc., Snell had been anticipated by T. Keightley (1847); also by R. Reitzenstein, *Epigramm und Skolion* (Giessen, 1893), 132 n.

[7] See 7. 4 n., G. Jachmann, 'L'Arcadia come paesaggio bucolico', *Maia*, 5 (1952), 171 = *Ausgewählte Schriften* (Königstein im Taurus, 1981), 359.

[8] G.–P. 19 (= *A.Pl.* A 231); available to Virgil in Meleager's *Garland*.

[9] See l. 1 n. (for Virgil, of course, the poem was authentic).

Virgil's Gallus is evidently imitating the 'real' Gallus, for similar passages or motifs occur later in the elegies of Propertius and Tibullus.[10] Like Propertius and Tibullus, Gallus takes a morbid pleasure in imagining his own death;[11] like Tibullus, another lovelorn soldier, he dreams of enjoying pastoral ease with his mistress, only to have his dream shattered by the harsh reality of war;[12] and, like Milanion of old, he seeks to alleviate his passion by hunting on Mt. Parthenius:

> Milanion nullos fugiendo, Tulle, labores
> saeuitiam durae contudit Iasidos.
> nam modo Partheniis amens errabat in antris,
> ibat et hirsutas ille uidere feras;
> ille etiam Hylaei percussus uulnere rami
> saucius Arcadiis rupibus ingemuit.[13]

Poetic compliment, subtle praise of poet by poet, may take the form of verbal reminiscence or quotation. Thus Catullus, in heralding the long-awaited publication of Cinna's *Zmyrna*, refers to the Satrachus, an exquisite small stream mentioned in his friend's poem.[14] And Ovid, in his elegy on the death of Tibullus, incorporates a line of Tibullus, 1. 1. 60 'te teneam moriens deficiente manu'; cf. Ov. *Am.* 3. 9. 58 'me tenuit moriens deficiente manu'.[15] In his note on l. 46 of the Tenth *Eclogue*, Servius remarks: 'hi autem omnes uersus Galli sunt, de ipsius translati carminibus'. Which lines does he mean? Quite possibly Servius did not know. For it is not to be supposed that he had read and collated Gallus with Virgil; he copied, rather, and perhaps not very carefully, the note of an earlier scholiast. Be that as it may, ll. 46–9 are apparently 'translated' from a *propempticon* and form a distinct unit, whose excited, abrupt rhythm resembles that of elegiac couplets:

> tu procul a patria (nec sit mihi credere tantum)
> Alpinas, a! dura niues et frigora Rheni
> me sine sola uides. a, te ne frigora laedant!
> a, tibi ne teneras glacies secet aspera plantas! (46–9)

10 See F. Skutsch (1901), 14–15, Ross (1975), 85.

11 Prop. 1. 17. 19–24, 1. 19, 2. 13. 17–58, Tib. 1. 1. 59–68, 3. 53–66.

12 Tib. 1. 3. 23–56, 10. 7–14. See Klingner 172–3.

13 Prop. 1. 1. 9–14, where see Fedeli; also Ross (1975), 90–1.

14 Catullus 95; see Clausen 6–7.

15 Similarly, Wordsworth, in his *Remembrance of Collins*, 19 'For him suspend the dashing oar', incorporates a line of Collins' own *Ode Occasioned by the Death of Mr. Thompson*, 15 'And oft suspend the dashing oar'.

The similarity of these lines to two couplets in Propertius was noticed by La Cerda:

> tune audire potes uesani murmura ponti
> fortis et in dura naue iacere potes?
> tu pedibus teneris positas fulcire pruinas,
> tu potes insolitas, Cynthia, ferre niues? (1. 8. 5–8)[16]

Imitation of Gallus is not, however, confined to ll. 46–9; it has been detected or surmised in other lines as well,[17] and may be presumed to exist undetected in still others.

'Haec sat erit . . .' (70). Virgil's poem, his song for Gallus, now concludes with an epilogue of eight lines (70–7), which balances the prologue of eight lines (1–8). The poet's love for Gallus, implied in the prologue by the phrase 'meo Gallo' and the iteration of Gallus' name, is now declared: the poet's love for Gallus—again his name is iterated, with greater feeling—grows hour by hour, grows like the green alder when spring is new. But suddenly it is evening; as the shadows deepen underneath the juniper, the singer rises and the poet takes his leave of pastoral song.

Bibliography

F. Skutsch, *Aus Vergils Frühzeit* (Leipzig, 1901), 2–27.

—— *Gallus und Vergil* (Leipzig, 1906), 155–90.

Klingner 166–74.

G. Williams, *Tradition and Originality in Roman Poetry* (Oxford, 1968), 233–9.

Ross (1975) 85–106.

The New Gallus.

G. B. Conte, *Virgilio: Il genere e i suoi confini* (Milan, 1984), 13–42 = C. Segal (ed.), *The Rhetoric of Imagination* (Ithaca, NY, 1986), 100–29.

D. F. Kennedy, '*Arcades ambo*: Virgil, Gallus and Arcadia', *Hermathena*, 143 (1987), 47–59.

R. Whitaker, 'Did Gallus Write "Pastoral" Elegies?', *CQ*, NS 38 (1988), 454–8.

[16] See Fedeli, pp. 203–6; also Ross (1975), 85.

[17] See notes on ll. 12, 22, 33, 38, 42, 48, 52, 55, 60, 66.

1. Extremum . . . laborem: see above, pp. xxiv–xxv. Cicero seems to have been the first to describe a literary composition as a *labor*, *De leg.* 1. 8 'intellego equidem a me istum laborem iam diu postulari' (*TLL* s.v. 794. 80). The concept, however, is Hellenistic: Callimachus 55. 1–2 G.–P. (of a poem on the capture of Oechalia) *Τοῦ Σαμίου πόνος εἰμὶ δόμῳ ποτὲ θεῖον ἀοιδόν | δεξαμένου*, 'I am the work of the Samian who once received the divine singer in his house', Asclepiades 28. 1 G.–P. (=*AP* 7. 11) *Ὁ γλυκὺς Ἠρίννας οὗτος πόνος*, 'This is the work in which Erinna delighted', where see note; also Gow on Theocr. 7. 51. Cf. *G.* 2. 39 'inceptumque una decurre laborem'.

1–2. Arethusa, mihi concede laborem: / pauca meo Gallo: cf. 7. 21–2 'Nymphae . . . Libethrides, aut mihi carmen, / quale meo Codro, concedite'.

1. Arethusa: the most renowned of the springs bearing this name; see Hoekstra on *Od.* 13. 408, *TLL* s.v. 511. 49. The story of Arethusa, alluded to in Pind. *Nem.* 1. 1–2 and in *A.* 3. 694–6, is told (by Arethusa) in Ov. *Met.* 5. 572–641. Pursued by the river-god Alpheus and rendered liquid by a compassionate Diana, the Arcadian Nymph Arethusa fled through the depths of the Ionian Sea to the great harbour of Syracuse, where she emerged on the island of Ortygia as a freshwater spring mingled with the river-god. (Sadly, the water of Arethusa, which Emerson found 'sweet and pure' (*Journal*, 23 Feb. 1833), is now impregnated with salt.) Arethusa was introduced into pastoral poetry by Theocritus, 1. 117 *χαῖρ', Ἀρέθοισα*, 'Farewell, Arethusa', where Daphnis, mortally sick of love, takes his leave of the pastoral world. Cf. also Theocr. 16. 102 *Σικελὴν Ἀρέθοισαν*, 'Sicilian Arethusa'. For [Moschus], Arethusa is the source of pastoral song and, as such, not inferior to Hippocrene, which he imagines as the source of Homeric song, *Epitaph. Bion.* 76–7 (Homer and Bion) *ὃς μὲν ἔπινε | Παγασίδος κράνας, ὃ δ' ἔχεν πόμα τᾶς Ἀρεθοίσας*, 'one ever drank of the Pegasean fount, but the other would drain a draught of Arethusa' (Lang). Her head, surrounded by plunging dolphins, appears on Cimon's tetradrachms; C. M. Kraay, *Archaic and Classical Greek Coins* (Berkeley, 1976), 222–3: 'among the most delightful creations of Greek coinage'.

2–3. Cf. Callim. *Hymn* 2. 30–1 *οὐδ' ὁ χορὸς τὸν Φοῖβον ἐφ' ἓν μόνον ἦμαρ ἀείσει, | ἔστι γὰρ εὔυμνος· τίς ἂν οὐ ῥέα Φοῖβον ἀείδοι;*, 'Nor will the choir sing of Phoebus for one day only. For he is much sung of; who would not sing readily of Phoebus?'

2. meo Gallo: 7. 22 n.

Lycoris: a pseudonym invented by Gallus, apparently modelled on 'Lycoreus', a cult-title of Apollo, and therefore suggesting Hellenistic elegance. For 'Lycoreus' is first attested in Callim. *Hymn* 2. 18–19 *ὅτε κλείουσιν ἀοιδοί | ἢ κίθαριν ἢ τόξα, Λυκωρέος ἔντεα Φοίβου*, 'when minstrels celebrate the cithara or the bow, the gear of Phoebus Lycoreus', where see F. Williams; cf. Euphorion, fr. 80. 3 Powell *Λυκωρέος οἰκία Φοίβου*, 'the abode of Phoebus Lycoreus'. Similarly, Propertius and Tibullus used cult-titles of Apollo when they called their mistresses 'Cynthia' and 'Delia'; see 6. 3 n., 7. 29 n.

Servius on 10. 1 identifies Lycoris with Cytheris, a freedwoman of Volumnius Eutrapelus, one of Antony's intimates. A *mima* or *meretrix* (in the Roman view there was no difference), Volumnia adopted 'Cytheris' as her stage-name, to which, by a convention of Latin love-poetry, 'Lycoris' is metrically equivalent. For the sentimental history of Volumnia Cytheris—she was Antony's mistress and aroused the jealousy of Cicero's wife—see the New Gallus, pp. 152–3, W. Kroll, *RE* xii. 218–19. This identification has been generally accepted, though R. Syme, *History in Ovid* (Oxford, 1978), 201, following Kroll, intrudes 'gentle dubitation'; see also R. F. Thomas, *CP* 83 (1988), 54 n. 1.

3. dicenda: 3. 55 n.

4. sic: followed by *incipe* (6), as in 9. 30–2, where see note. For *tibi* in the weak second position cf. Catull. 17. 5 'sic tibi' Hor. *Carm.* 1. 3. 1, Tib. 2. 5. 121, Prop. 3. 6. 2, Ov. *Her.* 4. 169, and see Wackernagel, *IF* 1 (1892), 411 = *Kleine Schriften* (Göttingen, 1955), i. 79.

subterlabere: cf. *G.* 2. 157.

Sicanos: first attested here in Latin, and found earlier in only two Greek poets: Callim. *Hymn* 3. 57 *Τρινακρίη, Σικανῶν ἕδος*, 'Trinacria, seat of the Sicanians', and Lycophron 870 *ἠδὲ Σικανῶν πλάκας*, 'and the plains of the Sicanians', 951, 1029. So scanned and, as elsewhere in V., placed at the end of the hexameter, *Sicanus* seems to be V.'s innovation, in which he was followed by Propertius, Ovid, and Statius, with the partial defection of the faithful Silius.

5. Doris: wife of Nereus and mother of the Nereids; here, by metonymy, the sea. There are several related metonymies in Hellenistic and Latin poetry. (1) *Nereus*: 6. 35, Callim. *Hymn* 1. 40, Lycophron 164, Opp. *Cyneg.* 2. 68, Nonnus, *Dionys.* 25. 51 *οὐ φονίῃ ῥαθάμιγγι Λίβυς φοινίσσετο Νηρεύς*, 'no Libyan Nereus turned red in a shower of blood', *A.* 10. 764 'medii . . . Nerei', Ov. *Met.* 1. 187, where see Bömer; see also O. Gross, *De metonymiis sermonis Latini a deorum nominibus petitis* (Diss. Philol. Halenses 19; Halle, 1911), 386–8. (2) *Thetis*: 4. 32, Lycophron 22, Stat. *Theb.* 1. 39, *Silu.* 3. 2. 74, Mart. 10. 30. 11 (Gross 396–7). (3) *Tethys*: Callim. *Aet.* 110. 70 Pf. (whence Catull. 66. 70), Lycophron 1069, Archias, *AP* 7. 214. 6, Luc. 5. 623, Stat. *Theb.* 3. 34 (Gross 393–4). (4) *Amphitrite*: Catull. 64. 11 (the Argo) 'illa rudem cursu prima imbuit Amphitriten', Ov. *Met.* 1. 14, *Fast.* 5. 731, and repeatedly in Opp. *Hal.* and Dionys. *Perieg.* Note that this metonymy, which must be Hellenistic, is first attested in Latin; see 7. 37 n. *Doris*, however, is special; the metonym does not occur in Greek poetry, either before or after V., and all instances in later Latin poetry are derived from V. (*TLL* Onomast. s.v.). It would seem, therefore, that V. invented the metonym *Doris*, as he later invented the metonym *Vesta* (the hearth), *G.* 4. 384 (Cyrene) 'ter liquido ardentem perfundit nectare Vestam' (Gross 406–7). V. was no doubt aware of the double meaning of *Ἑστία*; see H. Lloyd-Jones, *RhM* 103 (1960), 78–9 = *Academic Papers* (Oxford, 1990), i. 307–8. Cf. M. Haupt, *Opuscula* (Leipzig, 1876, repr. Hildesheim, 1967), ii. 74: 'neque Latinos poetas, quibus hae figurae paene magis quam Graecis placuisse videntur, multa in hoc genere ausos esse putamus sine Alexandrinorum poetarum exemplo'. See 6. 15 n.

amara: not, apparently, an adjective used of the sea before V., though Varro, *RR* 3. 17. 3, uses it of a saltwater fishpond. Cf. Stat. *Silu.* 2. 2. 18–19 'e terris occurrit dulcis amaro / nympha mari' and see *TLL* s.v. 1820. 65.

intermisceat: a curious and unusual verb, first attested in the *Bellum Hispaniense* 31. 6 'cum clamor esset intermixtus gemitu', where it is hardly more than a strengthened *misceo*. Here, however, as in a contemporary poem, Hor. *Serm.* 1. 10. 29–30 'patriis intermiscere petita / uerba foris', it means to taint or contaminate; see Fraenkel, *Horace*, 134–5. The source of the conceit may be Moschus, fr. 3. 4–5 Gow (Alpheus) *καὶ βαθὺς ἐμβαίνει τοῖς κύμασι τὰν δὲ θάλασσαν | νέρθεν ὑποτροχάει, κοὺ μίγνυται ὕδασιν ὕδωρ*, 'and deep into the waves he goes and runs beneath the sea, nor is water mingled with water'.

6. sollicitos Galli dicamus amores: cf. Theocr. 1. 19 *τὰ Δάφνιδος ἄλγε' ἀείδες*, 'you are accustomed to sing the woes of Daphnis'. The liaison of Gallus and Lycoris seems to have been a troubled one, troubled enough to inspire a quantity of passionate, self-regarding poetry. In the New Gallus, Gallus complains of her wantonness, 1 'tristia nequitia . . ., Lycori, tua'.

7. attondent: used figuratively for comic effect by Plautus; here first in its literal sense (*TLL* s.v.). V. uses the verb only once again, of pruning, in *G*. 2. 407.

simae . . . capellae: cf. [Theocr.] 8. 50 *σιμαὶ . . . ἔριφοι*, 'blunt-faced kids'. But the adjective occurs in the earliest Latin poetry, Livius Andronicus, *trag.* 5 R.[3] 'lasciuum Nerei simum pecus'.

uirgulta: young growth, trees or brush; see Mynors on *G*. 2. 3.

8. canimus surdis: a proverbial expression, e.g. Prop. 4. 8. 47 'cantabant surdo', Ov. *Am*. 3. 7. 61 'quid iuuet ad surdas si cantet Phemius aures'; see Otto, no. 1715.

respondent omnia siluae: see above, p. xxvi.

9–12. Cf. Theocr. 1. 66–9 *πᾷ ποκ' ἄρ' ἦσθ', ὅκα Δάφνις ἐτάκετο, πᾷ ποκα, Νύμφαι; | ἢ κατὰ Πηνειῶ καλὰ τέμπεα, ἢ κατὰ Πίνδω; | οὐ γὰρ δὴ ποταμοῖο μέγαν ῥόον εἴχετ' Ἀνάπω, | οὐδ' Αἴτνας σκοπιάν, οὐδ' Ἄκιδος ἱερὸν ὕδωρ*, 'Where ever were you when Daphnis was wasting, where ever were you, Nymphs? In the beautiful vales of the Peneus or of Pindus? For surely you kept not the great stream of the Anapus, nor Etna's peak, nor Acis' sacred water'. Theocritus imagines the Nymphs, minor divinities attached to a particular place, as goddesses who might be far away, in the vales of the Peneus or of Pindus; for had they been at home in Sicily, they would have come to the aid of their loved Daphnis. Adapting Theocritus to his own purpose, V. imagines the Nymphs or Muses (7. 21 n.) as being absent from their accustomed Greek haunts; for they could easily have crossed the Corinthian Gulf from Parnassus or Helicon—that is, Aonian Aganippe, of which V. was chiefly thinking; see below, 12 n.—to comfort Gallus where he lay dying of love in Arcadia. Thus, in Quintus of Smyrna 3. 594–6, the Muses leave Helicon and hasten across the Aegean Sea to Troy to grieve with Thetis for the death of Achilles (cf. Hom. *Od*. 24. 60–1,

Pind. *Isth.* 8. 57–8). And Ovid describes the Muses as coming from Helicon to share the anxieties of his flight, *Trist.* 4. 1. 49–52 'iure deas igitur ueneror mala nostra leuantes, / sollicitae comites ex Helicone fugae, / et partim pelago partim uestigia terra / uel rate dignatas uel pede nostra sequi'. Cf. Cic. *De nat. deor.* 2. 69 'concinneque, ut multa, Timaeus, qui cum in historia dixisset qua nocte natus Alexander esset eadem Dianae Ephesiae templum deflagrauisse, adiunxit minime id esse mirandum, quod Diana, cum in partu Olympiadis adesse uoluisset, afuisset domo'.

9. nemora: the scholiast on Theocr. 1. 67 remarks that *τέμπη* generally means *ἄλση*, 'groves' (Posch 64 n. 1). No doubt V. used a commentary of some sort on Theocritus; see 2. 9 n.

saltus: 6. 56 n.

9–10. puellae / Naides: 2. 46 n., 5. 59 n. For the position of the two nouns cf. Ap. Rhod. 4. 1398–9 *Νύμφαι* | *Ἑσπερίδες*. Water-nymphs, Naiads, seem especially appropriate here as Aganippe is the source of Gallus' love-poetry (6. 64 n.).

10. For the rhythm of this line see 6. 46 n.

indigno . . . amore: 8. 18 n.

11. nam neque Parnasi . . . nam neque Pindi: very rarely does *neque* occur in the fifth foot: here in a precise imitation of Theocr. 1. 67 *ἢ κατὰ Πηνειῶ . . . ἢ κατὰ Πίνδω*; in *G.* 2. 153 'per humum neque tanto', an expressive rhythm (see Mynors ad loc.); and in *A.* 9. 794, for no apparent reason.

Pindi: a high, windswept mountain range (Callim. *Hymn.* 4. 138–9) dividing Thessaly from Epirus; the Peneus rises on its eastern slope. Pindus was never a haunt of the Muses, but V., by associating it with Parnassus and Helicon, makes it seem so; see Nisbet–Hubbard on Hor. *Carm.* 1. 12. 6.

12. ulla: 5. 61 n.

moram fecere: an old idiom, e.g. Plaut. *Epid.* 691, *Most.* 75, Cic. *Pro Sulla* 58, *A.* 3. 473; varied by V., *A.* 1. 414 'moliriue moram', as by others (*TLL* s.v. *mora* 1470. 28).

Aonie Aganippe: for the prosody see 2. 24 n. (*Actaeo Aracyntho*); for Aganippe, 6. 64 n. The adjective *Ἀόνιος*, 'Boeotian', is not securely attested before Callimachus; see Pfeiffer on fr. 572. In Latin, adjective and spring first appear in Catull. 61. 27–30 'perge linquere Thespiae / rupis Aonios specus, / nympha quos super irrigat / frigerans Aganippe'. The only other reference to Aonian Helicon in the *E.*, 6. 65 'Aonas in montis', also involves Gallus; is 'Aonie Aganippe' a phrase borrowed from his poetry?

13–15. See 5. 27 n., 28 n.

13. lauri: for the hiatus see 3. 6 n.

fleuere, 15 fleuerunt: for this variation cf. Catull. 62. 28 *pepigere . . . pepigerunt*, Ov. *Met.* 3. 505 *planxere*, 507 *planxerunt*, 14. 563 *uiderunt*, 564 *uidere*.

myricae: 4. 2 n.

14. pinifer: first attested here; cf. *A.* 10. 708 'Vesulus . . . pinifer', 4. 249 'piniferum caput', *E.* 8. 22–3 'Maenalus argutumque nemus pinusque loquentis / semper habet'.

sola sub rupe: Catull. 64. 154.

iacentem: see Introduction, p. 288 n. 3.

15. Maenalus: 8. 22 n.

gelidi . . . Lycaei: no colder than Maenalus. Placed so, before the caesura, an adjective may have little more than a rhetorical function, to weight the second noun and balance the line.

Lycaei: a mountain in western Arcadia, reputed to be the birthplace of Pan, and, with Maenalus, one of his favourite haunts. Cf. Theocr. 1. 123–4 *ὦ Πὰν Πάν, εἴτ' ἐσσὶ κατ' ὤρεα μακρὰ Λυκαίω, | εἴτε τύγ' ἀμφιπολεῖς μέγα Μαίναλον*, 'O Pan, Pan, whether you are on the high hills of Lycaeus or range mighty Maenalus', *G.* 1. 16–17 'ipse nemus linquens patrium saltusque Lycaei / Pan, ouium custos, tua si tibi Maenala curae'.

17–18. nec te paeniteat pecoris, diuine poeta: / et formosus ouis ad flumina pauit Adonis: an admonition followed by a one-line *exemplum* or mythological paradigm. Commentators cite Theocr. 1. 109–10 *ὡραῖος χὤδωνις, ἐπεὶ καὶ μῆλα νομεύει | καὶ πτῶκας βάλλει καὶ θηρία πάντα διώκει*, 'Adonis, too, is in his bloom; he herds his sheep and kills hares and hunts all the wild beasts', and 3. 46 (Adonis) *ἐν ὤρεσι μῆλα νομεύων*, 'herding his sheep upon the hills', and no doubt V. had these passages, if vaguely, in mind. His formal model, however, was [Theocr.] 8. 51–2, where, as here, an admonition—Milon is warned not to scorn a goatherd lover—is followed by a one-line *exemplum*, *ἴθ', ὦ κόλε, καὶ λέγε, 'Μίλων, | ὁ Πρωτεὺς φώκας καὶ θεὸς ὢν ἔνεμεν'*, 'Go, stump-horn, and say, "Milon, Proteus herded seals even though he was a god"'.

17. nec te paeniteat: 2. 34.
diuine poeta: 5. 45; cf. also 6. 67.

18. formosus ... Adonis: cf. 4. 57 'formosus Apollo'.
ad flumina: 6. 64 n.

19–20. uenit et upilio, tardi uenere subulci, / uuidus hiberna uenit de glande Menalcas: cf. Theocr. 1. 80 *ἦνθον τοὶ βοῦται, τοὶ ποιμένες, ᾡπόλοι ἦνθον*, 'The cowherds came, the shepherds, the goatherds came', with spondaic delay before instead of after the caesura as in V. (G. Lee).

19. upilio: an unusual spelling, noticed in antiquity (*TLL* s.v. *opilio*); Ernout–Meillet, *Dict. étym.* s.v. *opilio*: 'la variation *o*/*u* est probablement d'origine dialectale'. The rustic *upilio* and earthy *subulci* are accommodated here by the elegance of V.'s rhetoric: ll. 19–20 form a tricolon crescendo, of which the third colon, l. 20, is 'golden'; see 2. 50 n.

subulci: swineherds have no place in the pastoral hierarchy of Theocritus (L. E. Rossi, *SIFC*, NS 43 (1971), 6–7) nor are they found elsewhere in V.'s pastoral landscape. La Cerda, therefore, with Ursinus and others, reads *bubulci*, which he defends in a very learned note; cf. Apul. *Flor.* 3 'nihil aliud plerique callebant quam Vergilianus upilio seu busequa' (see *TLL* s.v. *busequa*). Yet swine are essential to country living—'quis enim fundum colit nostrum quin sues habeat?' asks the aptly named Tremellius Scrofa in Varro, *RR* 2. 4. 3—and are referred to by V. in the *G.*, 2. 72 'glandemque sues fregere sub ulmis', 520 'glande sues laeti redeunt'. If V. had a particular landscape in mind, it was not (how could it have been?) Arcadia, as Voss suggests, but his native Cisalpina, where the oak forests produced a plentiful supply of acorns and swine abounded; according to Strabo 5. 1. 12, Rome was fed mainly on Cisalpine swine.

20. uuidus hiberna ... de glande Menalcas: Menalcas is wet through from handling the winter mast. Acorns were collected in late autumn and steeped in water for preservation during the winter; see Mynors on *G.* 1. 305.

21–3. omnes 'unde amor iste' rogant 'tibi?' uenit Apollo: / 'Galle, quid insanis?' inquit. 'tua cura Lycoris / perque niues alium perque horrida castra secuta est': cf. Theocr. 1. 81–5 *πάντες ἀνηρώτευν τί πάθοι κακόν. ἦνθ' ὁ Πρίηπος | κῆφα 'Δάφνι τάλαν,* *τί τὺ τάκεαι; ἁ δέ τυ κώρα | πάσας ἀνὰ κράνας, πάντ' ἄλσεα ποσσὶ φορεῖται | . . . | ζάτεισ'*, 'all asked what ailed him. Priapus came and said, "Poor Daphnis, why wasting away? For you the girl wanders by all the springs, through all the glades . . . searching"'.

21. uenit Apollo: as patron of herdsmen, Apollo Nomios, and poets. Heyne compares [Moschus], *Epitaph. Bion.* 26–8 *σεῖο, Βίων, ἔκλαυσε ταχὺν μόρον αὐτὸς Ἀπόλλων, | καὶ Σάτυροι μύροντο μελάγχλαινοί τε Πρίηποι· | καὶ Πᾶνες στοναχεῦντο τὸ σὸν μέλος*, 'Your sudden doom, Bion, Apollo himself lamented, and the Satyrs mourned and the Priapi clad in black, and the Pans sorrowed for your song'.

22. tua cura Lycoris: the phrase 'tua cura' occurs in 1. 57 (where see note), but here first with an erotic connotation (*TLL* s.v. 1475. 42), which O. Skutsch, *RhM* 99 (1956), 198–9, attributes to Gallus. Given the closeness with which V. imitates Theocritus, the bilingual 'echo', *κώρα Lycoris*—the questing, unnamed girl (*kora*) and the runaway Lycoris—must be intentional. For further, ingenious speculation see Ross (1975), 69.

23. perque . . . perqu(e): exactly reproducing the anaphora *πάσας . . . πάντ'* (Posch 70).

24–6. uenit et agresti capitis Siluanus honore, / florentis ferulas et grandia lilia quassans. / Pan deus Arcadiae uenit: cf. Lucr. 4. 586–7 'cum Pan / pinea semiferi capitis uelamina quassans', a passage imitated by V. elsewhere; see 1. 2 n.

24. agresti capitis Siluanus honore: a Virgilian ablative of description; cf. *A.* 3. 618 'domus sanie dapibusque cruentis', 5. 77 'mero . . . carchesia Baccho'. The 'rustic honour of his head' is the enormous garland worn by Silvanus.

Siluanus: an old Latin god of the woods hardly distinguishable from the Graeco-Latin Pan/Faunus, whom he seems to have displaced in popular imagination; later, as the woods receded, Silvanus became the protector of fields and flocks, Hor. *Epod.* 2. 22 'tutor finium', *A.* 8. 601 'aruorum pecorisque deo'. See Wissowa 212–16, Mynors on *G.* 1. 20.

25. ferulas: the giant fennel, *Ferula communis* L., which grows to a height of six feet or more (Sargeaunt 46). Without perfume, but used in garlands for its colour (Pliny, *NH* 21. 55); here its dark green leaves may be imagined as contrasting with the exceptional whiteness of the lilies (Pliny, *NH* 21.23 'candor . . . eximius').

lilia: the Madonna lily, *Lilium candidum* L.; Pliny, *NH* 21. 23 'nec ulli florum excelsitas maior, interdum cubitorum trium, languido semper collo et non sufficiente capitis oneri'.

26. Pan deus Arcadiae: *G*. 3. 392.

quem uidimus ipsi: surprisingly, the poet intervenes in his own fiction. The Ancients describe the experience of epiphany simply as 'seeing': Callim. *Hymn* 2. 10 ὅς μιν ἴδῃ, μέγας οὗτος, 'Whosoever seeth him (Apollo), great is he', Hor. *Carm*. 2. 19. 1–2 'Bacchum in remotis carmina rupibus / uidi docentem', where see Nisbet–Hubbard; see also A. Henrichs, *HSCP* 82 (1978), 209–10. Cf. 1. 42.

27. See 6. 20–2 n.

sanguineis . . . bacis: *G*. 2. 430.

ebuli: the dwarf-elder, *Sambucus ebulus* L., an invasive, mal-odorous weed with reddish-black berries.

minio: cinnabar or red sulphide of mercury imported from Spain, Prop. 2. 3. 11 'minio . . . Hibero'.

28. ecquis erit modus?: cf. 2. 68.

30. cytiso: 1. 78 n.; cf. Pliny, *NH* 13. 131 'apes quoque numquam defore cytisi pabulo contingente promittunt Democritus atque Aristomachus'.

31. In Homer and Hesiod, with a very few minor exceptions, and in Apollonius of Rhodes, speeches begin at the beginning of the line. The first poets to diverge from this traditional practice were Callimachus and Theocritus. In Latin hexameter poetry, from Ennius onwards, speeches often begin, as here, within the line. See J. Kvíčala, *Neue Beiträge zur Erklärung der Aeneis* (Prague, 1881), 265–74, Norden on *A*. 6. 45 ff., Hopkinson on Callim. *Hymn* 6. 41, O. Skutsch, *The Annals of Q. Ennius* (Oxford, 1985), 54–5.

tamen cantabitis: 'alii *tamen* superioribus iungunt, sed melius ut sic legamus *tamen cantabitis*, ut sit sensus: licet ego duro amore consumar, tamen erit solacium, quia meus amor erit uestra cantilena quandoque' (Serv.). Turcius Rufius Apronianus Asterius, who in AD 494 'revised and punctuated' his text of the *Eclogues*, now the codex Mediceus, agrees with Servius; see M. Geymonat, *Enc. Virg*. s.v. *Aproniano*. This, the modern punctuation, was established by Wagner in his revision of Heyne (1830). *Cantabitis*, as T. E. Page observes, 'is partly an assurance which Gallus utters as a solace to himself, partly a request'.

31–3. Arcades . . . / . . . / Arcades: 6. 21 n.

32. montibus: probably dative (Conington, Lee) rather than local ablative (Dryden, Coleman); like Corydon (2. 5), the Arcadians will sing to their hills. In a local sense, *montibus* is ordinarily accompanied with *in* or qualified, e.g. 6. 52 'in montibus', *A.* 7. 387 'frondosis montibus'.

cantare periti: 7. 5 n. *Peritus* occurs only here in V. and is rarely found in poetry; see Axelson 102.

33. molliter ossa quiescant: as would befit a love-poet, a writer of soft elegiac verse: Prop. 1. 7. 19 'mollem . . . uersum', 2. 1. 2 'mollis . . . liber', Ov. *Trist.* 2. 349 'mollia carmina', *Ex Pont.* 3. 4. 85 'molles elegi'. Although as old as Plautus, *molliter* is first used here (it does not occur elsewhere in V.) of the repose of death (*TLL* s.v. 1380. 83)—here, and by Ovid, *Her.* 7. 162 'et senis Anchisae molliter ossa cubent', *Am.* 1. 8. 108 'ut mea defunctae molliter ossa cubent', *Trist.* 3. 3. 76 'Nasonis molliter ossa cubent'. A. Barchiesi, *A&R* 26 (1981), 162–3, calls attention to several striking similarities between the language of Gallus, brief as the new fragment of his poetry is, and that of Ovid. May it not be that Gallus wrote 'molliter ossa cubent', which V. adapted by substituting *quiescant* for *cubent*? Cf. Prop. 1. 17. 22 'molliter et tenera poneret ossa rosa'.

ossa quiescant: later a pious commonplace, adumbrated in Ennius, *Sc.* 365 V.[2] = 299 J. 'ubi remissa humana uita corpus requiescat malis', and first expressed here; see J. E. Church, *ALL* 12 (1900–2), 226, Norden on *A.* 6. 328 'ossa quierunt'.

34. fistula: 2. 37n.

35–6. atque utinam . . .: cf. Theocr. 7. 86–8 (Lycidas wishes that he might have shared a goatherd's life with Comatas) *αἴθ' ἐπ' ἐμεῦ ζωοῖς ἐναρίθμιος ὤφελες ἦμεν, | ὥς τοι ἐγὼν ἐνόμευον ἀν' ὤρεα τὰς καλὰς αἶγας | φωνᾶς εἰσαΐων*, 'Oh that in my day you had been numbered among the living, that I might have herded your pretty she-goats upon the hills and listened to your voice'.

35. ex uobis: the only other instance of *ex* before a consonant in the *E.*, 6. 19 'ipsis ex uincula sertis', is not quite comparable because there, as in the similar phrase, 6. 33 'his ex omnia primis', *ex* does not cohere with the following word. *E* occurs only once in the *E.*, 7.13. See Axelson 119–20, Ross (1969), 46–9.

36. custos gregis: cf. 3.5, 5.44, *G.* 1. 17.

37. siue mihi Phyllis siue esset Amyntas: previously associated—Phyllis as the object of a heterosexual, Amyntas of a homosexual passion—in 3. 74, 76. There may be a reminiscence here of Theocr. 7. 105 εἴτ' ἔστ' ἆρα Φιλῖνος ὁ μαλθακὸς εἴτε τις ἄλλος, 'whether it be the pliant Philinus or another'.

38. furor: here first and virtually unique in this sense (*TLL* s.v. 1632. 80). Quite possibly an imitation of Gallus.

38–9. (quid tum, si fuscus Amyntas? / et nigrae uiolae sunt et uaccinia nigra): an imitation of Asclepiades conflated with an imitation of Theocritus: Asclep. 5. 3–4 G.–P. (= *AP* 5. 210) εἰ δὲ μέλαινα, τί τοῦτο; καὶ ἄνθρακες· ἀλλ' ὅτε κείνους | θάλψωμεν λάμπουσ' ὡς ῥόδεαι κάλυκες, 'If she is dark, what does it matter? So are coals, and yet when heated they glow like blown roses' (E. Krüger, *RhM* 67 (1912), 480, but already noticed by La Cerda), Theocr. 10. 28 καὶ τὸ ἴον μέλαν ἐστὶ καὶ ἁ γραπτὰ ὑάκινθος, 'Dark is the violet and the lettered hyacinth'. Cf. Longus 1. 16. 4 (Daphnis speaking of himself) μέλας, καὶ γὰρ ὁ ὑάκινθος, 'I am dark, but so is the hyacinth'.

39. et nigrae uiolae sunt et uaccinia nigra: the basic structure of the line is Theocritean (10. 28 καὶ . . . ἐστὶ καὶ, 'et . . . sunt et'), but the elaborate rhetoric—chiastic word-order emphasized by alliteration, rhyming pairs of adjectives and nouns, adjective repeated with shift of ictus—is Virgilian. See 2. 44 n. (*munera nostra*).

uiolae: 2. 47 n.

uaccinia nigra: 2. 18, where however *nigra* is not predicative.

40. inter salices lenta sub uite: the vine has been trained not on the usual elm but on the willow. Columella 5. 7. 1 briefly describes a kind of plantation or *arbustum* (1. 39 n.) found in Gaul which consists of dwarf trees, among them the willow; but (he says) the willow should be planted only in wet places, where other trees will hardly take root, because it spoils the flavour of the wine. A memory of V.'s youth, like the alder (6. 63 n.)? 'Vines . . . are, I am told, trained on willows in Lombardy in the present day' (Conington).

lenta: 1. 4n.

42–3. hic gelidi fontes, hic mollia prata . . ., / hic nemus; hic . . .: perhaps V. was thinking of Theocr. 5. 31–4 *ἅδιον ᾀσῇ | τεῖδ' ὑπὸ τὰν κότινον καὶ τἄλσεα ταῦτα καθίξας· | ψυχρὸν ὕδωρ τουτεὶ καταλείβεται· ὧδε πεφύκει | ποία, χἀ στιβὰς ἅδε, καὶ ἀκρίδες ὧδε λαλεῦντι*, 'More pleasantly will you sing seated here beneath the wild olive and these trees. Here drips cold water, here is grass and this our couch, and here the grasshoppers chatter'. See Posch 79.

hic . . . hic . . . / hic . . . hic: repeated three times in 7. 49–51, 9. 40–1 (a similar passage), 60–2.

42. gelidi fontes: cf. Cic. *De nat. deor.* 2. 98 'fontium gelidas perennitates'.

mollia prata: a phrase not found before V. and perhaps invented by him (*TLL* s.v. *mollis* 1370. 63) or by Gallus. Note the typically elegiac contrast between *mollis* and *durus* (47). Cf. *G.* 2. 384, 3. 520–1.

Lycori: suddenly and rather awkwardly intruded into his dream of pastoral felicity. Here, it may be, V. begins to imitate one of Gallus' elegies more consistently, with scant regard, however, to the preceding lines or, in ll. 44–5, to the imagined scene.

43. consumerer aeuo: a unique inversion of a phrase itself apparently unique, Lucr. 5. 1431 (the human race) 'in curis consumit inanibus aeuum'; see *TLL* s.v. *consumo* 615. 53.

44–5. nunc insanus amor duri me Martis in armis / . . . detinet: *nunc* in contrast with the idyllic existence he has imagined; cf. Tib. 1. 10. 11–13 'tunc mihi uita foret . . . / . . . / nunc ad bella trahor' and see Introduction, p. 291. *Amor* should probably be taken with *Martis*, although most commentators—Conington, Forbiger, and T. E. Page excepted—refer *amor* to Lycoris, with *Martis* dependent on *in armis*. Cf. above, 21 'amor iste', 22 'Galle, quid insanis?' But how can it be said that his mad passion for Lycoris keeps Gallus in the army? It is rather that his mad passion for war keeps Gallus from Lycoris. Cf. *A.* 7. 461 'saeuit amor ferri et scelerata insania belli', 550 'accendamque animos insani Martis amore' (Forbiger). Ancient readers, too, were puzzled: 'ex affectu amantis ibi se esse putat ubi amica est, ut *me* sit *meum animum*' (Serv.). Gallus seems to have forgotten for the moment that he is in Arcadia.

45. tela inter media: *A.* 10. 237; cf. *E.* 9. 12.
aduersos ... hostis: *A.* 12. 266, 456, 461.

46–9. See Introduction, p. 291.

46. (nec sit mihi credere tantum): Gallus would rather not believe what he knows to be the case; see T. C. W. Stinton, '*Si credere dignum est*: Some Expressions of Disbelief in Euripides and Others', *PCPS*, NS 22 (1976), 65 n. 12 = *Collected Papers* (Oxford, 1990), 241 n. 12. For the construction cf. *A.* 8. 676 'cernere erat' ('Graeca figura', Serv.), Hor. *Serm.* 1. 2. 101 'tibi paene uidere est', Tib. 1. 6. 24 'tunc mihi non oculis sit timuisse meis', and see E. Wölfflin, *ALL* 2 (1885), 135–6, Hofmann–Szantyr 349.

47. Alpinas ... Rheni: for this arrangement of proper nouns or adjectives enclosing the line cf. 3. 76, 84, 4. 57, 5. 52, 6. 42, 7. 26, 37, 8. 5, 56, 9. 28, 10. 15, 22, 68, 72. The higher frequency of this arrangement in *E.* 10 may be significant.

Alpinas ... niues et frigora Rheni: Gallic snows, 'Gallicana ... frigora' (Serv.), were proverbial; cf. Petronius 19. 3 'ego autem frigidior hieme Gallica factus', Philodemus 15. 4 (= *AP* 10. 21) *τὸν χιόσι ψυχὴν Κελτίσι νιφόμενον*, 'my soul blizzarded by Celtic snows'; see Gow–Page, *The Garland of Philip* (Cambridge, 1968), ad loc.

a: there are nine instances of this exclamation in the *E.*, two in the *G.* (2. 252, 4. 526), but none in the *A.*; on *A.* 4. 657, where *a* would be more idiomatic, see Clausen 60. Of the instances in the *E.*, 2. 69 is an imitation of Theocritus, 6. 47 and 52 quotations of Calvus, and the three in close proximity here (47, 48, 49), to which 1. 15 is similar, may be taken as representing the agitated, emotional style of Gallus. The two remaining instances, 2. 60 and 6. 77, both conform to standard elegiac practice; see A. Kershaw, *CP* 75 (1980), 71–2.

48. me sine sola uides: metrically equivalent to the second half of a pentameter.

me sine: the postposition of *sine* is very rare: Lucr. 1. 213 'nostro sine quaeque labore', 4. 1173 'hac sine', Hor. *Serm.* 1. 3. 68 'uitiis nemo sine', 5. 99 'flamma sine', *G.* 1. 161 'quis sine', 3. 42, *A.* 12. 883 'te sine', [Tib.] 4. 14. 3 'nostro sine facta dolore', and twice in Ovid. *Her.* 8. 80 'clamabam "sine me, me sine mater abis?"', *Ars* 2. 454 'quo sine'. See C. W. Weber, 'Tmesis, Sandhi, and the Two Caesurae of the Hexameter in Virgil's Aeneid' (Ph.D.

thesis, Univ. of California, Berkeley, 1975), nn. 543, 550, 564. Weber conjectures (p. 156) that 'me sine' is one of V.'s *translationes* from Gallus.

a, te ne frigora laedant!: cf. Cic. *De nat. deor* 2. 129 (mother birds with their chicks) 'ita tuentur ut et pinnis foueant ne frigore laedantur', Varro, *RR* 2. 1. 23 (the treatment of a sickly animal) 'et inicitur aliquid ne frigus laedat', *G.* 3. 298–9 'glacies ne frigida laedat / molle pecus'. The lover's concern lest his beloved should injure herself while fleeing from him may be a Hellenistic motif (here suitably modified); cf. Ov. *Met.* 1. 508–9 (Apollo to Daphne) 'me miserum, ne prona cadas indignaue laedi / crura notent sentes et sim tibi causa doloris', Nonnus, *Dionys.* 16. 91–3 (Dionysus to Nicaea) *παρθενικὴ ῥοδόεσσα, τί σοι τόσον εὔαδον ὗλαι; | σῶν ἐρατῶν μελέων περιφείδεο, μηδ' ἐπὶ πέτραις | ἀστορέες σέο νῶτα μετατρίψωσι χαμεῦναι*, 'Rosy maiden, why do you like the forest so much? Spare your lovely limbs, nor let the rough unstrown pallet upon the rocks chafe your back' (W. H. D. Rouse in the Loeb Classical Library).

50–1. ibo et . . . modulabor: colloquial in tone, e.g. Plaut. *Bacch.* 871 'ibo et faciam sedulo', *Cist.* 531 'ibo et persequar' (*TLL* s.v. *eo* 631. 20). Impetuously Gallus decides to change his life—to change, that is, his style of poetry: no more love-elegies, and the epyllion he composed in Chalcidic verse he will set to music on a Sicilian shepherd's pipe. The Grynean Grove, a *locus amoenus*, to judge from DServius' description (see 6. 72 n.), would easily lend itself to a pastoral rendering.

50. Chalcidico . . . uersu: Quintilian 10. 1. 56 'quid? Euphorionem transibimus? quem nisi probasset Vergilius idem, numquam certe conditorum Chalcidico uersu carminum fecisset in Bucolicis mentionem'. Gallus evidently wrote two kinds of poetry: epyllia (or at least one—*carmina* appears to be a poetic plural—on the Grynean Grove) in the style of Euphorion and love-elegies; cf. Serv. on 10. 1 'Gallus . . . fuit poeta eximius; nam et (*et* DServ.: *et* om. Serv.) Euphorionem, ut supra diximus (6. 72 n.), transtulit in Latinum sermonem et amorum suorum de Cytheride scripsit libros quattuor'. Where Servius and DServius all but agree, as here, DServius usually preserves the better or more complete text; it is far more likely, therefore, that Servius omitted *et*, whether accidentally or as seeming unnecessary, than that DServius added it. It has been established beyond reasonable doubt that Euphorion did not write elegiac poetry; see F. Skutsch, *RE* vi. 1176–8, Ross (1975), 40–3. With 'Chalcidico . . . uersu' cf. 6. 1 'Syracosio . . . uersu'.

condita: 6. 7 n.

51. modulabor: 5. 14 n.

auena: 1. 2 n.

52–3. certum est in siluis . . . / malle pati: Gallus is resolved to suffer in the woods rather than in the camp. For *pati* used absolutely cf. Ov. *Met.* 10. 25–6 (Orpheus in the Underworld) 'posse pati uolui nec me temptasse negabo: / uicit Amor'.

52. in siluis inter spelaea ferarum: cf. *A.* 3. 646 'in siluis inter deserta ferarum', 7. 404 'inter siluas, inter deserta ferarum'. *Deserta* as a noun is first attested in Lucr. 1. 163–4 'genus omne *ferarum*, / incerto partu culta ac *deserta* tenerent' (*TLL* s.v. *desero* 688. 33).

spelaea: σπήλαιον, the ordinary Greek word for cave, occurs in Plato, *Rep.* 7. 514 A, but nowhere in poetry. Norden, on *A.* 6. 10, suggests that *spelaeum*, which V. uses only here, and which he seems deliberately to avoid in the *A.* (see preceding note), was Gallus' innovation.

53–4. tenerisque meos incidere amores / arboribus: Acontius seems to have been the first to carve the name of his beloved on a tree, Callim. *Aet.* fr. 73 Pf. ἀλλ' ἐνὶ δὴ φλοιοῖσι κεκομμένα τόσσα φέροιτε | γράμματα, Κυδίππην ὅσσ' ἐρέουσι καλήν, 'but on your bark may you bear so many letters carved as will declare Cydippe beautiful', that is, the imagined inscription would read Κυδίππη καλή; see K. J. Dover, *Greek Homosexuality* (Cambridge, Mass., 1978), 111–15. Cf. Prop. 1. 18. 22 'scribitur et teneris Cynthia corticibus', Ov. *Her.* 5. 21–2 'incisae seruant a te mea nomina fagi, / et legor Oenone falce notata tua', and see 5. 13 n.

54. arboribus: crescent illae: a reminiscence of a pastoral passage in Lucretius, 1. 253 'arboribus, crescunt ipsae fetuque grauantur' (Wakefield).

crescent illae, crescetis, amores: 'hoc vere, si quid aliud, Virgiliana elegantia dignum' (Heyne). Cf. Ov. *Her.* 5. 23 'et quantum trunci, tantum mea nomina crescunt', Spenser, *Colin Clouts Come Home Againe*, 632–3 'Her name in every tree I will endosse, / That as the trees do grow, her name may grow'. For the anaphora, a feature of poetic technique as old as Homer, cf. *Il.* 24. 516, *Od.* 4. 149, Catull. 67. 1, Tib. 2. 1. 17, Hor. *Ars* 269.

55. interea: in the meantime, while waiting for the young trees to grow and his pastoral love to mature, Gallus will range over Maenalus with the Nymphs or take up hunting in the vain hope, as it proves to be, of assuaging his passion for Lycoris; see F. Skutsch (1906), 164–5. The notion of hunting as a cure for love derives ultimately from Euripides, the source of so much in Latin comedy and elegy; see 2. 4–5 n. In *Hipp.* 215–22, Phaedra, sick with love, longs to go to the mountain, to be among the pinewoods, where bloodthirsty hounds pursue the dappled deer, and there to halloo the hounds, hurl the javelin . . . The therapeutic value of hunting became a motif of Latin love-elegy, e.g. Ov. *Rem.* 199–206, and no doubt V. found something of the sort, a passage in an elegy or an elegy, in Gallus' *Amores*; see Introduction, pp. 290–1.

55–6. mixtis lustrabo Maenala Nymphis / aut acris uenabor apros: Conington compares *A.* 1. 322–4 'uidistis si quam hic errantem forte sororum / . . . / aut spumantis apri cursum clamore prementem' and says that *aut* 'merely distinguishes the actual chase from its preliminaries'. All things are possible in the pastoral world; still, it seems odd that Nymphs should be hunting the wild boar (see, however, 6. 56 n.) and l. 55 may suggest the lover wandering distractedly, like Minos in Callim. *Hymn* 3. 190–1 *ἧς ποτε Μίνως | πτοιηθεὶς ὑπ' ἔρωτι κατέδραμεν οὔρεα Κρήτης*, 'maddened with love of whom, Minos once roamed the hills of Crete', or Orion in Cic. *Arat.* 421 'excelsis errans in collibus amens', or Milanion in Prop. 1. 1. 11 'Partheniis amens errabat in antris'. See D. J. Kubiak, *CJ* 77 (1981–2), 16–18, Clausen, *HSCP* 90 (1986), 165–6; also 6. 52 n.

55. Maenala: 6. 30 n.

56. acris . . . apros: cf. Hor. *Epod.* 2. 31–2 'aut trudit acris hinc et hinc multa cane / apros in obstantis plagas'.

57. Parthenios: see Introduction, p. 289.

58–9. lucosque sonantis / ire: *G.* 4. 364–5.

59–60. Partho torquere Cydonia cornu / spicula: cf. *A.* 12. 856–8 'non secus ac neruo per nubem impulsa sagitta, / armatam saeui Parthus quam felle ueneni, / Parthus siue Cydon, telum immedicabile, torsit', and see R. M. Rosen and J. Farrell, *TAPA* 116 (1986), 251. The Cydonians, *Κύδωνες*, were an ancient people inhabiting north-western Crete (*Od.* 3. 292, 19. 176), but Callimachus, in keeping with Hellenistic practice, extended the local name to Cretans in general; see 9. 13 n., McLennan on Callim. *Hymn* 1. 45, Bornmann on *Hymn* 3. 197. Although the adjective *Κυδώνιος* occurs in early Greek poetry (see D. L. Page, *Poetae Melici Graeci* (Oxford, 1962), index verborum, s.v.), V. probably regarded it as Callimachean: *Aet.* fr. 560 Pf. *τόξου ... Κυδωνίου*, 'a Cydonian bow', *Hymn* 3. 81 *Κυδώνιον ... τόξον*; here first in Latin, and here only in V. The Cretans enjoyed a reputation for archery reaching back to Homer; see Nisbet–Hubbard on Hor. *Carm.* 1. 15. 17. Of Parthian archery the Romans had had bitter experience.

torquere ... / spicula: again of shooting arrows in *A.* 11. 773 'spicula torquebat Lycio Gortynia cornu'; but cf. *A.* 7. 164–5 'aut acris tendunt arcus aut lenta lacertis / spicula contorquent', where *(con)torqueo* has its usual sense of hurling spears: *A.* 7. 741, 9. 402, 665 'intendunt acris arcus amentaque torquent', 10. 585, 11. 676, 12. 578.

59. cornu: a metrical variant for *arcu*; cf. *A.* 11. 654 'spicula conuerso fugientia derigit arcu' with *A.* 7. 497 'Ascanius curuo derexit spicula cornu'. See *TLL* s.v. 969. 13, Lyne on *Ciris* 299.

60. spicula: 'arrows', as in *A.* 7. 497 (see preceding note), 9. 606 'spicula tendere cornu', 11. 773 (see above, 59–60 n.). More frequently of spears: *A.* 5. 307, 586, 7. 186, 626, 687, 10. 888, 11. 606, 676, 12. 403, 563.

medicina furoris: no doubt 'Gallus' is quoting Gallus here; cf. Prop. 2. 1. 57–8 'omnis humanos sanat medicina dolores: / solus amor morbi non amat artificem', 1. 5. 27–8 'non ego tum potero solacia ferre roganti, / cum mihi nulla mei sit medicina mali', and see Tränkle 22–3, Ross (1975), 66–8.

62. Hamadryades: only here in V.; found earlier in Catull. 61. 23. See 6. 55–6 n.

65–8. Cf. Theocr. 7. 111–13 *εἴης δ' Ἠδωνῶν μὲν ἐν ὤρεσι χείματι μέσσῳ | Ἕβρον πὰρ ποταμὸν τετραμμένος ἐγγύθεν Ἄρκτω, | ἐν δὲ θέρει πυμάτοισι παρ' Αἰθιόπεσσι νομεύοις*, 'and in midwinter (mayst thou) find thyself on the mountains of the Edonians, turned towards the river Hebrus, hard by the pole. And in summer mayst thou herd thy flock among the furthest Ethiopians' (Gow). These lines are from the song of Simichidas, in which he asks Pan to further the love-affair of his friend Aratus; if Pan declines to do so, may he sleep on nettles and experience winter in the mountains of Thrace and summer in the African desert.

65. Hebrum: the principal river of Thrace, now the Marica (Maritsa) / Meric / Evros; see G. Bonamente, *Enc. Virg.* s.v.

66. Sithonias: a rare toponym, with a short *o* as in Lycophron 1357 *Σιθόνων*, Euphorion, fr. 58. 2 Powell *Σιθονίῃ*, and Parthenius, *Ἐρωτ. παθ.* 11. 4 (= *Suppl. Hell.* 646. 3) *Σιθονίῳ κούρῳ*. Originally the middle peninsula of Chalcidice (Herod. 7. 122 *Σιθωνίῃ*), but later extended to the whole of Thrace (see above, 59–60 n.). Here first in Latin (Gallus?), and here only in V.

hiemis . . . aquosae: depends on *frigoribus mediis*, 'the middle frosts / Of watery winter' (Lee); so Housman on Manil. 1. 455. Strictly *aquosae* is inconsistent with *frigoribus* and *niues*, but V. was thinking of an Italian winter (Wagner); cf. *A.* 9. 671 'aquosam hiemem'.

67. cum moriens alta liber aret in ulmo: Conington's conjecture *aret Liber*, which Housman commends, while insulting Conington, on Manil. 3. 662 'tum Liber grauida descendit plenus ab ulmo', destroys the bucolic diaeresis.

68. uersemus: 'quoniam qui pascit, huc et illuc agit pecus' (DServ.).

sub sidere Cancri: the sign of the Crab, a midsummer sign; cf. Manil. 4. 758 'ardent Aethiopes Cancro, cui plurimus ignis'.

69. E. Fraenkel, *JRS* 45 (1955), 6 = *Kleine Beiträge* (Rome, 1964), ii. 95–6: 'There existed, we may almost say from time immemorial, a manner of denoting the transition from a paraenetic tale or a general maxim to its application to the case in hand by means of the phrase *καὶ σύ* or *οὕτω καὶ σύ*, 'and so you (too) . . .'; n. 1: 'The famous line . . . "omnia vincit Amor; et nos cedamus Amori", is a perfect instance of the pattern with which we are concerned'.

omnia uincit Amor: metrically equivalent to the second half of a pentameter.

Amor: W. S. Allen, *Vox Latina*[2] (Cambridge, 1978), 128: 'a natural pause permits the second syllable of *amŏr* to be heavy'. Cf. *A.* 11. 323 *amor, et*, 12. 422 *dolor, omnis*, *G.* 3. 118 *labor, aeque*.

70. diuae: only here in the *E.*; cf. Theocr. 10. 24–5 (below, 72 n.) and see *E.* 7. 25–6 n.

71. sedet: 3. 55 n.

gracili fiscellam texit hibisco: 'allegoricos autem significat se composuisse hunc libellum tenuissimo stilo' (Serv.)—a literary comment of some value.

fiscellam: 2. 72 n.

hibisco: 2. 30 n.; also Mynors on *G.* 1. 165.

72. Pierides: uos haec facietis maxima Gallo: cf. Theocr. 10. 24–5 *Μοῖσαι Πιερίδες, συναείσατε τὰν ῥαδινάν μοι | παῖδ'· ὧν γάρ χ' ἅψησθε, θεαί, καλὰ πάντα ποεῖτε*, 'Pierian Muses, celebrate with me the slender girl, for whatever you touch, goddesses, you make beautiful'. For the last time ever, V. invokes the Pierian Muses; see 3. 85 n.

72–3. Gallo, / Gallo, cuius amor: 9. 47–8 n.

74. se subicit: cf. *G.* 2. 18–19 'etiam Parnasia laurus / parua sub ingenti matris se subicit umbra'.

alnus: 6. 63 n.; also Abbe 77.

75–7. See above, pp. xxv–xxvi.

75–7. Cf. Lucr. 6. 783–5 'arboribus primum certis grauis umbra tributa / usque adeo, capitis faciant ut saepe dolores, / si quis eas subter iacuit prostratus in herbis'. But why the juniper tree? Poetry required a particular tree and V.'s choice of the juniper may be owing to personal observation; cf. C. Fantazzi and C. W. Querbach, *Phoenix*, 39 (1985), 364 n. 18: 'In the most frequent type of pasture land encountered in Italy, a rather rough maquis, juniper trees are usually found grouped densely together in what is called a *ginepraio*, and their leaves are a favorite food of goats'.

GENERAL INDEX

INDEX OF NAMES

INDEX OF LATIN WORDS